Physical Modifications of Starch

Zhongquan Sui • Xiangli Kong
Editors

Physical Modifications of Starch

Second Edition

Springer

Editors
Zhongquan Sui
Department of Food Science & Engineering
Shanghai Jiao Tong University
Shanghai, Shanghai, China

Xiangli Kong
College of Agriculture and Biotechnology
Zhejiang University
Hangzhou, Zhejiang, China

ISBN 978-981-99-5389-9 ISBN 978-981-99-5390-5 (eBook)
https://doi.org/10.1007/978-981-99-5390-5

This Springer imprint is published by the registered company Springer Nature Singapore Pte Ltd.
The registered company address is: 152 Beach Road, #21-01/04 Gateway East, Singapore 189721, Singapore

Contents

Chapter 1
Molecular Structure of Starch

Xiangli Kong

Abstract The physicochemical characteristics of starches from enormous sources show a very wide range, resulting in various functional properties, which are attributed to the distinctive molecular structure and biosynthesis pathway of starch macromolecules. Three major components, amylose, amylopectin and occasionally intermediate material, are present in starch granules. Amylopectin and amylose have distinctively different functional properties. Hence the ratio of amylose/amylopectin was observed to have significant effects on rheological and functional attributes to starch. Moreover, the molecular structure of amylopectin and amylose was also shown to have considerable effects on the functional properties of starch. This chapter serves as a point of basic information on starch molecular structure and stimulates further understanding of the topic in every chapter for the readers of this book.

Keywords Starch · Structure · Amylose · Amylopectin

1.1 Introduction

Starch, one of the most abundant biopolymers on earth, is biosynthesized by green plants, algae, or some cyanobacteria. Starch provides the major calories for human diet and the primary feedstock for bioindustry. There is a very wide range in the physicochemical properties of starches from enormous origins, attributed to the differences in starch molecular structure and biosynthesis pathway, resulting in various functionalities. Even so, the native starches still have some limitations when facing huge kinds of applications due to their intrinsic structures. In terms of chemical composition, starch is a very simple molecule, builting from glucose only. However, its structure is more complex, although normally being comprised of three polymers, amylose and amylopectin and occasionally intermediate material in some

X. Kong (✉)
College of Agriculture and Biotechnology, Zhejiang University, Hangzhou, Zhejiang, China
e-mail: xlkong@zju.edu.cn

Z. Sui, X. Kong (eds.), *Physical Modifications of Starch*,
https://doi.org/10.1007/978-981-99-5390-5_1

oat or high amylose mutant maize starches (Kong et al. 2009a). This chapter serves as a point of basic information on starch molecular structure and stimulates further understanding of the topic in every chapter for the readers of this book.

The starch structure can be described in terms of physicochemical properties of the constituent molecules, compositional variation, the multi-scale level structure, and architecture of starch granule (Tester et al. 2004). The compositional variation, minor components, molecular structure of amylose and amylopectin, and granular structure all play key roles in starch physicochemical properties (Zhu 2018). Especially, the enzymatic characteristic of gelatinized starch is mainly attributed to starch molecular structure; moreover, the molecular structure contributes to the physical structure of native starch granule. In recent years, starch molecular structures are also observed to be essentially important in determining brewing quality (Gous and Fox 2017; Yu et al. 2020). In this chapter, we mainly focus on the molecular structure of starch. The physical or granular structure and the physicochemical properties of starch will be introduced in the following chapters.

1.2 Starch Components

For most starches, amylopectin is the major component therein and constitutes up to 65–85% of the total starch weight for most wild-type plants, although some mutant starches contain amylopectin solely (waxy type) or very high amylose (up to ~70%). The structural framework of starch granule is based on amylopectin, which underlies the starch semi-crystalline nature and has profound effect on the physical and biological characteristics, and amylose is believed to fill spaces in the semi-crystalline matrix (Pérez and Bertoft 2010; Pfister and Zeeman 2016). Amylopectin and amylose have distinctively different functional properties. Gelatinized amylopectin is stable in water and makes soft gels and weak films, whereas amylose retrogrades with high tendency and forms tough gels and strong films (Pérez and Bertoft 2010). Therefore, the ratio of amylose/amylopectin was observed to have a significant effect on rheological and functional attributes of starch (Kong et al. 2009b, 2015a). Furthermore, the molecular structure of amylopectin and amylose, depending on planting conditions and genetic factors, has been shown to have considerable effects on the physicochemical properties of starch (Kong et al. 2008, 2010, 2015b, 2016; Syahariza et al. 2013; Li et al. 2016; Ni et al. 2022), which are important for the diverse applications of starch.

Meanwhile, some minor components, such as proteins, lipids, phosphorus, and minerals, are also present in starch granules and have influences on properties (Vamadevan and Bertoft 2015). The content of proteins ranges between 0.06% and 0.7% (Debet and Gidley 2006; Pérez and Bertoft 2010), which depends on the starch sources and the degree of purification during starch isolation (Dhital et al. 2019). Starch proteins can be categorized into surface protein and granule-associated protein. The surface protein can be removed by using surfactants or salt solutions. However, removal of the granule-associated protein needs strong detergents and swelling of granule. Amylose content was observed to be positively correlated with granule-associated protein in rice (Ye et al. 2019). The physicochemical properties and microstructure of starch were altered after granule-associated proteins removal (Ma et al. 2021, 2022). Although lipids are present both on surface and in internal parts of the starch granule, only the latter is considered as the true starch lipids (Dhital et al. 2019). In cereal starches, the amounts of lipids are up to 1.5%, and there is a positive correlation between amylose content and lipid content (Morrison 1995). Furthermore, some portions of amylose are complexed with lipids (Pérez and Bertoft 2010). The amylose-lipid complex plays key roles in functional properties of starch, therefore affecting the quality of starch-based foods. The phosphorus was present in starch granules with three major forms (phosphate monoesters, phospholipids, and inorganic phosphates). The phosphorus content in cereal starches is very low, while it is relatively higher in tuber starches. For example, the amount is approximately 0.5% in potato starch. A relatively high amount of phosphate, even as a minor component, tends to increase granule hydration, lower crystallinity, and yield starch pastes with higher viscosity, freeze–thaw stability, and transparency (Pfister and Zeeman 2016). Other minerals present in starch have subtle effects on functional properties (Tester et al. 2004).

1.3 Molecular Structure of Amylose

Amyloses are composed of long α-1,4-linked linear chains with only a few α-1,6-linked branched points, making them almost linear molecules. Both linear and branched amylose molecules are presented in starch, and more than ten branches exist in those amyloses with large molecular weights. A relatively higher DP (degree of polymerization) range was observed for branched amylose molecules comparing with that of linear ones, and larger branched amylose molecules have a higher number of side chains (Hanashiro 2015). The molecular weight of amylose molecules is estimated to be ~10^6 Da. However, the actual size cannot be measured because of shear scission after dispersion (Seung 2020). Atomic force microscopy revealed that pea starch contained single-branched molecules with long side chains and multiple-branched molecules with shorter side chains (Gunning et al. 2003). The size of amylose molecules shows variations among starch sources, the amyloses from tuber or root such as potato and sweet potato starches were observed to be much larger than those from cereal starches. Furthermore, the molar fraction of branched amylose from sweet potato is higher than those from cereals (Bertoft 2017).

The molecular structure of amylose was characterized by analyzing the chain length profile after debranching, showing that amylose molecules consist of shorter chains with DP 100–700 and longer chains with DP 700–40,000 (Wang et al. 2014). It should be noted that the chain length profile in their work is produced from a mixture of linear and branched amyloses; therefore, the chain length distribution of sole branched amylose remains unclear. In another report, very short chains with DP

< 100 were also identified, despite their weight-based amount (Hanashiro et al. 2013). Furthermore, because of steric hindrance or retrogradation during the debranching process, it was verified that the amylose molecules could not be debranched by bacterial enzymes (isoamylase or pullulanase) completely in many reports (Mercier 1973; Takeda et al. 1989; Hanashiro et al. 2013).

The branched structure of amylose can be described by the value of β-amylolysis limit, which is the percentage of amylose converted to maltose upon exo-acting β-amylase hydrolysis. During hydrolysis, the β-amylase cleaves linear amyloses into maltose completely from non-reducing ends and leaves maltose or maltotriose; however, it cannot bypass the branched points, and consequently produces residues of so-called β-amylase limit dextrin. Therefore, the β-amylolysis limit is influenced by both the contents and α-(1,6)-branch linkage amounts of branched amylose molecules, and also the position of the outermost branch point (Hanashiro 2015). The actual molar ratios of linear and branched amylose molecules can be usually measured by determining the β-amylase limit dextrin that is labeled at its reducing end before β-amylolysis. The parameter of molar percentage for branched amylose molecules is vital to understand the structural indices of amylose deeply, since these indices are normally measured by using a mixture of linear and branched amyloses. Other parameters, such as average chain length (CL) including molar-based CL and weight-based CL, number of chains (NC), and extent of debranching, are also employed to describe the branched structure of amylose molecules.

Amylose molecules can entangle with hydrophobic molecules to form amylose-hydrophobic complex, for example, amylose-lipid complex exhibiting V-type crystal (Chi et al. 2021), which will limit the complexion of iodine with amylose and under-estimate the value of amylose content measured by the iodine binding method. The complexes can contribute to positive and negative attributes for starch-based foods, such as stickiness reducing, retrogradation minimizing, or lipids oxidation, etc. (Dhital et al. 2019).

1.4 Molecular Structure of Amylopectin

Amylopectin is usually the major component in starch and plays the key role in starch functional properties (Apriyanto et al. 2022). Compared with amylose, the molecular structure of amylopectin is much more complicated. It is a highly branched biopolymer and comprised of relatively large amounts of short linear α-(1,4)-D-glucosyl chains interconnected each other through branches of α-(1,6) linkages, which account up to ~5% of total glycosidic linkages and make the amylopectin molecule highly branched with a molecular weight of ~10^7 to 10^8 Da (Tetlow and Bertoft 2020). Therefore, the size of amylopectin is approximately one or two orders of magnitude higher than amylose. Three size fractions were observed to be present in amylopectin component from various starch sources, number-average DP (DP_n) of the largest size fraction ranges from 13,400 to 26,500, DP_n

ranges between 4400 and 8400 for intermediate size fraction, and between 700 and 2100 for the smallest fractions, respectively (Bertoft 2015).

1.4.1 Chain Length Profile of Amylopectin

One of the most common methods used to characterize the structural parameter of amylopectin is unit chain length profile, which is obtained by debranching amylopectin with enzymes (isoamylase and pullulanase), and analyzed by gel permeation chromatography (GPC), fluorophore-assisted carbohydrate electrophoresis (FACE), and high performance anion exchange chromatography with pulsed amperometric detector (HPAEC-PAD). Both FACE and HPAEC-PAD could give high-resolution profile characterizations. The unit chains in amylopectin can be categorized into three classes: A-chains, B-chains, and C-chain. Among them, the A-chains do not carry any chains and are attached to other chains by α-(1,6) linkages. The B-chains carry at least one other chain; meanwhile, they are attached to other chains (B-chains or C-chain) by α-(1,6) linkages. The C-chain carries other types of chains (A-chains or B-chains) and have free reducing end; therefore, the C-chain is not attached to any chain. However, there is a much more common classification method in the literature, that is, the unit chains are categorized into four fractions (Fig. 1.1), fa

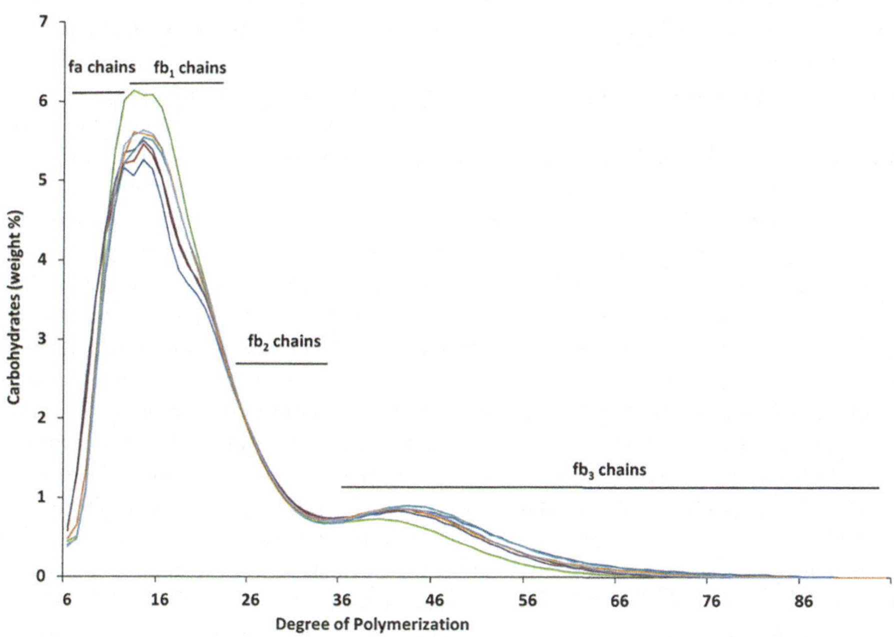

Fig. 1.1 Typical weight-based chain length profile of various rice amylopectins (data adopted from Kong et al. 2015b). (Reused with permission from *ACS Publications*)

(DP 6–12), fb_1 (DP 13–24), fb_2 (DP 25–36), and fb_3 (DP > 36) according to the length of chains (Hanashiro et al. 1996). The ratios of these fractions were observed to be correlated with physicochemical, functional, or digestion properties significantly (Jane et al. 1999; Benmoussa et al. 2007; Kong et al. 2008, 2010, 2015b, 2016; Li et al. 2016; Ni et al. 2022), although it should be noted that these fractions of chains are not real A-chains or B-chains.

1.4.2 Internal Structure of Amylopectin

The amylopectin internal structure attracted more and more attention in recent years, and it was observed that the physicochemical properties of starch were determined by the internal structure of amylopectin to a large extent (Zhu 2018). Bertoft et al. (2016) found that small differences in the internal structure of amylopectin could explain great variations in functional properties of starch, since only tiny differences in the internal chain lengths were observed and the external chain lengths were similar. External chains represent molecular segments from the outermost branches to the non-reducing ends of the chains, while the remaining segments are regarded as internal.

The internal structural parameters of amylopectin can be analyzed after removing the external chains using exo-acting enzymes from non-reducing ends. β-amylase is the most commonly used exo-acting enzyme; it can cleave α-D-(1 → 4) glucosidic linkage from non-reducing ends of amylopectin molecules but cannot bypass the branch points, eventually produces the small molecule—maltose, and leaves a relatively large molecule—β-limit dextrin. The β-limit dextrin can be purified from the hydrolysate mixture of maltose and β-limit dextrin by alcohol precipitation or tangential flow filtration system. All A-chains in the produced β-limit dextrin are cleaved into maltosyl or maltotriosyl residues, which depended on the DP number of original chains (even or odd). Another exo-acting enzyme usually employed to explore the internal structure of amylopectin is phosphorylase *a*, which can remove glucose from non-reducing ends of amylopectin and cleave all A-chains into DP 4 residues; however, all B-chains have external segments of DP 3 and are longer than DP 4 in the final ϕ-limit dextrin product. The ϕ-limit dextrin can be further hydrolyzed by β-amylase, thereafter, every external chain in the ϕ-limit dextrin releases one maltose molecule (Fig. 1.2) and ϕ, β-limit dextrin is produced. In the ϕ, β-limit dextrin, all A-chains are in the form of DP 2 residues. The limit values representing percentages of external segments in amylopectin molecules are normally calculated through the total amounts of small molecules, such as maltose and glucose, released from amylopectin exo-acting enzymatic reactions.

The resulting limit dextrins with different types can be debranched by a combination of isoamylase and pullulanase to calculate the ratios of real A-chains and B-chains. Based on ϕ, β-limit dextrin debranching, Bertoft et al. (2008) further categorized the real B-chains into B_{fp}-chains (fingerprint B-chains, DP 3-7), BS_{major}-chains (DP 8-23), and BL-chains (DP > 23); the authors found that the

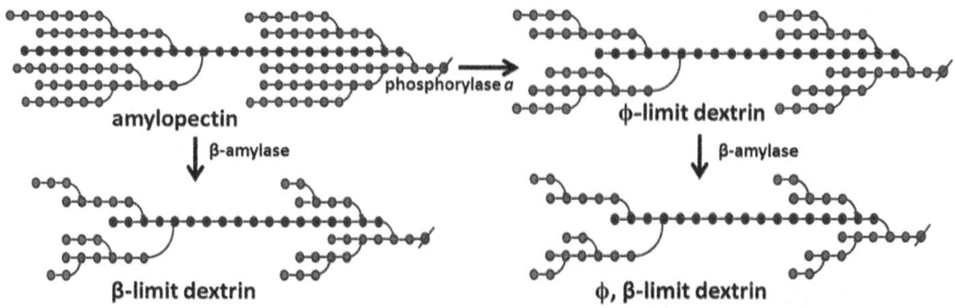

Fig. 1.2 General scheme of β-limit dextrin, ϕ-limit dextrin, and ϕ, β-limit dextrin produced from amylopectin by phosphorylase *a* and β-amylase

B_{fp}-chains were special to a plant source and can be possibly used for starch source identification (Fig. 1.3).

1.4.3 Organization of Chains in Amylopectin

The organization of unit chains in amylopectin molecules has not been fully clarified yet due to the limitations of current techniques; however, it is of great importance for understanding the entire macromolecular structure and predicting how it will behave (Vamadevan and Bertoft 2015; Hamaker 2021; Nakamura and Kainuma 2022). Several structural models were raised to explain the organization of chains in amylopectin, since the amylopectin was identified. In fact, the structural model evolved continuously with discoveries promoted by the application of new analytical techniques in the past 80 years. The most widely accepted structural model is cluster model, which was proposed by Nikuni and French independently (Nikuni 1969; French 1972), and later, quantitatively improved by Hizukuri (1986). The basic concept of the traditional cluster is that a group of short linear chains constitute a densely branched cluster, the clusters are then connected to each other by long B-chains tandemly, and the B-chains are classified into B1, B2, B3, and B4 chains according to the number of clusters they connect, respectively. However, as the new experimental data on amylopectin structure were produced recently, the traditional cluster model was challenged since the actual clusters have never been isolated and characterized (Tetlow and Bertoft 2020; Zhong et al. 2022). Currently, in order to explore the organization of chains in amylopectin, the branched units should be isolated and then analyzed. In the way of exploration, the endo-acting enzyme of *B. amyloliquefaciens* α-amylase was commonly used in conjunction with some exo-acting enzymes, such as β-amylase and phosphorylase *a*, to produce branched limit dextrins with branch points close to each other. These resistant branched limit dextrins were named as building blocks, which were regarded as the elemental structural units in amylopectin molecules for the building block backbone model raised recently (Tetlow and Bertoft 2020). The major difference between the cluster

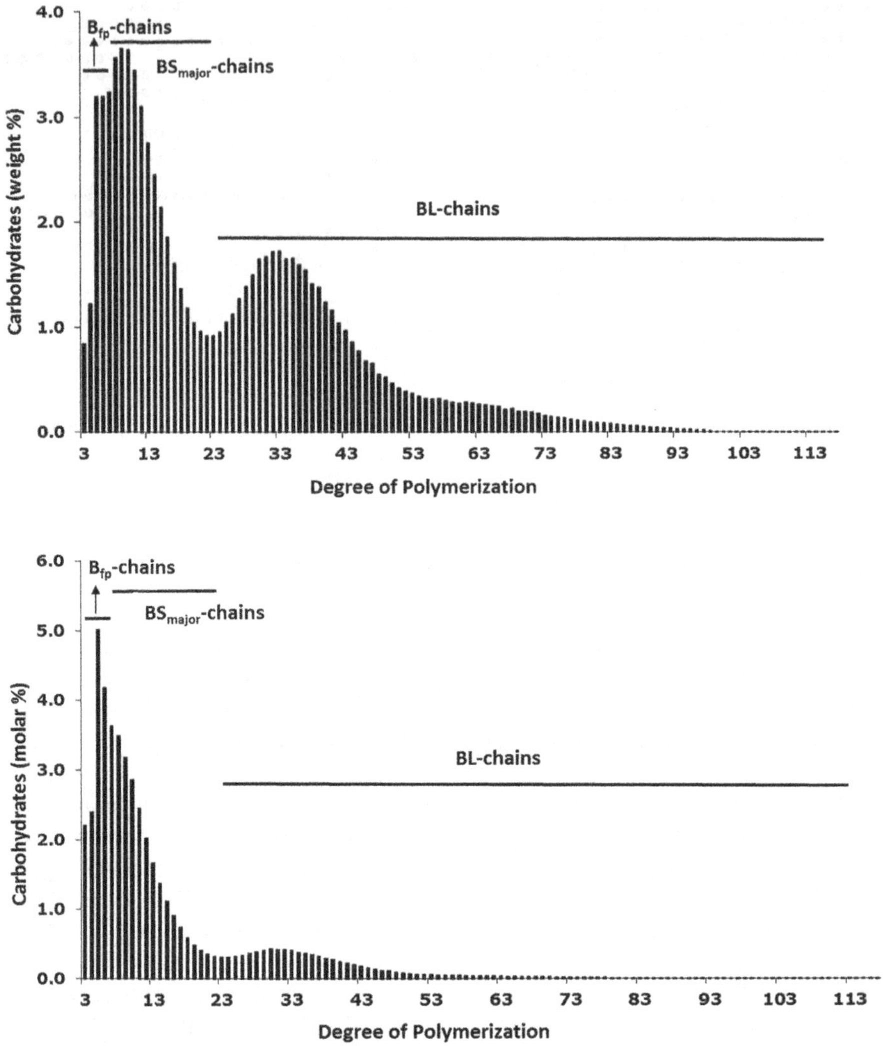

Fig. 1.3 Typical internal chain length profiles (B-chains only) of ϕ,β-limit dextrin produced from *Amaranthus* amylopectin (left, weight-based; right, molar-based)

model and building block backbone model is the orientation of long chains. The long chains have essentially the same orientation as the double helices and penetrate amorphous and crystalline lamella; however, in building block backbone model, the long chains are figured to be mainly in amorphous lamella and perpendicularly oriented to the double helices (Fig. 1.4). These two representative models are still in debate; however, the debate will progress the approaching of the actual fine structure of amylopectin.

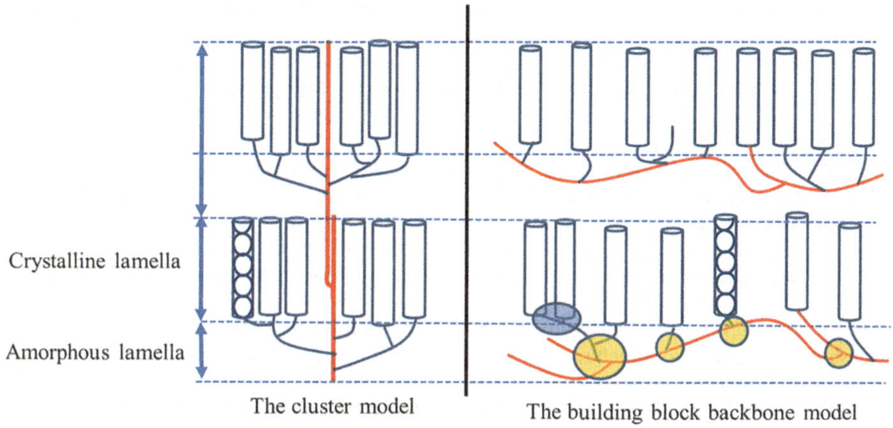

Fig. 1.4 The cluster model and the building block backbone model (red lines represent long B-chains in amylopectin)

1.5 Conclusion Remarks

The functional properties of starch are determined by the molecular structure of starch components, mainly amylose and amylopectin, so advances in molecular structure characterization of these macromolecules will improve our oriented design of starch for applications. Much progress in molecular structure of starch has been made in last decades; however, the inner organization and structure-functionality relationship need further investigation. Moreover, the molecular structure changes in starch modification process should also be given attention to make the oriented design of starch for specific application possible.

References

Apriyanto A, Compart J, Fettke J (2022) A review of starch, a unique biopolymer – structure, metabolism and in planta modifications. Plant Sci 318:111223

Benmoussa M, Moldenhauer KAK, Hamaker BR (2007) Rice amylopectin fine structure variability affects starch digestion properties. J Agric Food Chem 55:1475–1479

Bertoft E (2015) Fine structure of amylopectin. In: Nakamura Y (ed) Starch: metabolism and structure. Springer, Tokyo

Bertoft E (2017) Understanding starch structure: recent progress. Agronomy 7:56

Bertoft E, Piyachomkwan K, Chatakanonda P, Sriroth K (2008) Internal unit chain composition in amylopectins. Carbohydr Polym 74:527–543

Bertoft E, Annor GA, Shen X, Rumpagaporn P, Seetharaman K, Hamaker BR (2016) Small differences in amylopectin fine structure may explain large functional differences of starch. Carbohydr Polym 140:113–121

Chi C, Li X, Zhang Y, Chen L, Li L, Miao S (2021) Progress in tailoring starch intrinsic structures to improve its nutritional value. Food Hydrocoll 113:106447

Debet MR, Gidley MJ (2006) Three classes of starch granule swelling: influence of surface proteins and lipids. Carbohydr Polym 64:452–465

Dhital S, Brennan C, Gidley MJ (2019) Location and interactions of starches in planta: effects on food and nutritional functionality. Trends Food Sci Technol 93:158–166

French D (1972) Fine structure of starch and its relationship to the organization of starch granules. J Jpn Soc Starch Sci 19:8–25

Gous PW, Fox GP (2017) Review: Amylopectin synthesis and hydrolysis – understanding isoamylase and limit dextrinase and their impact on starch structure on barley (Hordeum vulgare) quality. Trends Food Sci Technol 62:23–32

Gunning AP, Giardina TP, Faulds CB, Nathalie J, Ring SG, Williamson G, Morris VJ (2003) Surfactant-mediated solubilisation of amylose and visualisation by atomic force microscopy. Carbohydr Polym 51:177–182

Hamaker BR (2021) Current and future challenges in starch research. Curr Opin Food Sci 40:46–50

Hanashiro I (2015) Fine structure of amylose. In: Nakamura Y (ed) Starch: metabolism and structure. Springer, Tokyo

Hanashiro I, Abe J-I, Hizukuri S (1996) A periodic distribution of the chain length of amylopectin as revealed by high-performance anion-exchange chromatography. Carbohydr Res 283:151–159

Hanashiro I, Sakaguchi I, Yamashita H (2013) Branched structures of rice amylose examined by differential fluorescence detection of side-chain distribution. J Appl Glycosci 60:79–85

Hizukuri S (1986) Polymodal distribution of the chain lengths of amylopectins, and its significances. Carbohydr Res 147:342–347

Jane J, Chen YY, Lee LF, Mcpherson AE, Wong KS, Radosavljevic M, Kasemsuwan T (1999) Effects of amylopectin branch chain length and amylose content on the gelatinization and pasting properties of starch. Cereal Chem 76:629–637

Kong X, Bertoft E, Bao J, Corke H (2008) Molecular structure of amylopectin from amaranth starch and its effect on physicochemical properties. Int J Biol Macromol 43:377–382

Kong X, Corke H, Bertoft E (2009a) Fine structure characterization of amylopectins from grain amaranth starch. Carbohydr Res 344:1701–1708

Kong X, Bao J, Corke H (2009b) Physical properties of Amaranthus starch. Food Chem 113:371–376

Kong X, Kasapis S, Bertoft E, Corke H (2010) Rheological properties of starches from grain amaranth and their relationship to starch structure. Starch 62:302–308

Kong X, Zhu P, Sui Z, Bao J (2015a) Physicochemical properties of starches from diverse rice cultivars varying in apparent amylose content and gelatinisation temperature combinations. Food Chem 172:433–440

Kong X, Chen Y, Zhu P, Sui Z, Corke H, Bao J (2015b) Relationships among genetic, structural, and functional properties of rice starch. J Agric Food Chem 63:6241–6248

Kong X, Kasapis S, Zhu P, Sui Z, Bao J, Corke H (2016) Physicochemical and structural characteristics of starches from Chinese hull-less barley cultivars. Int J Food Sci Technol 51: 509–518

Li H, Prakash S, Nicholson TM, Fitzgerald MA, Gilbert RG (2016) The importance of amylose and amylopectin fine structure for textural properties of cooked rice grains. Food Chem 196:702–711

Ma M, Chen X, Zhou R, Li H, Sui Z, Corke H (2021) Surface microstructure of rice starch is altered by removal of granule-associated proteins. Food Hydrocoll 121:107038

Ma M, Zhu H, Liu Z, Sui Z, Corke H (2022) Removal of starch granule-associated proteins alters the physicochemical properties of diverse small granule starches. Food Hydrocoll 124:107318

Mercier C (1973) The fine structure of corn starches of various amylose-percentage: waxy, normal and amylomaize. Starch 25:78–83

Morrison WR (1995) Starch lipids and how they relate to starch granule structure and functionality. Cereal Foods World 40:437–446

Nakamura Y, Kainuma K (2022) On the cluster structure of amylopectin. Plant Mol Biol 108:291–306

Ni D, Yang F, Lin L, Sun C, Ye X, Wang L, Kong X (2022) Interrelating grain hardness index of wheat with physicochemical and structural properties of starch extracted therefrom. Foods 11: 1087

Nikuni Z (1969) Starch and cooking (in Japanese). Sci Cook 2:6–14

Pérez S, Bertoft E (2010) The molecular structures of starch components and their contribution to the architecture of starch granules: a comprehensive review. Starch 62:389–420

Pfister B, Zeeman SC (2016) Formation of starch in plant cells. Cell Mol Life Sci 73:2781–2807

Seung D (2020) Amylose in starch: towards an understanding of biosynthesis, structure and function. New Phytol 228:1490–1504

Syahariza ZA, Sar S, Hasjim J, Tizzotti MJ, Gilbert RG (2013) The importance of amylose and amylopectin fine structures for starch digestibility in cooked rice grains. Food Chem 136:742–749

Takeda C, Takeda Y, Hizukuri S (1989) Structure of amylomaize amylose. Cereal Chem 66:22–25

Tester RF, Karkalas J, Qi X (2004) Starch—composition, fine structure and architecture. J Cereal Sci 39:151–165

Tetlow IJ, Bertoft E (2020) A review of starch biosynthesis in relation to the building block-backbone model. Int J Mol Sci 21:1–37

Vamadevan V, Bertoft E (2015) Structure-function relationships of starch components. Starch 67: 55–68

Wang K, Hasjim J, Wu AC, Henry RJ, Gilbert RG (2014) Variation in amylose fine structure of starches from different botanical sources. J Agric Food Chem 62:4443–4453

Ye X, Zhang Y, Qiu C, Corke H, Sui Z (2019) Extraction and characterization of starch granule-associated proteins from rice that affect in vitro starch digestibility. Food Chem 276:754–760

Yu WW, Zhai HL, Xia GB, Tao KY, Li C, Yang XQ, Li LH (2020) Starch fine molecular structures as a significant controller of the malting, mashing, and fermentation performance during beer production. Trends Food Sci Technol 105:296–307

Zhong Y, Qu JZ, Liu X, Ding L, Liu Y, Bertoft E, Petersen BL, Hamaker BR, Hebelstrup KH, Blennow A (2022) Different genetic strategies to generate high amylose starch mutants by engineering the starch biosynthetic pathways. Carbohydr Polym 287:119327

Zhu F (2018) Relationships between amylopectin internal molecular structure and physicochemical properties of starch. Trends Food Sci Technol 78:234–242

Chapter 2
Granular Structure of Starch

Mengting Ma, Matteo Bordiga, and Zhongquan Sui

Abstract Granule shape and size are one of the most important morphological differentiating factors between different sources of starch. This chapter summarizes an overview of aspects related to the shape and size of various plant starch granules, as well as the multi-scale structure of starch. The chapter also provides the microstructural features (including surface pores, channels, cavities) and non-starch components (proteins/lipids), thus revealing the distribution of starch granule-associated proteins/lipids and their influence on starch properties. Consequently, the presence of these microstructural features or proteins/lipids may be of potential significance in the enzymatic hydrolysis and chemical reactions of starch by facilitating or preventing direct access of enzymes or chemical reagents to the internal granule matrix.

Keywords Starch granule · Structure · Pores/channels · Starch granule-associated proteins/lipids

2.1 Introduction

In the endosperm cells of plants, native starch mainly exists in the form of granule. Starch is the second abundant natural polymer in nature after cellulose and chitin. The difference is that starch can be hydrolyzed by enzymes in the human digestive system, providing 70–80% of energy and playing a pivotal role in human life. In addition to its nutritional value, starch can also influence the physicochemical properties of food and is widely used in the food processing industry as an additive for adhesives, powder sprays, and preservatives. At the same time, as a renewable,

M. Ma · Z. Sui (✉)
Department of Food Science & Technology, School of Agriculture and Biology, Shanghai Jiao Tong University, Shanghai, China
e-mail: zsui@sjtu.edu.cn

M. Bordiga
Department of Pharmaceutical Sciences, Università degli Studi del Piemonte Orientale, Novara, Italy

Z. Sui, X. Kong (eds.), *Physical Modifications of Starch*,
https://doi.org/10.1007/978-981-99-5390-5_2

degradable, and cheap natural resource, starch is also an important industrial raw material. The application of starch is determined by its properties, and the properties are determined by the structure of the starch. The structure analysis of starch has been widely concerned by researchers. It is well known that the structure of starch granules is organized at different length scales, namely whole granule (μm), growth rings (~0.1 μm), blocklets (20–500 nm), lamellar structure (8–11 nm), and molecular scale (~0.1 nm) (Pérez and Bertoft 2010).

Starches isolated from different plant sources show characteristic granule morphology (shape, size, and surface features). For example, potato starch is spherical or ellipsoid, rice starch and maize starch are irregular polygons (Dhital et al. 2010). Even starches of the same plant origin often have differences in their granule morphology. Some starch granules have special pores/channels or crack structures on the surface. The surface characteristics of starch granules may directly influence the physicochemical, enzymatic, and chemical modification of starch through influencing the interaction between starch and water, chemicals, and enzymes. Maize starch has pores which are connected to channels large enough to allow enzymes to diffuse within the granules, resulting in an "inside-out" hydrolysis pattern (Dhital et al. 2014). The surface of potato starch granules is relatively smooth and dense, so that the enzyme molecules can only slowly hydrolyze starch from the surface to the interior of the granule, showing a hydrolysis pattern of "outside-in" (Dhital et al. 2010). Likewise, the abovementioned differences in starch surface structure can also significantly affect the chemical modification for starch. In addition, it is worth noting that there are some endogenous proteins and lipid molecules in the interior, surface or channels of starch granules (Ma et al. 2022a, b). Although the protein/lipid content is very low, they would have a profound effect on the physicochemical properties of starch.

2.2 Granular Structure

2.2.1 Size and Shape

Different types of starch granules have different shapes. There are mainly three types, namely round, oval, and polygon. As shown in Fig. 2.1a, the granule shape of cereal starch can be flat, plump, oval, rod-shaped, or irregular polygonal shape. Both oat and B-type wheat starches are oval. However, the surface of oats is clean and smooth, while the surface of type B wheat is rough. Amaranth, quinoa, rice, and waxy maize starch granules were polygonal with sharp edges (Yao et al. 2020). For legume starches (Fig. 2.1b), most granules were characterized with smooth surface, round, ellipsoid, or spherical shapes (Ma et al. 2017). Additionally, most of the tuber and root starch granules are oval. Turmeric and poria starch granules were flat triangular shape (Leonel 2007). Some nut starches exhibited unusual granule morphology of half-spheres, although most were oval or spherical in shape with smooth surfaces (Miao et al. 2012).

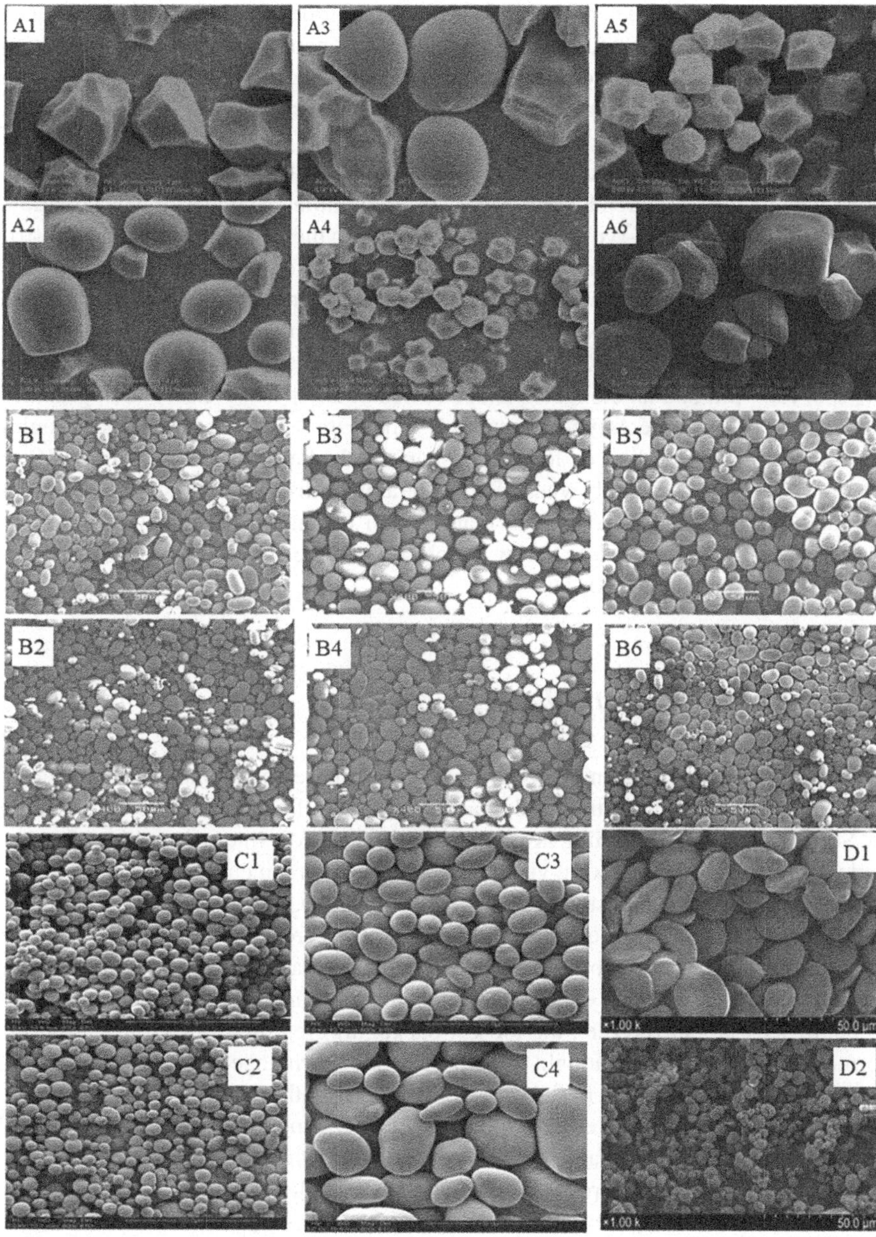

Fig. 2.1 SEM images of cereal starches (**A1**, rice starch; **A2**, wheat B-type starch; **A3**, oat starch; **A4**, quinoa starch; **A5**, amaranth starch; **A6**, waxy maize starch), legume starch (**B1**, black eye bean starch; **B2**, chick pea starch; **B3**, baby lima bean starch; **B4**, small red bean starch; **B5**, lentil starch; **B6**, mung bean starch), potato starch with different granular size (**C**) and wheat A-type (**D1**) and B-type wheat starch (**D2**). Reproduced from Yao et al. (2020), Ma et al. (2017), Chen et al. (2019), and Zhang et al. (2016). (Reused with permission from Elsevier Publications)

Granule size can be determined by various techniques such as scanning electron microscopy (SEM), sieving, and laser scattering. In general, granule size may vary from less than 1 μm to more than 110 μm (Hoover 2001), which was divided into large (diameter >25 μm), medium (10 μm < diameter <25 μm), small (diameter <10 μm), and very small granules (diameter <5 μm) (Yao et al. 2020). The sources of small granular starch include rice, wheat, oats, and millet. At the same time, amaranth and quinoa starch belong to very small granules. Starches of wheat, barley, rye, and triticale (wheat-rye hybrids) are known for their bimodal grain size distribution. Starches of the same plant origin often have differences in their granule size and morphology. As illustrated in Fig. 2.1c, the smaller starch fractions with $D_{(4,3)}$ (volume equivalent mean diameter) of 13.7 and 20.5 μm had a spherical shape, the medium ones with $D_{(4,3)}$ of 25.0 μm had an elliptical shape, and the large ones with $D_{(4,3)}$ of 55.9 μm had an elongated shape (Chen et al. 2019). Wheat starch granules contain two types: large A-type and small B-type (Fig. 2.1d). Wheat A-type granules possess a typical disc-like or lenticular shape with diameters ranging from 10 to 30 μm. On the other hand, wheat B-type granules display both spherical and/or irregular shape with diameters less than 10 μm (Zhang et al. 2016). A-type and B-type granules from hull-less barley starch showed the similar shape and size with wheat starch (Chen et al. 2021).

2.2.2 *Microstructural Characteristics*

Starch granules can have unique microstructural features including pores, channels, and cavities. These microstructural characteristics are large enough to be observed by SEM and confocal scanning laser microscopy (CSLM). Internal cavities of potato and wheat starch granules were visualized and the enlargement of these cavities due to dehydration was demonstrated (Baldwin et al. 1994). Huber and BeMiller (1997, 2000) further elucidated the presence of cavities in cereal starch (maize, sorghum, and millet) and confirmed that cavities are a practical feature of dried starch granules (which may be enlarged by dehydration of starch granules). The hilum is considered to be the least organized (i.e., the loosest packaging) region of the granule, and the formation of the cavity is thought to be the result of shrinkage of granules following water loss (from the inside to the outside). Cavities within granules can directly be connected to the external environment through channels. In addition, equatorial grooves on the granule surfaces could be observed in wheat, barley, hull-less barley, and rye starch. Equatorial grooves were more susceptible to hydrolysis by α-amylase and/or amyloglucosidase than the broad or flat granule surfaces (Li et al. 2012).

Pores were visualized on the granule surface of cereal starch (maize, sorghum, and millet) and on the equatorial groove of wheat, barley, and rye starch granules (Huber and BeMiller 1997, 2000, 2001; Kim and Huber 2008; Fannon et al. 1992). Fannon et al. (1992) verified that the pores were characteristic of native starch and not an artifact produced by enzymatic attack or dehydration of starch granules. Pores were randomly distributed to varying degrees on each granule, and often occurring in clusters on granules (Glaring et al. 2006). Moreover, pores might be openings leading to the granule interior. Subsequently, the channels were identified in maize and sorghum starch granules by Huber and BeMiller (1997), and it was observed that each granule was radially oriented in different numbers, with a diameter of about 0.07–0.10 mm. Therefore, they proposed a model of the starch channel. Starch granules have structural channels that connect the surface of the starch granules with the internal cavity through microtubules. Some of the channels can run through the entire starch granules, and some of the channels are distributed inside the starch granules. Generally, cereal starch granules with A-type crystalline have pore/channel structures, meanwhile, B-type (tuber) starch granules have smooth surfaces without pores/channels (Kim and Huber 2008). Tang et al. (2006) hypothesized that the natural peripheral pore structures on the surface of starch granules were due to the loose aggregation of "defective blocklets." The phosphate attached to the surface of potato starch blocklets increased the degree of binding among blocklets and formed a more compact molecular structure (Tang et al. 2006). However, the latest study found that potato starch also has some extremely small nano-scale pore structures by nitrogen adsorption technology, but no pore or channel structure could be observed (Chen et al. 2019). Although a lot of starch granules possess some channels, the reason for their occurrence and their specific feature is not yet understood.

Additionally, pores or channels would affect the interaction between the starch granules and foreign materials, such as water, chemicals, and enzymes. Observations on phosphorylated potato and sorghum starches showed that the granule reaction pattern was influenced by the presence of pores/channels. In sorghum starch, reagents flow from channels (laterally) and cavities (from the inside out) into the granular matrix. In contrast, the reagent diffused inward through the surface of the outer potato starch granules (Huber and Bemiller 2001). In general, enzymatic hydrolysis of starch granules include enzyme diffusion onto the surface of the granules, enzyme adsorption onto the surface, and hydrolysis of starch molecules. Maize starches were usually rapidly hydrolyzed due to their channels/pores connecting to the center of the granule, so enzymes easily penetrated the granule interior in an "inside-out" pattern, i.e., hydrolyzing starch from inside to outside. In contrast, the hydrolysis process of potato starch was slower, and its pattern was an "outside-in" mechanism because potato starch granules lack channels/pores (Dhital et al. 2010).

2.3 Growth Ring Structure

Growth ring, also known as shell structure, refers to concentric ring-like structure that alternates between soft and hard (alternating semi-crystalline and amorphous layers) around the hilum (Pérez and Bertoft 2010). Some large starch granules can be directly observed under a light microscope (LM) as a growth ring structure. The

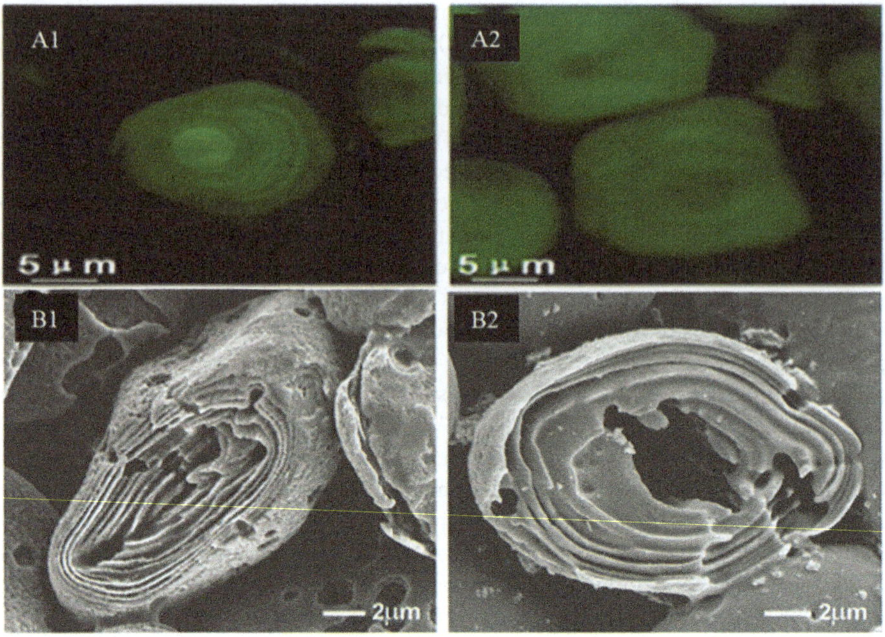

Fig. 2.2 CLSM images of HCL-hydrolyzed maize starches and SEM images of α-amylase-hydrolyzed hull-less barley starches. Reproduced from Chen et al. (2009) and Li et al. (2004). (Reused with permission from Elsevier Publications)

growth rings start at hilum, and their number and size depend on their botanical origin and the amount of sugar biosynthesis. The alternating rotation of day and night on earth is responsible for the unique periodic growth ring of starch granules enriched in plant tissues, as the alternation of day and night results in a strong and weak cycle of photosynthesis in the plant (Whistler and Paschall 2009). In general, the shell structure can be observed under a LM after staining or cytochemical labeling of the granules. The amorphous growth rings of starch disrupted to highlight the semi-crystalline growth rings by etching (e.g., acid, alkaline, or enzymatic methods), and then observed by SEM and transmission electron microscopy (TEM) with a growth ring (shell structure) of 120–400 nm thickness (Whistler and Paschall 2009). Chen et al. (2009) removed the amorphous domains of maize starches through acid hydrolysis, as shown in Fig. 2.2, and showed that maize starches exhibited a distinct growth ring structure. The semi-crystalline growth rings of waxy and normal hull-less barley starches after α-amylase hydrolysis treatments were observed by SEM (Li et al. 2004). In the future, the use of cutting techniques in conjunction with electron microscopy to in situ observe the growth ring structure of starch could be considered.

2.4 Blocklet Structure

The blocklets could be observed in both the surface and inside of starch granules through SEM and atomic force microscopy (AFM), which were spherical or ellipsoidal in size between 20 and 500 nm, between the growth ring and lamellar structures (Jiranuntakul et al. 2013; Ridout et al. 2002). Each blocklet may correspond to an amylopectin molecule or a cluster structure in amylopectin, while amylose may be involved in the composition of blocklets or play a linking role among blocklets (Herrera et al. 2016). The blocklets structure is the bridge between the molecular, lamellar structure and the growth ring and granule structure. Based on previous studies on the blocklets structure, the blocklet structure-shell structure-granule structure model of starch was proposed (Fig. 2.3). Tang et al. (2006) proposed model 1, suggesting that the blocklets forming the hard shell are "normal" and consist of amylopectin molecules, the reduced ends of which may be close to the hilum of starch granule, whereas the blocklets forming the soft shell are "defective" and contain some amylose molecules, which are loosely arranged in the soft shell. The pores on the surface of the starch granule are thought to be composed of "defective" blocklets. A schematic structure of the blocklets is presented based solely on the main components (Fig. 2.3), lacking an assessment of the characteristics of the amylopectin molecules, such as the length of the short amylopectin chains (thickness of the crystalline lamellae), the ratio of long to short amylopectin chains, and the degree of polymerization (Tang et al. 2006).

There are differences in the size and number of blocklet structures in starch from different plant sources. The size of the blocklets is generally larger in diameter in

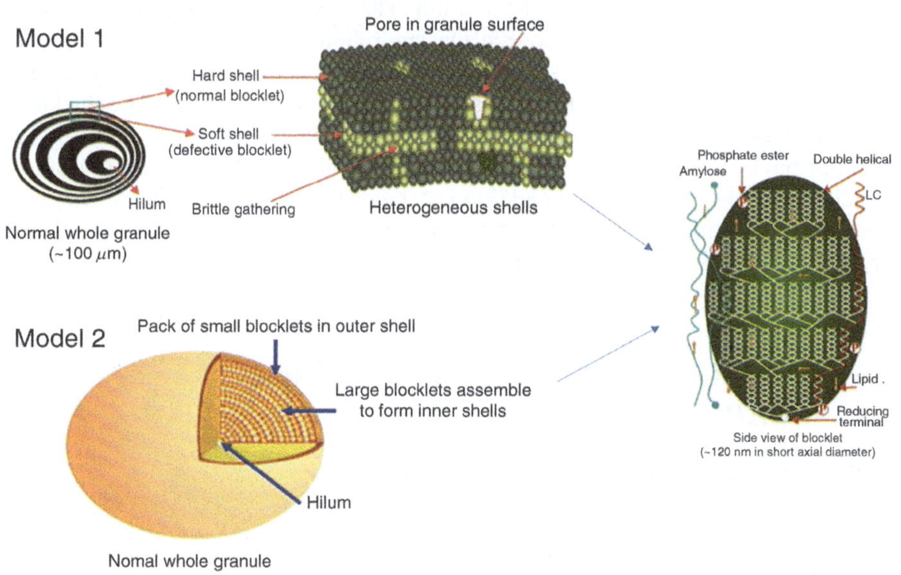

Fig. 2.3 Starch granule model with blocklets structure. Reproduced from Tang et al. (2006) and Huang et al. (2014). (Reused with permission from Elsevier Publications)

B-type and C-type starch than in A-type starch. Observations of the surfaces of potato and wheat starch granules revealed that potato starch had a more pronounced blocklets structure (Pérez and Bertoft 2010). This may be related to the phenomenon that wheat starch granules have higher protein content on the surface than potato starch. The surface blocklets of wheat starch granules become less visible due to the coverage of the surface protein, while the surface blocklets are more visible in potato starch. For wheat starch, blocklets of larger size (100 nm) are mainly in the crystalline region, while those of smaller size (25 nm) are mainly found in the amorphous region. The surface shell of a potato starch granule has a large number of blocklets measuring 400–500 nm, while the interior of the granule consists of mainly small blocklets (Le Corre et al. 2010). Model 2 is also presented in Fig. 2.3, in which the larger blocklets were loosely assembled inside, but smaller blocklets are densely packed in the outer shell of the granule (Huang et al. 2014).

2.5 Lamellar Structure

Amylopectin molecules are highly branched and have a high molecular weight, forming the main backbone of the starch granule (Kossmann and Lloyd 2000). Branched points of amylopectin molecules and some disordered branched amylopectin molecules form an amorphous lamellar structure, possibly interspersed with amylose chains (Morrison 1990). The short-branched chains of amylopectin molecules are stacked to form ordered double helix amylopectin clusters to form crystalline lamellae. The semi-crystalline layer is composed of an amorphous lamella of 2–5 nm and a crystalline lamella of 5–6 nm. The thickness of the semi-crystalline layer (the repetition distance between the amorphous and crystalline lamellae) is approximately 9 nm, and each semi-crystalline growth ring (120–400 nm) consists of approximately 16 alternating repeats of the semi-crystalline layer. The thickness of the semi-crystalline layer also depends on the type and origin of the starch. A-type starches have shorter branched chains, which contribute to the thinner crystalline lamellae than B-type starches. In addition, A-type starch is more susceptible to acid and enzyme treatment, presumably due to the presence of vulnerable structure in the crystalline lamellae (Jane et al. 1997).

2.6 Starch Granule-Associated Proteins/Lipids (SGAPs/SGALs)

2.6.1 The Distribution of SGAPs/SGALs

During growth and development, starch and protein are formed simultaneously in the endosperm or cotyledons. The storage proteins in the grain form as discrete proteinoplast during grain filling and fuse into a continuous protein matrix around

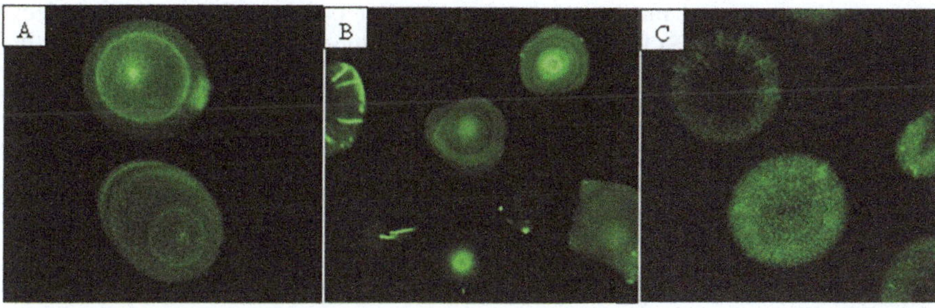

Fig. 2.4 Confocal laser scanning microscopy of potato (**a**), maize (**b**), and wheat (**c**) starch granules with CBQCA stained. Reproduced from Han and Hamaker (2002). (Reused with permission from Elsevier Publications)

the starch granule later in development, so the extraction of starch requires repeated removing of protein to improve the purity of the starch. At later stages of grain filling, the amyloplast has completed its biological function and its peripheral biofilm degrades, with some of the lipids/proteins from the biofilm migrating and attaching to the surface of the starch granule (Baldwin et al. 1997). Most proteins are easily removed by repeated washing of starch in water, but some proteins are bound to the starch granule and cannot be removed. Schofield and Greenwell firstly defined starch granule-associated proteins (SGAPs) in 1987, which are located on the surface or inside the starch granule and cannot be removed by washing. SGAPs can be divided into surface and internal proteins (Schofield and Greenwell 1987). As shown in Fig. 2.4, Han and Hamaker highlighted SGAPs in granule surface from potato starch and granule channels from cereal (maize and wheat) starch and using CBQCA (Han and Hamaker 2002). Thus, SGAPs for cereal starches are divided into three detailed types: surface proteins and channel proteins are proteins embedded on the granule surface or channels; internal proteins refer to proteins located in the granule matrix.

More recent reports revealed the distribution of SGAPs in starch granule. Naguleswaran et al. (2011) stained the starch granules with fluorescamine and revealed that surface proteins were uniformly distributed in the peripheral layer of triticale starch granules, forming a thin coating or film with strong fluorescence intensity. Radially distributed channel proteins extending inward from the granule surface were often observed in the Pronghorn starch, whereas only a few channel proteins were observed in the starch from ultima triticale (Naguleswaran et al. 2011). Han and Hamaker (2002) also observed radially oriented channel proteins in some larger wheat starch granules. The very weak fluorescence intensity of triticale starch granule interior may be due to the limited access of closely packed peripheral regions near the granule surface (Naguleswaran et al. 2011). Channel proteins were observed more frequently in waxy maize starch than in normal maize starch, but normal maize starch exhibited stronger internal fluorescence intensity than waxy (Bae et al. 2020). The fluorescence signal consistent with channel proteins and surface proteins disappeared in rice and maize starch granules after protease, 2% sodium dodecyl sulfate (SDS) or 0.2% SO_2 treatment, while the fluorescence associated with the granule matrix remained (Bae et al. 2020; Ma et al. 2022b; Naguleswaran et al. 2011). Merbromin was mainly confined to the outermost surface of the native starch granule and its channels and cavity while merbromin was able to penetrate more deeply into the granule matrix following protease treatment (Bae et al. 2020). This suggested that the elimination of SGAPs facilitated the penetration of merbromin into the granular matrix.

Surface proteins range in molecular weight (Mw) from 5 to 60 kDa that depend on the Mw of the surrounding storage proteins. Internal proteins are mainly biosynthetic or degradative enzymes involved in starch, ranging in MW from 60 to 150 kDa, and these proteins can be divided into five categories: (a) 140–145 kDa of starch branching enzymes; (b) 100–115 kDa of starch synthases; (c) 87 kDa of starch branching enzymes; (d) 75–77 kDa of starch synthase I; and (e) 57–63 kDa of GBSS (Baldwin 2001). The SGAPs were specific to different plant sources, such as 22 kDa for maize starch, 160 and 98 kDa for potato starch, and 140, 115, 90, and 80 kDa for wheat starch (Yoon et al. 2010). Therefore, the presence of the SGAPs could be used to identify the source of starch in foods such as noodle.

The distribution of starch granule-associated lipids (SGALs) is similar to SGAPs, which includes surface lipids, channel lipids and internal lipids. Compared to cereal starches, tuber (cassava, potato) and legume (pea, kidney bean) starches have lower proportion of total and internal lipid content. Wheat starch contains 0.7% total lipids, of which 0.64% is internal lipids and the remainder is in the form of surface lipids (Vasanthan and Hoover 1992). There is a positive correlation between the amylose and SGALs content in starch. High-amylose starch has higher SGALs content than normal maize starch, meanwhile SGALs are few in waxy maize starch (~0.08%). In addition, smaller starch granules contained more SGALs than larger starch granules (Dhital et al. 2011). This may be due to the larger surface area and volume of the smaller starch granules, allowing more lipids to adhere to the surface. The SGALs content of soft wheat starch is higher than that of durum wheat starch, which is also reported that wheat B-type granule contains more SGALs than wheat A-type granule of the same genotype (Tao et al. 2016). The environment also influences SGALs content of starch granules. Higher environmental temperatures during the maturation of wheat and barley starch granules can result in higher lipids (Tester et al. 1991). SGALs are attached in situ to the surface and channels of the starch granule and are obtained from the surrounding amyloplast and endoplasmic reticulum (Ma et al. 2022a). The surface lipids of the starch granules of wheat are mainly polar lipids, consisting of triacylglycerides, free fatty acids, phosphatidylcholine, and lysophosphatidylcholine. The internal lipids of starch consist of monoacyl lipids, free fatty acids, and lysophospholipids (Rosicka-Kaczmarek et al. 2016). Both surface lipids and internal lipids are remnants of the biological origin of starch, but whether they are enzymatically active and their role in starch synthesis remains unknown.

2.6.2 The Effect of SGAPs/SGALs on Starch Properties

There are various intermolecular forces such as covalent, hydrogen, ionic, and disulfide bonds, as well as molecular entanglement between SGAPs and starch molecules (Dhital et al. 2019). Surface proteins and channel proteins could be removed by alkali, surfactant, or protease treatment from starch (Li et al. 2016; Tester et al. 2008; Sun et al. 2021). SDS may affect the noncovalent bonds between the starch molecules and forms complexes with the amylose/amylopectin, thereby influencing starch properties (Debet and Gidley 2006). Surface proteins and channel proteins have an important effect on the surface chemistry of starch, thereby affecting starch properties (swelling, gelatinization, pasting, gel, retrogradation properties, and enzyme modification). Removal of surface proteins and channel proteins from rice starch could enlarge channel diameter, pore volume, and increase the surface area of granules (Ma et al. 2021). Removal of SGAPs on surface and in channels decreased gelatinization temperature, pasting viscosities of small granule starches while SGAPs removal could lead to stronger gel networks in small granule starches due to increased amylose leaching (Ma et al. 2022b). Previous studies proved that SGAPs removal changed the influence of annealing on starches, and annealing showed different effect on waxy and normal rice or maize starch with SGAPs removal (Xu et al. 2022; Sun et al. 2021). Meanwhile, removal of SGAPs on surface and in channels greatly increased hydrolysis rate of α-amylase as well as amyloglucosidase for rice starches, and these hydrolyzed granules incurred a greater number and size of pores on their surfaces after SGAPs removal (Ma et al. 2020a, b).

The SGALs including surface lipids and channel lipids can be extracted by cold solvents, but the total extraction of SGALs requires the destruction of the granule structure by acid hydrolysis or thermal gelation (Morrison 1988). SGALs have a significant impact on the functional properties of starch due to their ability to form complexes with starch molecules during heating. Amylose molecules are helical in shape, with the hydroxyl groups of glucose residues present on the outer surface of the helix, while the inner space is hydrophobic and the hydrophobic tails of lipids can enter this space to form complexes (Putseys et al. 2010). When determining the apparent amylose content of starch, the complex of lipids with amylose can limit the amylose-iodine complex. It has been suggested that the lipid must be completely removed for an accurate determination of the amylose content using the iodine-binding method (Morrison and Laignelet 1983). Amylose-lipid complexes can inhibit the solubility and mobility of amylose molecules and therefore affect the functional properties of starch. It was reported that the presence of SGALs reduced the viscosity and retrogradation of starchy foods because amylose-lipid complexes reduced amylose leaching. Thus, SGALs could prevent uneven moisture distribution and maintain product quality, e.g., by delaying aging in bread (Vasanthan and Hoover 1992). SGALs also has a profound role in aggregation of small granule starches. The structure of the swollen granules became weaker after SGALs were removed, but removing SGALs result in the higher gel hardness (Ma et al. 2022a). For the SGALs on starch, more physical, chemical, and enzymic modification research is still needed. Moreover, it is still unclear what the interactions between SGAPs and SGALs in starch granule are.

References

Bae JE, Hong JS, Baik MY, Choi HD, Choi HW, Kim HS (2020) Impact of starch granule-associated surface and channel proteins on physicochemical properties of corn and rice starches. Carbohydr Polym 250:116908

Baldwin PM (2001) Starch granule-associated proteins and polypeptides: a review. Starch 53:475–503

Baldwin PM, Adler J, Davies MC, Melia CD (1994) Holes in starch granules: confocal, SEM and light microscopy studies of starch granule structure. Starch 46:341–346

Baldwin PM, Melia CD, Davies MC (1997) The surface chemistry of starch granules studied by time-of-flight secondary ion mass spectrometry. J Cereal Sci 26:329–346

Chen P, Yu L, Simon G, Petinakis E, Dean K, Chen L (2009) Morphologies and microstructures of cornstarches with different amylose–amylopectin ratios studied by confocal laser scanning microscope. J Cereal Sci 50:241–247

Chen L, Ma RR, Zhang ZP, Huang MG, Cai CX, Zhang RJ, McClements DJ, Tian YQ, Jin ZY (2019) Comprehensive investigation and comparison of surface microstructure of fractionated potato starches. Food Hydrocoll 89:11–19

Chen X, Ma M, Liu X, Zhang C, Xu Z, Li H, Sui Z, Corke H (2021) Multi-scale structure of A- and B-type granules of normal and waxy hull-less barley starch. Int J Biol Macromol 200:42–49

Debet MR, Gidley MJ (2006) Three classes of starch granule swelling: influence of surface proteins and lipids. Carbohydr Polym 64:452–465

Dhital S, Shrestha AK, Gidley MJ (2010) Relationship between granule size and in vitro digestibility of maize and potato starches. Carbohydr Polym 82:480–488

Dhital SAK, Shrestha JH, Gidley MJ (2011) Physicochemical and structural properties of maize and potato starches as a function of granule size. J Agric Food Chem 59:10151–10161

Dhital S, Warren FJ, Zhang B, Gidley MJ (2014) Amylase binding to starch granules under hydrolysing and non-hydrolysing conditions. Carbohydr Polym 113:97–107

Dhital S, Brennan C, Gidley MJ (2019) Location and interactions of starches in planta: effects on food and nutritional functionality. Trends Food Sci Technol 93:158–166

Fannon JE, Hauber RJ, BeMiller JN (1992) Surface pores of starch granules. Cereal Chem 69:284–288

Glaring MA, Koch CB, Blennow A (2006) Genotype-specific spatial distribution of starch molecules in the starch granule: a combined CLSM and SEM approach. Biomacromolecules 7:2310–2320

Han XZ, Hamaker BR (2002) Location of starch granule-associated proteins revealed by confocal laser scanning microscopy. J Cereal Sci 35:109–116

Herrera MP, Vasanthan T, Hoover R (2016) Characterization of maize starch nanoparticles prepared by acid hydrolysis. Cereal Chem 93:323–330

Hoover R (2001) Composition, molecular structure, and physicochemical properties of tuber and root starches: a review. Carbohydr Polym 45:253–267

Huang JR, Wei NG, Li HL, Liu SX, Yang DQ (2014) Outer shell, inner blocklets, and granule architecture of potato starch. Carbohydr Polym 103:355–358

Huber KC, BeMiller JN (1997) Visualization of channels and cavities of corn and sorghum starch granules. Cereal Chem 74:537–541

Huber KC, BeMiller JN (2000) Channels of maize and sorghum starch granules. Carbohydr Polym 41:269–276

Huber KC, Bemiller JN (2001) Location of sites of reaction within starch granules. Cereal Chem 78: 173–180

Jane JL, Wong KS, McPherson AE (1997) Branch-structure difference in starches of A- and B-type x-ray patterns revealed by their Naegeli dextrins. Carbohydr Res 300:219–227

Jiranuntakul W, Sugiyama S, Tsukamoto K, Puttanlek C, Rungsardthong V, Puncha-Arnon S, Uttapap D (2013) Nano-structure of heat-moisture treated waxy and normal starches. Carbohydr Polym 97:1–8

Kim HS, Huber KC (2008) Channels within soft wheat starch A- and B-type granules. J Cereal Sci 48:159–172

Kossmann J, Lloyd J (2000) Understanding and influencing starch biochemistry. Crit Rev Plant Sci 19:171–226

Le Corre D, Bras J, Dufresne A (2010) Starch nanoparticles: a review. Biomacromolecules 11: 1139–1153

Leonel M (2007) Analysis of the shape and size of starch grains from different botanical species. Cien Tecnol Aliment 27:579–588

Li JH, Vasanthan T, Hoover R, Rossnagel BG (2004) Starch from hull-less barley: V. In-vitro susceptibility of waxy, normal, and high-amylose starches towards hydrolysis by alpha-amylases and amyloglucosidase. Food Chem 84(4):621–632

Li CY, Cheng L, Lu ZX, Li WH, Cao LP (2012) Morphological changes of starch granules during grain filling and seed germination in wheat. Starch 64:166–170

Li WH, Wu GL, Luo QG, Jiang H, Zheng JM, Ouyang SH, Zhang GQ (2016) Effects of removal of surface proteins on physicochemical and structural properties of A- and B-starch isolated from normal and waxy wheat. J Food Sci Technol 53:2673–2685

Ma MT, Wang YJ, Wang MX, Jane JL, Du SK (2017) Physicochemical properties and in vitro digestibility of legume starches. Food Hydrocoll 63:249–255

Ma M, Xu Y, Liu Z, Sui Z, Corke H (2020a) Removal of starch granule-associated proteins promotes α-amylase hydrolysis of rice starch granule. Food Chem 330:127313

Ma M, Xu Z, Li P, Sui Z, Corke H (2020b) Removal of starch granule-associated proteins affects amyloglucosidase hydrolysis of rice starch granules. Carbohydr Polym 247:116674

Ma MT, Chen XJ, Zhou RZ, Li HT, Sui ZQ, Corke H (2021) Surface microstructure of rice starch is altered by removal of granule-associated proteins. Food Hydrocoll 121:107038

Ma MT, Wen YD, Zhang CC, Xu ZK, Li HT, Sui ZQ, Corke H (2022a) Extraction and characterization of starch granule-associated surface and channel lipids from small-granule starches that affect physicochemical properties. Food Hydrocoll 126:107370

Ma MT, Zhu HX, Liu ZY, Sui ZQ, Corke H (2022b) Removal of starch granule-associated proteins alters the physicochemical properties of diverse small granule starches. Food Hydrocoll 124: 107318

Miao M, Jiang H, Jiang B, Cui SW, Jin Z, Zhang T (2012) Structure and functional properties of starches from Chinese ginkgo (Ginkgo biloba L.) nut. Food Res Int 49:303–310

Morrison WR (1988) Lipids in cereal starches - a review. J Cereal Sci 8:1–15

Morrison WRK (1990) Methods in plant biochemistry. Academic, New York

Morrison WR, Laignelet B (1983) An improved colorimetric procedure for determining apparent and total amylose in cereal and other starches. J Cereal Sci 1:9–20

Naguleswaran S, Li JH, Vasanthan T, Bressler D (2011) Distribution of granule channels, protein, and phospholipid in triticale and corn starches as revealed by confocal laser scanning microscopy. Cereal Chem 88:87–94

Pérez S, Bertoft E (2010) The molecular structures of starch components and their contribution to the architecture of starch granules: a comprehensive review. Starch 62:389–420

Putseys JA, Lamberts L, Delcour JA (2010) Amylose-inclusion complexes: formation, identity and physico-chemical properties. J Cereal Sci 51:238–247

Ridout MJ, Gunning AP, Parker ML, Wilson RH, Morris VJ (2002) Using AFM to image the internal structure of starch granules. Carbohydr Polym 50:123–132

Rosicka-Kaczmarek J, Makowski B, Nebesny E et al (2016) Composition and thermodynamic properties of starches from facultative wheat varieties. Food Hydrocoll 54:66–76

Schofield JD, Greenwell P (1987) Wheat starch granule proteins and their technological significance. Elsevier, Amsterdam

Sun L, Xu Z, Song L, Ma M, Zhang C, Chen X, Xu X, Sui Z, Corke H (2021) Removal of starch granule associated proteins alters the physicochemical properties of annealed rice starches. Int J Biol Macromol 185:412–418

Tang HJ, Mitsunaga TH, Kawamura Y (2006) Molecular arrangement in blocklets and starch granule architecture. Carbohydr Polym 63:555–560

Tao H, Wang P, Wu F, Jin Z, Xu X (2016) Particle size distribution of wheat starch granules in relation to baking properties of frozen dough. Carbohydr Polym 137:147–153

Tester RF, South JB, Morrison WR, Ellis RP (1991) The effects of ambient temperature during the grain-filling period on the composition and properties of starch from four barley genotypes. J Cereal Sci 13:113–127

Tester RF, Yousuf R, Karkalas J, Kettlitz B, Roper H (2008) Properties of protease-treated maize starches. Food Chem 109:257–263

Vasanthan T, Hoover R (1992) Effect of defatting on starch structure and physicochemical properties. Food Chem 45:337–347

Whistler RL, Paschall EF (2009) Starch: chemistry and technology. Elsevier, Amsterdam

Xu Z, Song L, Ming S, Zhang C, Li Z, Wu Y, Sui Z, Corke H (2022) Removal of starch granule associated proteins affects annealing of normal and waxy maize starches. Food Hydrocoll 131: 107695

Yao TM, Wen YD, Xu ZK, Ma MT, Li P, Brennan C, Sui ZQ, Corke H (2020) Octenylsuccinylation differentially modifies the physicochemical properties and digestibility of small granule starches. Int J Biol Macromol 144:705–714

Yoon JW, Jung JY, Chung HJ, Kim MR, Kim CW, Lim ST (2010) Identification of botanical origin of starches by SDS-PAGE analysis of starch granule-associated proteins. J Cereal Sci 52: 321–326

Zhang Y, Guo Q, Feng N, Wang JR, Wang SJ, He ZH (2016) Characterization of A- and B-type starch granules in Chinese wheat cultivars. J Integr Agric 15:2203–2214

Chapter 3
Physicochemical Properties of Starch

Binjia Zhang, Yabin Guo, Zihang Cheng, and Dongling Qiao

Abstract Starch is a major biopolymer in many flour-based foods consumed worldwide and a crucial material in various industrial areas. The starches from different botanical origins often show differences in the composition and structure, and thus in the physicochemical properties (e.g., thermal, rheological, solubility, swelling, hydrolysis, and degradation features), which stands at the core of the design and production of starch-based products. This chapter summarizes the physicochemical properties of starch and related influencing factors, providing a foundational information for other chapters.

Keywords Thermal · Rheological · Hydrolysis

3.1 Introduction

Starch, typically a naturally occurring carbohydrate in many plants (e.g., cereals, tubers, roots, legumes, and some immature fruits like mangos or bananas) (Alcázar-Alay and Meireles 2015), is a major food component providing energy for the human body and has been used in various industrial areas (Cai et al. 2021). Starch is composed of two D-glucan polymers, including amylose, a mostly linear 1,4-α-D-glucan, and amylopectin, mainly 1,4-α-D-glucan having 1,6-α linkages at branching points (Zhang et al. 2013; Pratiwi et al. 2018). These two kinds of biopolymers can form multi-scale structures inside the starch granule mainly through hydrogen bonding, involving the granule (<1–100 μm), alternating semicrystalline and amorphous shells (100–400 nm), amorphous-crystalline lamellae (9–10 nm), and macromolecular chains (~nm) (Oates 1997; Xie et al. 2012).

B. Zhang (✉) · D. Qiao
Chongqing Key Laboratory of Speciality Food Co-Built by Sichuan and Chongqing, College of Food Science, Southwest University, Chongqing, China

Y. Guo · Z. Cheng
Group for Cereals and Oils Processing, College of Food Science and Technology, Huazhong Agricultural University, Wuhan, China

Z. Sui, X. Kong (eds.), *Physical Modifications of Starch*,
https://doi.org/10.1007/978-981-99-5390-5_3

The composition (e.g., amylose content) and structure of starch have important influences on its physicochemical properties such as thermal (gelatinization and pasting) and rheological (e.g., steady state, dynamic, and textural) features, which are closely related to the processing performance of starch-based products. For instance, potato and tapioca starches, with low gelatinization temperature, a quick swelling rate and a high viscosity, can endow noodles with increased elasticity and chewiness (Obadi and Xu 2021). Starch hydrogels with different amylose contents and crystalline types display different mechanical performances, and a higher amylose content can produce a higher hardness value (Cui et al. 2022). Apart from the abovementioned physicochemical properties, other properties such as starch digestibility also show close links to the starch structure (Zhu 2018). To widen the application of starch in food and industry, it is essential to understand the physicochemical features of starch. This chapter summarizes the current knowledge on the starch physicochemical properties (including thermal, rheological, and other properties), which provides fundamental information for the rational usage of starch resources.

3.2 Thermal Properties

Starch usually needs to be processed before it is turned into usable forms. Among them, the thermal processing (e.g., steaming, cooking, and extrusion) for aqueous starch systems has been widely practiced. While heated in water, the semi-crystalline architecture of starch could be gradually disrupted, resulting in disordered starch matrices (sol/gel forms) (Fig. 3.1a). This process is well known as "gelatinization", which is an irreversible process that involves starch granule swelling, crystallite melting, and molecule solubilization (Sullivan and Johnson 1964).

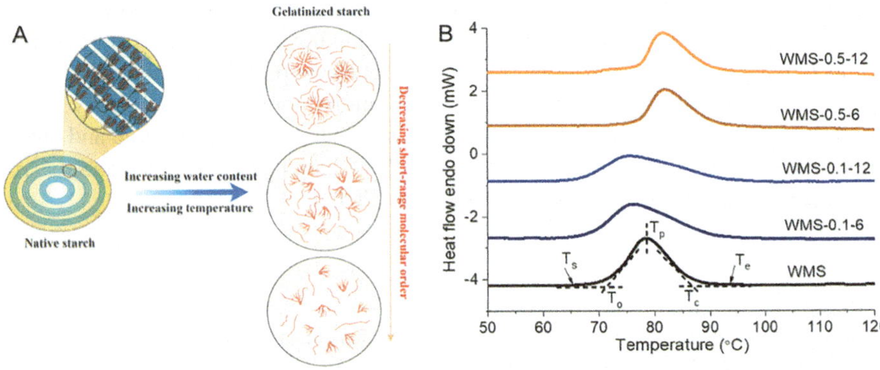

Fig. 3.1 Thermal transition of starch during hydrothermal treatment (**a**), and DSC thermograms of waxy maize starch (WMS) and its alkali-treated samples (**b**). (**a**) is reprinted from Huang et al. (2021) with permission from Elsevier, Copyright 2021; (**b**) is adapted from Qiao et al. (2017) with permission from Elsevier, Copyright 2017

The thermal properties of starch are affected by a series of factors such as starch source, moisture content, additives, heating time, and temperature. Understanding the gelatinization and pasting properties of starch can help us in better regulate the thermal processing of starch-based products and eventually their quality attributes (Wang et al. 2021).

3.2.1 Gelatinization

The gelatinization of starch can be divided into three stages (Copeland et al. 2009): (1) at the reversible water absorption stage, the amorphous part of starch granules absorbs a small amount of water and produces a limited expansion; (2) at the irreversible water absorption stage, a large amount of water enters the interior of the starch granules, the volume of the starch granules expands significantly, some amylose overflows on the surface, and the viscosity of the system increases rapidly; (3) at the final granule disintegration stage, the starch granules break and lose their original shape, the viscosity begins to decrease, and the starch molecules overflow and disperse in the solution. The gelatinization of starches plays a key role in their food and non-food applications (Genkina et al. 2014). The gelatinization process can be inspected using the differential scanning calorimeter (DSC). Several investigations also characterized the degree of starch gelatinization using optical microscopy, synchrotron X-ray scattering, and nuclear magnetic resonance (Gonera and Cornillon 2002; Carlstedt et al. 2015).

The DSC enables us to acquire the gelatinization temperatures (starting (T_s), onset (T_o), peak (T_p), conclusion (T_c), and end (T_e)), enthalpy (ΔH), and gelatinization temperature range (R) of starch (Fig. 3.1b) (Qiao et al. 2017). In general, starch gelatinization occurs at a temperature range of 60–80 °C (Liu et al. 2019), which depends on the composition, starch molecular weight and structure, moisture content, and the addition of substances (Schirmer et al. 2015). The starch granule size also affects the gelatinization behaviors (Puncha-arnon et al. 2008). Amylopectin with a high degree of polymerization (DP) could exhibit high crystallinity, thus providing a more stabilized starch structure and limiting the hydration of amorphous regions; this could enhance the resistance of the granules to gelatinization (Shevkani et al. 2016). The gelatinization temperature normally increases as the amylose content increases (Genkina et al. 2014); T_o and T_p are positively correlated with the relative length of amylopectin short chains, and T_c is related to the relative length of the medium chains of amylopectin (Li and Gong 2020).

The starch gelatinization parameters can be changed by various treatments. For instance, annealing treatment at 50 °C increased the starch gelatinization temperature by stabilizing the starch crystals and increasing the amylose-amylopectin interactions, but had little effect on the enthalpy (Wang et al. 2017a, b). The pressure treatment disrupted the starch structure and promoted water migration during gelatinization, decreasing gelatinization enthalpy but not changing it linearly (Liu et al. 2009).

The reduction of water content for starch can allow increases in T_o, T_p, and T_c and a decrease in ΔH. As indicated by previous findings, T_p increased and then decreased with the increase of NaCl concentration (Li et al. 2014), since the presence of a small amount of NaCl could protect the hydrogen bonding between starch molecules and delay the destruction of starch crystallization; the addition of protein and protein fibrils enhanced the gelatinization temperature and reduced the enthalpy (Chen et al. 2020; Wang et al. 2020); the lipids on the surface of starch affect starch gelatinization, and the removal of lipids could decrease the crystallinity, gelatinization temperature, and enthalpy (Li et al. 2016); sugars and sugar alcohols could form intermolecular interactions with starch in the amorphous regions, suppressing starch gelatinization (Allan et al. 2018).

3.2.2 Pasting Properties

The pasting properties represent changes in the viscosity of starch during pasting. The pasting process is usually studied using an amylograph, Rapid Visco Analyzer (RVA), a Brabender Amylograph, an Ottawa Starch Viscometer, or a dynamic Rheometer. These instruments are used to record the viscosity changes of the starch pastes/suspensions under shear with temperature. Among these, RVA has been used extensively to study the pasting properties of starches and related foods, due to its fast testing speed and the small sample size required (Sneh et al. 2020). A typical pasting profile with common parameters measured by RVA is included in Fig. 3.2a. The temperature at which the viscosity starts to increase indicates the pasting temperature (PT). The peak time or maximum viscosity time is the heating time at which the maximum paste viscosity occurs. The maximum viscosity reached during heating is defined as the peak viscosity (PV). The breakdown viscosity (BD) refers to the difference between the peak viscosity and the minimum viscosity. The difference between the final viscosity and the minimum viscosity represents the setback viscosity (SB), closely related to the retrogradation of gelatinized starch. The final viscosity (FV) corresponds to the viscosity value at the end of a cooling stage (Balet et al. 2019).

The pasting properties of thermally processed starches or related food matrices are determined by several factors, such as the moisture content, the starch source, additives, and the temperature and time of thermal treatment (Xia et al. 2021). For instance, the paste viscosity of wheat and rice flours apparently depends on the water content, attributed to the gradual disruption of starch granules followed by interactions between starch, protein, and lipid (Guo et al. 2018). The relationship between rice starch structure and pasting properties after citric acid treatment could be understood in Fig. 3.2b (Huo et al. 2018). The effect of thermal processing on the paste viscosity of starchy foods is also associated with the starch source. The paste viscosities (PV, FV, TV, SB, and BD) of cooked japonica flour or indica hybrid flour were lower than those of its raw flour, but waxy flour showed an opposite phenomenon (Wang et al. 2016). The paste viscosities of waxy rice flour decreased

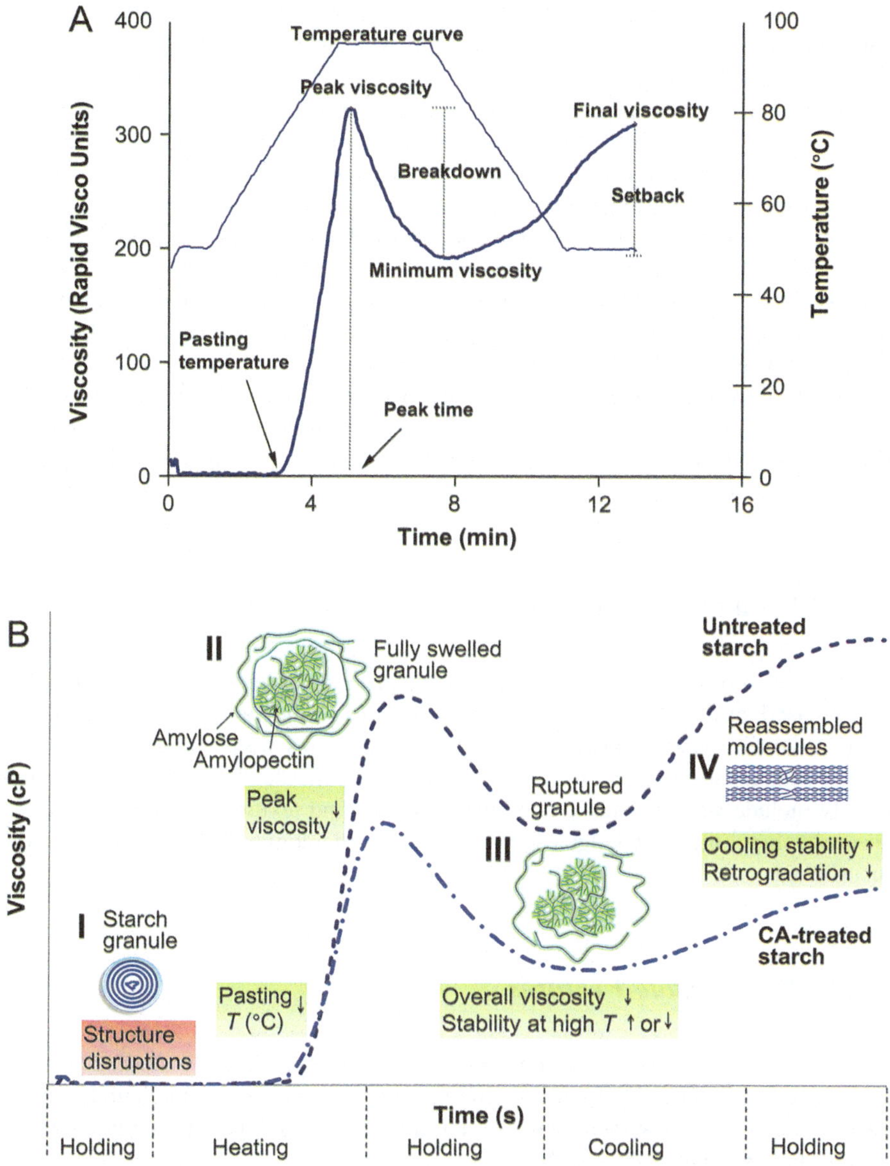

Fig. 3.2 Pasting profile of rice starch measured by RVA (**a**), and effect of citric acid treatment on rice starch pasting (**b**). (**a**) is reprinted with permission of the Royal Society of Chemistry (Copeland et al. 2009), permission conveyed through Copyright Clearance Center, Inc; (**b**) is adapted from Huo et al. (2018) with permission from Elsevier, Copyright 2018

significantly as the oven-heating temperature increased from 140 to 180 °C, attributed to the degradation of starch molecules at high temperatures (a decrease in the capacity of the starch to swell) (Xla et al. 2020). The peak viscosity and breakdown viscosity of the steamed noodles were significantly lower than those of the non-steamed noodles, and this phenomenon was more obvious as the steaming time increased (Luo et al. 2015).

Physical treatments also affect starch gelatinization, such as high hydrostatic pressure treatment (HHP) and heat-moisture treatment (HMT) modification. Research has shown that HHP affected the pasting behavior of various types of starch differently. This treatment could induce a transformation of the crystalline type of corn starch, and alter the pasting temperature and the paste viscosity (Katopo et al. 2002). The effect on rice starch pasting was not significant at low-pressure values (120 and 480 MPa), and the peak viscosity decreased significantly at 600 MPa (Li et al. 2012). Investigations on the pasting properties of HMT-treated starches showed that HMT increased the pasting temperature and thermal stability of the starch but decreased the peak viscosity and breakdown (Gunaratne and Corke 2007). The pasting temperature for normal and waxy maize starches increased as the HMT moisture level increased, but remained constant with the increase of HMT time (Ai et al. 2015).

3.3 Rheological Properties

Rheological properties play vital roles in the processing and quality regulation of starch-based products. The rheological properties could be used to predict and explain the flow and deformation of starch systems, as well as the conditions of texture changes for different starch-based foods. The rheological properties of starch mainly include steady state, dynamic rheological, and texture features. Relative to shear rheology, the knowledge on the extensional rheological behaviors of starch is insufficient.

3.3.1 Steady-State Features

The rheological features of starch under steady state are usually studied using a rotational rheometer to discuss the variations of flow behaviors of starch under shear, and the methods mainly include rotational flow shear and dynamic oscillation shear (Evans and Haisman 1980) (Table 3.1).

The flow behavior describes the rotationally shearing features of starch under the geometry, and the shear rate, shear stress, and apparent viscosity can be recorded. Flow shear mode is often conducted under steady shear state. During flow shear, the viscosity of starch paste decreases when shear rate increases (a shear-thinning feature), indicating that starch paste belongs to a pseudoplastic fluid, as the network of entangled starch molecules is disrupted during shear process (Ahuja et al. 2020). Usually, the viscosity of starch paste decreases in a wide range of sections of increasing shear rate but keeps almost constant when the shear rate is very low or

Table 3.1 Rheological modes and parameters of starch

Modes		Parameters	Models	Parameters via fitting
Flow shear	Steady flow shear	γ, σ, η	Ostwald-de Waele model	K, n
			Herschel-Bulkley model	σ_0, K, n
			Carreau model	η_0
	Transient flow ramp	γ, σ	Ostwald-de Waele model	S
Oscillation shear	Frequency sweep	ω, G', G''		
	Temperature sweep	T, G', G''		

Note: γ shear rate, σ shear stress, η apparent viscosity, ω angular frequency, T temperature, G' storage modulus, G'' loss modulus, K consistency coefficient, n flow index, σ_0 yield stress, η_0 zero shear viscosity, S area of hysteresis hoop

very high, since the deformation and recombination of network structure keep a dynamic equilibrium at very low or high shear rate.

To better understand the starch flow behaviors, the Ostwald-de Waele (Power law) model (Eq. (3.1)) and the Herschel-Bulkley model (Eq. (3.2)) are widely used in data analyses (Silva and Lucas 2018):

$$\sigma = K \times \gamma^n \tag{3.1}$$

$$\sigma = \sigma_0 + K \times \gamma^n \tag{3.2}$$

In these equations, σ is shear stress (Pa), σ_0 is yield stress (Pa), γ is shear rate (s^{-1}), K is consistency coefficient (Pa s^{-1}), and n is flow index. The coefficient K reflects the average viscosity of a non-Newtonian fluid (e.g., starch paste), and the flow index n indicates the deviation degree of the fluid from Newtonian flow. While the pseudoplastic fluid (e.g., starch paste) shows a shear-thinning behavior ($n < 1$), the dilatant fluid (e.g., starch suspension) has a shear-thickening behavior ($n > 1$) (Pang et al. 2020). For a starch paste, the n value ranges from 0 and 1, and the lower the n value the trend of shear-thickening is stronger. The K and n values are commonly determined by the solid content of starch paste, the amylose content, and the temperature of the environment. The addition of exogenous additives (e.g., hydrocolloids) also influences the flow behavior of starch pastes. In addition, the yield stress σ_0 is defined as the minimum shear stress required to initiate flow (Genovese and Rao 2003), and it is usually achieved via increasing solid content or adding polysaccharides into starch pastes.

Previous investigations also discussed the zero shear viscosity and infinite shear viscosity of starch pastes, as deduced from models such as the Carreau model:

$$\eta = \eta_\infty + (\eta_0 - \eta_\infty)\left[1 + (\lambda\gamma)^2\right]^{\frac{n-1}{2}} \tag{3.3}$$

In which, η is apparent viscosity (Pa s), η_0 is zero shear viscosity (Pa s), η_∞ is infinite shear viscosity (Pa s), λ is relaxation time (s), γ is shear rate (s^{-1}), and n is power-law index. Among these parameters, the zero shear viscosity η_0 is used to represent the linear region of Newtonian compliance in which the units flow past one another for the rapture of bonds, and this value normally increases as the average molar mass of unlinked flowing chains increases.

Thixotropy is another important rheological parameter of starch paste as it simultaneously has time-dependent and shear-thinning characteristics. The thixotropy is usually measured via a two-stage flow ramp test, including a stage of increasing shear rate and the other stage of decreasing shear rate through transient continuous variation. The area of hysteresis hoop (S) of up curve and down curve in the shear stress-shear rate graph reflects the thixotropic degree of starch systems (pastes), and the larger the area the stronger thixotropic behavior starch possesses (Liu et al. 2021). In fact, the thixotropy reflects the deformation and recombination of starch chains and network structures, significantly depending on factors such as the botanical source, starch concentration, amylose/amylopectin ratio, exogenous additives, and temperature (Ma et al. 2019; Zhang et al. 2019). Generally, increasing the concentration of starch pastes enhances the thixotropy.

3.3.2 Dynamic Rheological Features

The dynamic rheological analysis is used to determine the viscoelasticity of materials (Rao and Cooley 1992). The dynamic viscoelastic nature of starch involves the mechanical response law of starch pastes or gels under the action of periodic stress (strain). Such dynamic behavior is frequency-dependent and temperature-dependent. The dynamic viscoelasticity test can be conducted by fixing two of control variables (e.g., oscillation frequency, oscillation amplitude, test temperature, and test time) and changing the third one. The basic test modes include strain sweep, frequency sweep, temperature sweep, and event sweep. During a sweep mode, the related parameter is continuously changed in selected steps. Using dynamic measurements, we can obtain several key rheological parameters of starch matrices such as storage modulus G' (Pa) and loss modulus G'' (Pa), loss tangent ($\tan \delta = G''/G'$), and complex viscosity (η^*) (Kamweru 2020).

G' indicates the ability of starch material to recover during a deformation period, representing its elastic behavior. G'' shows the amount of energy dissipated during a deformation period, which is used to represent its viscous behavior (Tabilo-Munizaga and Barbosa-Cánovas 2005). The $\tan \delta$ indicates the relative effect of elasticity or viscosity in the system viscoelastic behavior (Gunasekaran and Ak 2000). The intersection of the storage modulus and the loss modulus curves generally is used to show the gel point of starch ($G' = G''$ and tan $\delta = 1$). The starch system has occurred at this point from viscosity to elasticity. The frequency and modulus corresponding to the intersection reflect the structure information inside starch matrices (Whistler et al. 2012). When the complex viscosity (η^*) is high (e.g., 39.8–97.5 Pa s), the starch gel is suitable as a good thickener or stabilizer (Yousefi and Razavi 2015).

The dynamic rheological properties of starch pastes and gels are affected by different factors, including starch particle size, concentration, gelatinization conditions (e.g., temperature and heating rate), and storage conditions (e.g., temperature and time) (Punia et al. 2020). In particular, when the starch was heated in water, the viscoelastic change of the starch system normally involves several stages (Fonseca-Florido et al. 2018). For instance, (1) at an early stage of heating, the G' and tan δ of the starch slurry are small, and amylose molecules leach from starch granules; (2) at or above the gelatinization temperature, the amylose matrix interacts with the low molecular weight amylopectin to strengthen the continuous network, resulting in a sharp increase in G' and G''. The tan δ value decreases at the same time (Wu et al. 2016).

3.3.3 Texture Features

The texture is an important quality aspect of starch-based products, since it is closely related to the practical performance (e.g., sensory and other attributes). There are some basic testing modes, including compression, puncture, shear, bending, and tensile analyses, in addition to texture profile analysis (TPA).

TPA and puncture tests are important for investigating starch texture. TPA simulates the chewing process by compressing the sample (mainly gel form) two times (Yu et al. 2020). Several texture parameters of starch matrices could be obtained using TPA, including hardness, springiness, gumminess, chewiness, cohesiveness, and adhesiveness (Wu et al. 2017). Among these parameters, (1) hardness is positively correlated to starch retrogradation and springiness shows a opposite trend (Yang et al. 2021); (2) cohesiveness is a parameter related to elasticity and intramolecular interactions, which determines the internal resistance of the starch under compression (Nishinari et al. 2019); (3) adhesiveness has been reported to correlate with sensory of many starch-based matrices such as rice (Nishinari et al. 2019); (4) gumminess (product of hardness and cohesiveness) is the energy required to disintegrate a gel; (5) chewiness (product of hardness, cohesiveness, and springiness) is the energy required to masticate a starch matrix (Mehboob et al. 2015). In addition, the strength, breaking stress, brittleness, and breaking distortion rate of starch could be obtained from puncture test and compression test.

These texture properties can be varied by multiple factors, such as amylose content, molecular weight, and additives (Ong and Blanshard 1995; Tao et al. 2019; Zhao et al. 2021). Tao et al. (2019) reported a positive correlation between the amylose content and the hardness of cooked rice. Starches of different sources (such as rice, quinoa, corn, and potato) could show significant differences in the texture features (Adawiyah et al. 2013; Wu et al. 2017). Adawiyah et al. (2013) confirmed that the gel from arenga starch was more rigid than that of sago starch at high concentrations, despite having a similar amylose content. The same results were seen for rice starch due to the molecular weight of amylopectin (Peng et al. 2021). The additives (e.g., sugars and tea derivatives) could affect the retrogradation of starch matrix and thus its hardness, springiness, and cohesiveness (Zhang et al. 2015, 2017).

3.4 Solubility and Swelling Power

Solubility and swelling power are key indicators to present the interaction between starch and water molecules. Solubility is the percentage of dissolved starch molecules at a certain temperature, corresponding to the degree of dissolution of starch molecular chains. Swelling power is the amount of water absorbed by per unit of starch, when it is heated in excess water at a certain temperature. This hydration (water-holding) property of starch is generally expressed as swelling power (*SP*) (Eq. (3.4)) and solubility (S_o) (Eq. (3.5)) in water at a certain temperature:

$$SP\ (\mathrm{g/g\ dry\ starch}) = m_1/(m_0 - m_2) \tag{3.4}$$

$$S_o\ (\mathrm{g/100\ g\ dry\ starch}) = (m_2/m_0) \times 100 \tag{3.5}$$

Here, m_0 means the weight of starch in dry basis, m_1 is the weight of swollen starch, and m_2 is the weight of dissolved starch.

The *SP* and S_o values of starch are greatly influenced by the temperature. More specifically, starch granules are highly robust and impermeable to water at ambient temperature, which prevents starch-water interaction to absorb water and increase viscosity. In contrast, when heated in excess water above the melting point of starch crystallites, gelatinization occurs to sharply increase the solubility and swelling power. Other influencing factors include the amylose and amylopectin features (e.g., ratio, molecular weight, and chain length), the presence of other components (lipids and phosphates), and the granule structure. Also, processing (e.g., shearing and cooking) and storage can change starch solubility and swelling power (Majzoobi and Farahnaky 2021; Li et al. 2022). For instance, it was shown that the swelling power of wheat starch was correlated negatively with amylose content. An analysis of the amylopectin structure showed that starch with a higher swelling power tended to contain a higher proportion of long chains with a degree of polymerization $\geq$35 (Ge et al. 2022).

3.5 Hydrolysis and Degradation

Starch may be hydrolyzed and degraded under numerous conditions. This section presents the knowledge on the hydrolysis (by enzyme, acid, or alcohol acid) of starch and its degradations during processing (plasma, dry heat, heat-moisture, or thermomechanical treatment).

3.5.1 *Enzyme Hydrolysis*

In recent years, the enzyme hydrolysis (in vitro digestion) of starch has been widely investigated. The hydrolysis features of starch can be quantitatively studied based on the first-order kinetic functions (Eqs. (3.6) and (3.7)) and the associated logarithm of the slope (LOS) plot (Fig. 3.3) (Li et al. 2020):

$$C_t = C_\infty \left(1 - e^{-kt}\right) \tag{3.6}$$

$$\ln \frac{dC_t}{dt} = -kt + \ln(C_\infty \times k) \tag{3.7}$$

In which, C_t (%) represents the amount of starch hydrolyzed at a time point t (min); C_∞ (%) represents the estimated percentage of starch hydrolyzed at the end of a hydrolysis (digestion) period; k (min^{-1}) stands for the rate coefficient of the hydrolysis (digestion). Such hydrolysis (digestion) parameters of starch are often varied by influencing factors. For instance, the enzyme hydrolysis of starch granules tends to be largely affected by the polymorph type; the A-type starch can show a higher enzyme susceptibility than that of B-type starch, and an intermediate susceptibility between A- and B-type starches could be seen for C-type starch (Zhu 2018). Additionally, the presence of lipids and proteins could mitigate the enzyme hydrolysis (digestion) of starch associated with the complexation between starch chains and lipid/protein molecules (Zhang et al. 2022).

The enzymes commonly used for starch hydrolysis include α-amylase, glucoamylase, and pullulanase. The presence of ingredients like lipid, protein, and polyphenol usually has a negative influence on starch hydrolysis (Yu et al. 2018; Zhang et al. 2022). The protein especially the water-insoluble protein could retard the hydrolysis (digestion) of starch by α-amylase due to protein-enzyme binding events (Yu et al. 2018). Zhang et al. (2022) studied the effects of non-starch components and starch molecular structure on the α-amylase hydrolysis of starch; they found that the differences in the structure of amylose did not show significant effects on the hydrolysis, and a higher molecular density of corn amylopectin resulted in a lower hydrolysis degree, as compared to the counterpart in cassava and potato starch. In addition, the removal of protein had a significant promotion effect on starch hydrolysis but lipid removal only showed a slight effect (Zhang et al. 2022). However, the traditional hydrolysis of starch by enzymes may have

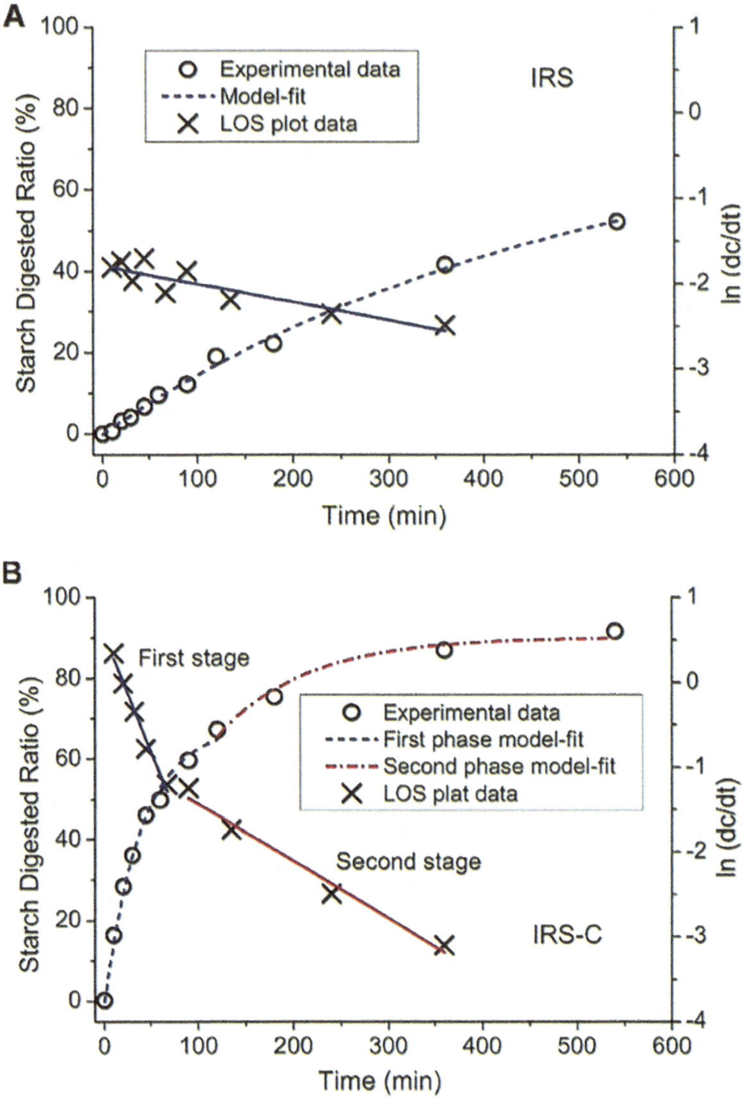

Fig. 3.3 Enzyme hydrolysis (in vitro digestion) data and related LOS plots of native indica rice starch (IRS) (**a**) and its cooked counterpart (**b**). Adapted from Li et al. (2020) with permission from Elsevier, Copyright 2020

disadvantages, such as high cost and slow kinetics; researchers tend to combine other techniques (such as ultrasound) with enzyme hydrolysis to make the starch hydrolysis cost-effective and efficient (Wang et al. 2017a, b). Activation of amylase with ultrasonication and Ca^{2+} ions could significantly increase the enzyme

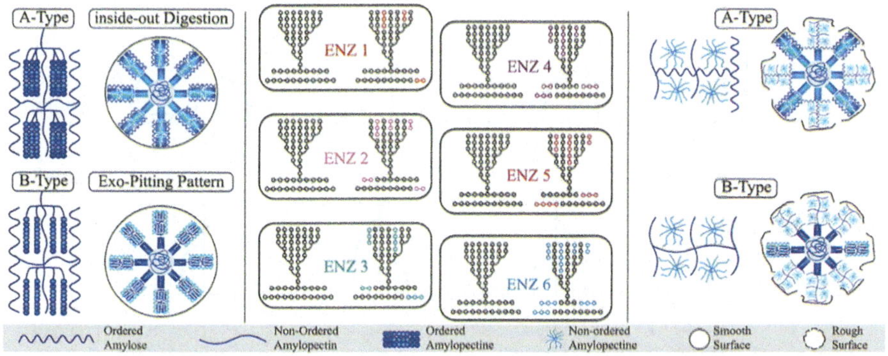

Fig. 3.4 Schematic representation of how activated amylase affects the enzyme hydrolysis of starch. Reprinted from Abedi et al. (2022) with permission from Elsevier, Copyright 2022

hydrolysis rate toward starch suffered from freezing-thawing pre-treatment (Fig. 3.4) (Abedi et al. 2022).

3.5.2 Acid Hydrolysis

To overcome the shortages of native starch (e.g., poor solubility and low shear resistance), acid hydrolysis has been widely practiced to improve the performance of starch and thus widen its industrial applications. The acid-treated starch shows changes in the features of gelatinization, retrogradation, and digestion (enzyme hydrolysis) (Li and Hu 2021). The acid treatment is conducted by exposing starch to mineral acids such as HCl, H_2SO_4, HNO_3, and H_3PO_4 at low temperature below the gelatinization point, and the degree of hydrolysis varies as a function of treatment time (Pratiwi et al. 2018). The acid hydrolysis of starch preferably starts from amorphous growth rings, followed by slow hydrolysis of semi-crystalline growth rings (Li and Hu 2021). Size-exclusion chromatography (SEC) can be used to characterize the chain lengths and molecule sizes of starch suffering from acid hydrolysis.

To inspect the hydrolysis patterns of amylose and amylopectin molecules, a first-order reaction model (Eq. (3.8)) and a linear model (Eq. (3.9)) are used to describe the acid hydrolysis data and the multiple steps during hydrolysis (Pinto et al. 2021):

$$C_p = C_0\left(1 - e^{-kt}\right) \tag{3.8}$$

$$\ln \frac{C_0}{C_0 - C_p} = kt \tag{3.9}$$

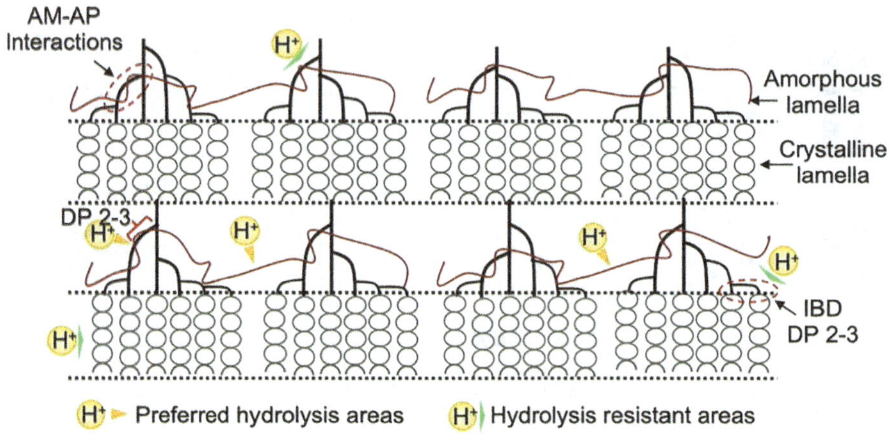

Fig. 3.5 Schematic representation for the hydrolysis pattern of hydrochloric acid on the amylose and amylopectin molecules in starch semi-crystalline lamellae. Reprinted from Li and Hu (2021) with permission from Elsevier, Copyright 2021

In these equations, C_p reflects the starch hydrolyzed at a time point; C_0 represents the initial percentage of starch hydrolysates; t and k stand for the reaction time (d) and the first-order rate constant (d^{-1}), respectively. Using the HCl hydrolysis pattern of amylose and amylopectin molecules as an example (Fig. 3.5), H^+ tends to hydrolyze the amylose long chains with DP > ~300 and the 2 or 3 glucose units away from the branching points of amylopectin long intra-cluster branches (Li and Hu 2021). The acid type, concentration, and hydrolysis time could affect the structure and functional properties of starch (Chen et al. 2017). The hydrolysis process exhibits two stages, including a first rapid stage (ascribed to the hydrolysis of amorphous layers of starch granules) and a second slower stage (related to the degradation of crystallites in starch granules) (Genkina et al. 2009). The acid hydrolysis pattern of starch is largely affected by amylose content; an exo-corrosion pattern for high amylose starch (G80) and an endo-corrosion pattern for waxy starch could be seen (Chen et al. 2017).

3.5.3 Processing-Induced Degradation

Starch may undergo degradation during processing, which plays a crucial role in determining the product processibility and quality. The starch chains have been found to be degraded by specific treatments such as extrusion, plasma treatment, and heat-moisture treatment.

It was found that starch underwent significant degradation during typical thermomechanical processing (extrusion) (Liu et al. 2010). While being extruded, the presence of high shear stress and temperature causes gelatinization and degradation of starch, while the rigid amylopectin crystallites are more susceptible to shear degradation than the flexible amorphous amylose (Liu et al. 2010). From a previous finding, the molecule degradation of starch during extrusion possessed the following features: (1) the size distribution of starch molecules gradually narrowed down as degradation proceeded; (2) the degradation was rapid at the beginning of extrusion and then slowed down with the size of the starch molecules decreased; (3) the sizes of amylopectin molecules (mainly larger than amylose molecules in size) became smaller and thus presented an apparent overlapping region with amylose molecules (Fig. 3.6a) (Liu et al. 2010). The molecule degradation of starch during extrusion predominantly occurred at the center of the chains, rather than the case that the outer chains were cleaved from the parent chain (Fig. 3.6b). For the same size, amylose was more stable than amylopectin with a given shear environment (Liu et al. 2010).

Plasma (the fourth state of matter) contains a variety of active substances (e.g., ions, electrons, and free radicals) (Wu et al. 2022), and has attracted great attention in the modification of starch. Plasma treatment (cold plasma) could alter the multi-scale structures (e.g., molecular structure, morphology, and crystallinity) of starch, and its properties (e.g., gelatinization and rheological features) (Gao et al. 2021; Okyere et al. 2022). In particular, the active species of plasma could not only endow starch with various chemical groups such as carboxyl, carbonyl, and peroxide groups (Gao et al. 2021), but also cause an evident degradation of starch chains (Okyere et al. 2022). Accompanying the starch degradation, plasma treatment could allow an increase in the proportion of linear chain fragments (Zhu 2015; Gao et al. 2019).

Heat-moisture treatment (HMT) is a typical modification method for starch, which is normally conducted at limited water levels (typically $<35\%$ w/w) and controlled temperature (below gelatinization point) and for 15 min to 16 h (Pratiwi et al. 2018). HMT could result in degradation of starch molecules, and the treatment effectiveness largely related to the starch features (e.g., botanical source, structure, and amylose content) and processing parameters (such as water content, temperature, and heating time) (Fonseca et al. 2021). In addition, dry heat treatment is a green method for starch physical modification and is conducted by heating starch at high temperatures (>100 °C) and low moisture content ($<10\%$) for several hours, which can degrade starch chains into smaller parts and increased the flexibility of starch branches (Chi et al. 2019).

3.6 Summary

To rationally use starch resources, great efforts have been made to understand their physicochemical properties, mainly including gelatinization, pasting, solubility and swelling power, as well as hydrolysis and degradation characteristics. Multiple techniques (methods) (e.g., DSC, RVA, XRD, and rheometer) and theoretical models (e.g., Herschel-Bulkley model and first-order kinetic function) have been applied to inspect the starch properties mentioned above. Such physicochemical

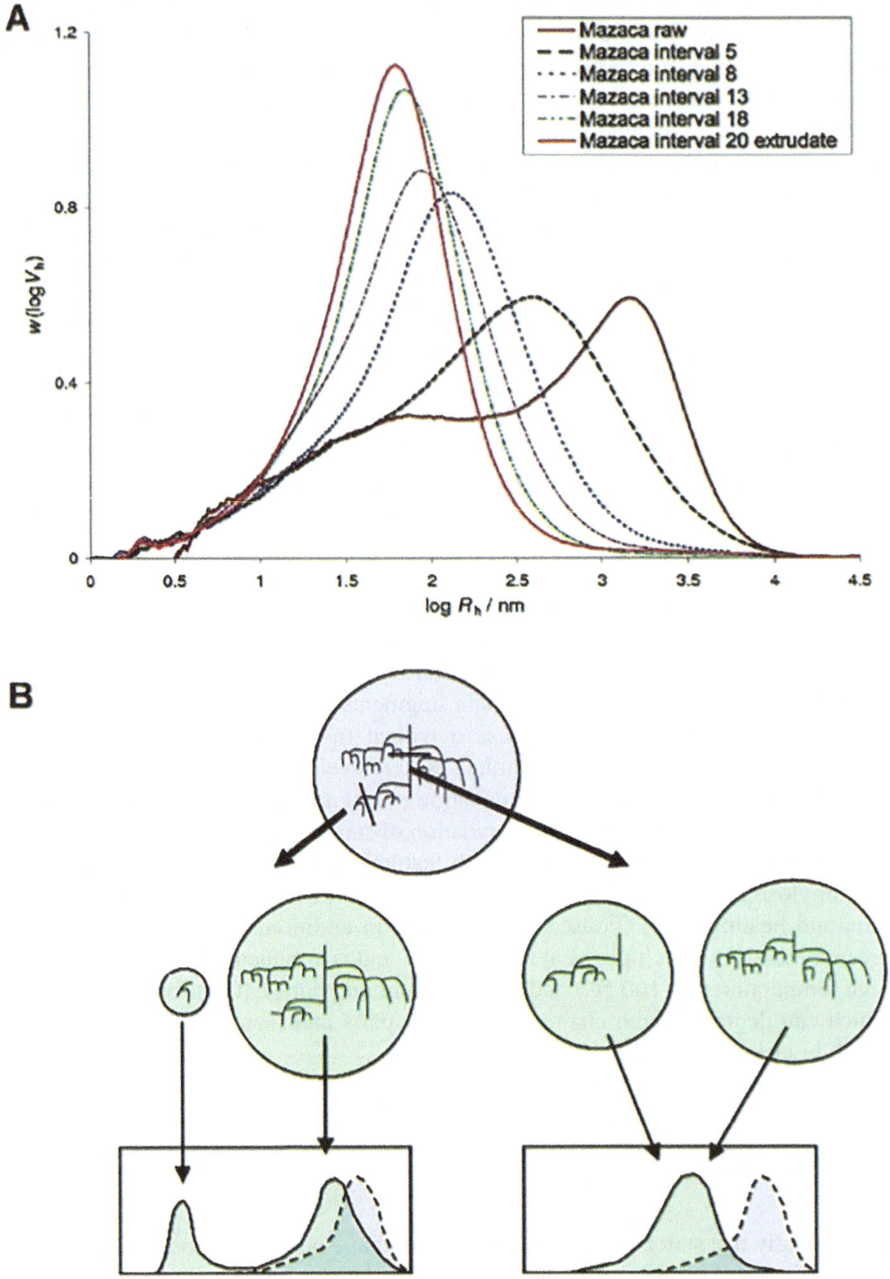

Fig. 3.6 Evolution of the size distribution of fully branched Mazaca starch that collected along the barrel of the extruder (**a**), and the different size distributions caused by the extrusion of intact starch molecules with different bond breaking positions (**b**). Reprinted with permission from Liu et al. (2010). Copyright (2010) American Chemical Society

features have close links to starch composition (e.g., amylose content) and structure (e.g., crystalline structure and granule structure), processing conditions (e.g., water content, temperature, and shear stress), and additives (e.g., alkalis, hydrocolloids, and lipids). Despite these advances, it is still highly necessary to disclose more correlations among processing conditions, starch-component interaction, starch-based matrix structure, and the resulting physicochemical properties.

References

Abedi E, Sayadi M, Pourmohammadi K (2022) Effect of freezing-thawing pre-treatment on enzymatic modification of corn and potato starch treated with activated α-amylase: investigation of functional properties. Food Hydrocoll 129:107676

Adawiyah DR, Sasaki T, Kohyama K (2013) Characterization of arenga starch in comparison with sago starch. Carbohydr Polym 92(2):2306–2313

Ahuja A, Lee R, Latshaw A, Foster P (2020) Rheology of starch dispersions at high temperatures. J Texture Stud 51(4):575–584

Ai L, Yao T, Sui Z, Zhao Y, Kong, Xiangli (2015) Effects of heat-moisture treatment reaction conditions on the physicochemical and structural properties of maize starch: moisture and length of heating. Food Chem 173:1125–1132

Alcázar-Alay SC, Meireles MAA (2015) Physicochemical properties, modifications and applications of starches from different botanical sources. Food Sci Technol 35:215–236

Allan MC, Rajwa B, Mauer LJ (2018) Effects of sugars and sugar alcohols on the gelatinization temperature of wheat starch. Food Hydrocoll 84:593–607

Balet S, Guelpa A, Fox G, Manley M (2019) Rapid visco analyser (RVA) as a tool for measuring starch-related physiochemical properties in cereals: a review. Food Anal Methods 12(10): 2344–2360

Cai T, Sun H, Qiao J, Zhu L, Zhang F, Zhang J, Tang Z, Wei X, Yang J, Yuan Q, Wang W, Yang X, Chu H, Wang Q, You C, Ma H, Sun Y, Li Y, Li C, Jiang H, Wang Q, Ma Y (2021) Cell-free chemoenzymatic starch synthesis from carbon dioxide. Science 373(6562):1523–1527

Carlstedt J, Wojtasz J, Fyhr P, Kocherbitov V (2015) Understanding starch gelatinization: the phase diagram approach. Carbohydr Polym 129:62–69

Chen D, Fang F, Federici E, Campanella O, Jones OG (2020) Rheology, microstructure and phase behavior of potato starch-protein fibril mixed gel. Carbohydr Polym 239:116247

Chen P, Xie F, Zhao L, Qiao Q, Liu X (2017) Effect of acid hydrolysis on the multi-scale structure change of starch with different amylose content. Food Hydrocoll 69:359–368

Chi C, Li X, Lu P, Miao S, Zhang Y, Chen L (2019) Dry heating and annealing treatment synergistically modulate starch structure and digestibility. Int J Biol Macromol 137:554–561

Copeland L, Blazek J, Salman H, Tang MC (2009) Form and functionality of starch. Food Hydrocoll 23(6):1527–1534

Cui C, Jia Y, Sun Q, Yu M, Ji N, Dai L, Wang Y, Qin Y, Xiong L, Sun Q (2022) Recent advances in the preparation, characterization, and food application of starch-based hydrogels. Carbohydr Polym 291:119624

Evans ID, Haisman DR (1980) Rheology of gelatinised starch suspensions. J Texture Stud 10(4): 347–370

Fonseca-Florido HA, Gómez-Aldapa CA, López-Echevarría G, Velazquez G, Morales-Sánchez E, Castro-Rosas J, Méndez-Montealvo G (2018) Effect of granular disorganization and the water content on the rheological properties of amaranth and achira starch blends. LWT 87:280–286

Fonseca LM, Halal SLME, Dias ARG, Zavareze EDR (2021) Physical modification of starch by heat-moisture treatment and annealing and their applications: a review. Carbohydr Polym 274: 118665

Gao S, Liu H, Sun L, Cao J, Yang J, Lu M, Wang M (2021) Rheological, thermal and in vitro digestibility properties on complex of plasma modified Tartary buckwheat starches with quercetin. Food Hydrocoll 110:106209

Gao S, Liu H, Sun L, Liu N, Wang J, Huang Y, Wang F, Cao J, Fan R, Zhang X, Wang M (2019) The effects of dielectric barrier discharge plasma on physicochemical and digestion properties of starch. Int J Biol Macromol 138:819–830

Ge X, Shen H, Sun X, Liang W, Zhang X, Sun Z, Lu Y, Li W (2022) Insight into the improving effect on multi-scale structure, physicochemical and rheology properties of granular cold water soluble rice starch by dielectric barrier discharge cold plasma processing. Food Hydrocoll 130: 107732

Genkina NK, Kiseleva VI, Noda T (2009) Comparative investigation on acid hydrolysis of sweet potato starches with different amylopectin chain-length. Starch 61(6):321–325

Genkina NK, Kozlov SS, Martirosyan VV, Kiseleva VI (2014) Thermal behavior of maize starches with different amylose/amylopectin ratio studied by DSC analysis. Starch 66(7-8):700–706

Genovese DB, Rao MA (2003) Vane yield stress of starch dispersions. J Food Sci 68(7):2295–2301

Gonera A, Cornillon P (2002) Gelatinization of starch/gum/sugar systems studied by using DSC, NMR, and CSLM. Starch 54(11):508–516

Gunaratne A, Corke H (2007) Effect of hydroxypropylation and alkaline treatment in hydroxypropylation on some structural and physicochemical properties of heat-moisture treated wheat, potato and waxy maize starches. Carbohydr Polym 68(2):305–313

Gunasekaran S, Ak MM (2000) Dynamic oscillatory shear testing of foods—selected applications. Trends Food Sci Technol 11(3):115–127

Guo P, Yu J, Copeland L, Wang S, Wang S (2018) Mechanisms of starch gelatinization during heating of wheat flour and its effect on invitro starch digestibility. Food Hydrocoll 82:370–378

Huang S, Chao C, Yu J, Copeland L, Wang S (2021) New insight into starch retrogradation: the effect of short-range molecular order in gelatinized starch. Food Hydrocoll 120:106921

Huo Y, Zhang B, Niu M, Jia C, Zhao S, Huang Q, Du H (2018) An insight into the multi-scale structures and pasting behaviors of starch following citric acid treatment. Int J Biol Macromol 116:793–800

Kamweru P (2020) Food rheology using dynamic mechanical analysis. A short review. IJRP, Dhaka

Katopo H, Song Y, Jane JL (2002) Effect and mechanism of ultrahigh hydrostatic pressure on the structure and properties of starches. Carbohydr Polym 47(3):233–244

Li C, Gong B (2020) Insights into chain-length distributions of amylopectin and amylose molecules on the gelatinization property of rice starches. Int J Biol Macromol 155:721–729

Li C, Hu Y (2021) Effects of acid hydrolysis on the evolution of starch fine molecular structures and gelatinization properties. Food Chem 353:129449

Li N, Wang L, Zhao S, Qiao D, Jia C, Niu M, Lin Q, Zhang B (2020) An insight into starch slowly digestible features enhanced by microwave treatment. Food Hydrocoll 103:105690

Li Q, Xie Q, Yu S, Gao Q (2014) Application of digital image analysis method to study the gelatinization process of starch/sodium chloride solution systems. Food Hydrocoll 35:392–402

Li W, Bai Y, Mousaa S, Zhang Q, Shen Q (2012) Effect of high hydrostatic pressure on physicochemical and structural properties of rice starch. Food Bioprocess Technol 5(6): 2233–2241

Li W, Gao J, Wu G, Zheng J, Ouyang S, Luo Q, Zhang G (2016) Physicochemical and structural properties of A- and B-starch isolated from normal and waxy wheat: effects of lipids removal. Food Hydrocoll 60:364-373-373

Li Y, Qi Y, Li H, Chen Z, Xu B (2022) Improving the cold water swelling properties of oat starch by subcritical ethanol-water treatment. Int J Biol Macromol 194:594–601

Liu H, Yu L, Dean K, Simon G, Petinakis E, Chen L (2009) Starch gelatinization under pressure studied by high pressure DSC. Carbohydr Polym 75(3):395–400

Liu W-C, Halley PJ, Gilbert RG (2010) Mechanism of degradation of starch, a highly branched polymer, during extrusion. Macromolecules 43(6):2855–2864

Liu Y, Chen X, Xu Y, Xu Z, Li H, Sui Z, Corke H (2021) Gel texture and rheological properties of normal amylose and waxy potato starch blends with rice starches differing in amylose content. Int J Food Sci Technol 56(4):1946–1958

Liu Y, Yu J, Copeland L, Wang S, Wang S (2019) Gelatinization behavior of starch: reflecting beyond the endotherm measured by differential scanning calorimetry. Food Chem 284:53-59-59

Luo LJ, Guo XN, Zhu KX (2015) Effect of steaming on the quality characteristics of frozen cooked noodles. LWT Food Sci Technol 62(2):1134–1140

Ma S, Zhu P, Wang M (2019) Effects of konjac glucomannan on pasting and rheological properties of corn starch. Food Hydrocoll 89:234–240

Majzoobi M, Farahnaky A (2021) Granular cold-water swelling starch; properties, preparation and applications, a review. Food Hydrocoll 111:106393

Mehboob S, Ali TM, Alam F, Hasnain A (2015) Dual modification of native white sorghum (Sorghum bicolor) starch via acid hydrolysis and succinylation. LWT Food Sci Technol 64(1):459–467

Nishinari K, Fang Y, Rosenthal A (2019) Human oral processing and texture profile analysis parameters: bridging the gap between the sensory evaluation and the instrumental measurements. J Texture Stud 50(5):369–380

Oates CG (1997) Towards an understanding of starch granule structure and hydrolysis. Trends Food Sci Technol 8(11):375–382

Obadi M, Xu B (2021) Review on the physicochemical properties, modifications, and applications of starches and its common modified forms used in noodle products. Food Hydrocoll 112: 106286

Okyere AY, Rajendran S, Annor GA (2022) Cold plasma technologies: their effect on starch properties and industrial scale-up for starch modification. Curr Res Food Sci 5:451–463

Ong MH, Blanshard JMV (1995) Texture determinants in cooked, parboiled rice. I: rice starch amylose and the fine structure of amylopectin. J Cereal Sci 21(3):251–260

Pang B, Wang S, Chen W, Hassan M, Lu H (2020) Effects of flow behavior index and consistency coefficient on hydrodynamics of power-law fluids and particles in fluidized beds. Powder Technol 366:249–260

Peng Y, Mao B, Zhang C, Shao Y, Wu T, Hu L, Hu Y, Tang L, Li Y, Tang W, Xiao Y, Zhao B (2021) Influence of physicochemical properties and starch fine structure on the eating quality of hybrid rice with similar apparent amylose content. Food Chem 353:129461

Pinto VZ, Moomand K, Deon VG, Biduski B, Zavareze EDR, Lenhani GC, Fidelis dos Santos GH, Lim L-T, Dias ARG (2021) Effect of physical pretreatments on the hydrolysis kinetic, structural, and thermal properties of Pinhão starch nanocrystals. Starch 73(7-8):2000008

Pratiwi M, Faridah DN, Lioe HN (2018) Structural changes to starch after acid hydrolysis, debranching, autoclaving-cooling cycles, and heat moisture treatment (HMT): a review. Starch 70(1-2):1700028

Puncha-arnon S, Pathipanawat W, Puttanlek C, Rungsardthong V, Uttapap D (2008) Effects of relative granule size and gelatinization temperature on paste and gel properties of starch blends. Food Res Int 41(5):552–561

Punia S, Sandhu KS, Dhull SB, Siroha AK, Purewal SS, Kaur M, Kidwai MK (2020) Oat starch: physico-chemical, morphological, rheological characteristics and its applications - A review. Int J Biol Macromol 154:493–498

Qiao D, Xie F, Zhang B, Zou W, Zhao S, Niu M, Lv R, Cheng Q, Jiang F, Zhu J (2017) A further understanding of the multi-scale supramolecular structure and digestion rate of waxy starch. Food Hydrocoll 65:24–34

Rao MA, Cooley HJ (1992) Rheological behavior of tomato pastes in steady and dynamic shear. J Texture Stud 23(4):415–425

Schirmer M, Jekle M, Becker T (2015) Starch gelatinization and its complexity for analysis. Starch 67(1-2):30–41

Shevkani K, Singh N, Bajaj R, Kaur A (2016) Wheat starch production, structure, functionality and applications-a review. Int J Food Sci Technol 52(1):38–58

Silva ÍGM, Lucas EF (2018) Rheological properties of xanthan gum, hydroxypropyl starch, cashew gum and their binary mixtures in aqueous solutions. Macromol Symp 380(1):1800070
Sneh P et al (2020) Oat starch: physico-chemical, morphological, rheological characteristics and its applications - a review. Int J Biol Macromol 154:493–498
Sullivan JW, Johnson JA (1964) Measurement of starch gelatinization by enzyme susceptibility. Cereal Chem 41(2):73–77
Tabilo-Munizaga G, Barbosa-Cánovas GV (2005) Rheology for the food industry. J Food Eng 67(1):147–156
Tao K, Yu W, Prakash S, Gilbert RG (2019) High-amylose rice: Starch molecular structural features controlling cooked rice texture and preference. Carbohydr Polym 219:251–260
Wang D, Ma X, Yan L, Chantapakul T, Wang W, Ding T, Ye X, Liu D (2017a) Ultrasound assisted enzymatic hydrolysis of starch catalyzed by glucoamylase: Investigation on starch properties and degradation kinetics. Carbohydr Polym 175:47–54
Wang L, Zhang L, Wang H, Ai L, Xiong W (2020) Insight into protein-starch ratio on the gelatinization and retrogradation characteristics of reconstituted rice flour. Int J Biol Macromol 146:524–529
Wang S, Li P, Zhang T, Wang S, Copeland L (2016) Trypsin and chymotrypsin are necessary for in vitro enzymatic digestion of rice starch. RSC Adv 7(7):3660–3666
Wang S, Wang J, Wang S, Wang S (2017b) Annealing improves paste viscosity and stability of starch. Food Hydrocoll 62:203–211
Wang Z, Ma S, Sun B, Wang F, Huang J, Wang X, Bao Q (2021) Effects of thermal properties and behavior of wheat starch and gluten on their interaction: a review. Int J Biol Macromol 177:474–484
Whistler RL, BeMiller JN, Paschall EF (2012) Starch: chemistry and technology. Academic, New York
Wu G, Morris CF, Murphy KM (2017) Quinoa starch characteristics and their correlations with the texture profile analysis (TPA) of cooked quinoa. J Food Sci 82(10):2387–2395
Wu K, Dai S, Gan R, Corke H, Zhu F (2016) Thermal and rheological properties of mung bean starch blends with potato, sweet potato, rice, and sorghum starches. Food Bioprocess Technol 9(8):1408–1421
Wu Z, Qiao D, Zhao S, Lin Q, Zhang B, Xie F (2022) Nonthermal physical modification of starch: an overview of recent research into structure and property alterations. Int J Biol Macromol 203: 153–175
Xia L, Sha B, Chen C, Jy A, Lc C, Swa B (2021) Changes of starch during thermal processing of foods: current status and future directions. Trends Food Sci Technol 119:320–337
Xie F, Halley PJ, Avérous L (2012) Rheology to understand and optimize processibility, structures and properties of starch polymeric materials. Prog Polym Sci 37(4):595–623
Xla C, Rui X, Jza B, Long CC, Zja C, Yta C (2020) Pasting, rheology, and fine structure of starch for waxy rice powder with high-temperature baking. Int J Biol Macromol 146:620–626
Yang H, Tang M, Wu W, Ding W, Ding B, Wang X (2021) Study on inhibition effects and mechanism of wheat starch retrogradation by polyols. Food Hydrocoll 121:106996
Yousefi AR, Razavi SMA (2015) Dynamic rheological properties of wheat starch gels as affected by chemical modification and concentration. Starch 67(7-8):567–576
Yu B, Ren F, Zhao H, Cui B, Liu P (2020) Effects of native starch and modified starches on the textural, rheological and microstructural characteristics of soybean protein gel. Int J Biol Macromol 142:237–243
Yu W, Zou W, Dhital S, Wu P, Gidley MJ, Fox GP, Gilbert RG (2018) The adsorption of α-amylase on barley proteins affects the in vitro digestion of starch in barley flour. Food Chem 241:493–501
Zhang B, Li X, Liu J, Xie F, Chen L (2013) Supramolecular structure of A- and B-type granules of wheat starch. Food Hydrocoll 31(1):68–73
Zhang H, Sun B, Zhang S, Zhu Y, Tian Y (2015) Inhibition of wheat starch retrogradation by tea derivatives. Carbohydr Polym 134:413–417

Zhang X, Li R, Kang H, Luo D, Fan J, Zhu W, Liu X, Tong Q (2017) Effects of low molecular sugars on the retrogradation of tapioca starch gels during storage. PLoS One 12:e0190180

Zhang X, Wang L, Xu J, Yuan J, Fan X (2022) Effect of starch chain structure and non-starch components on the hydrolysis of starch by α-amylase. Starch 74:2100107

Zhang Y-F, Li J-B, Zhang Z-Y, Wei Q-S, Fang K (2019) Rheological law of change and conformation of potato starch paste in an ultrasound field. J Food Measure Charact 13(3): 1695–1704

Zhao Q, Tian H, Chen L, Zeng M, Qin F, Wang Z, He Z, Chen J (2021) Interactions between soluble soybean polysaccharide and starch during the gelatinization and retrogradation: effects of selected starch varieties. Food Hydrocoll 118:106765

Zhu F (2015) Interactions between starch and phenolic compound. Trends Food Sci Technol 43(2): 129–143

Zhu F (2018) Relationships between amylopectin internal molecular structure and physicochemical properties of starch. Trends Food Sci Technol 78:234–242

Chapter 4
Heat-Moisture Treatment of Starch

Anil Gunaratne

Abstract Heat-moisture treatment (HMT) is a physical modification method that can be employed to modify the starch structure and functional properties without destroying the starch granular structure. In HMT, starch is exposed to a high temperature, commonly above the gelatinization temperature with insufficient water to gelatinize. Changes in X-ray pattern, crystallinity, starch chain interactions, swelling, amylose leaching, gelatinizing, pasting, gelling, and digestible properties of starch have been shown to occur during HMT. These changes have been found to vary with the source of starch, moisture content, incubation temperature, and time. During HMT, structural arrangement that occurred in starch chains within the crystalline and amorphous domains of starch granule modifies the physicochemical properties of starch.

Keywords Heat-moisture treatment · Starch properties

4.1 Introduction

Starch is a polysaccharide stored in plant tissues such as tubers, roots, stems, seeds, and grains and is the second largest biomass produced on earth next to cellulose. Starch is widely used in food and non-food industry. Granules are the basic physical structural units of starch. Functionality is the key to utilizing starches in a wide range of food applications such as specific viscosity, mouth feel, freeze-thaw stability, clarity, emulsion stability, color, and film coating properties. These properties have shown to be influenced by composition, morphology, and molecular arrangement of amylose and amylopectin (Hoover 2010). The granules are partially crystalline and insoluble in cold water and their size, shape, and composition are essentially genetical. Starch granule may be spherical, oval, polygonal, disk-shaped, kidney-shaped, or elongated (Jane et al. 1994). In general cereal starch granules are small and polyhedric, whereas tuber and root starch granules are large and spherical or

A. Gunaratne (✉)
Sabaragamuwa University of Sri Lanka, Belihuloya, Sri Lanka

Z. Sui, X. Kong (eds.), *Physical Modifications of Starch*,
https://doi.org/10.1007/978-981-99-5390-5_4

ellipsoid. Native starches usually do not possess desirable functional properties such as thermal, shear, and acid stability and solubility in clod aqueous liquids and resistant to phase separation (retrogradation) and thus need some sort of value addition through the modification. Modified starch plays a major role in food applications as functional ingredients. Starch is mainly modified either chemically or physically to improve functionality. But most starches presently used in food applications are chemically modified. Physical modification is favored with the growing demand generated for nontoxic consumer safety foods. Hydrothermal treatments, mainly heat-moisture treatment and annealing, can be modified functional properties of starch without destroying the starch granule structure. In heat-moisture treatment, starch is incubated at low moisture level (<35% water w/w) to a certain period of time, at temperature above the glass transition temperature but below the gelatinization temperature and thus, heat-moisture treatment occurs below the onset of starch gelatinization.

4.2 Properties of Amylose and Amylopectin and Starch Structure

Amylose and amylopectin are the major constituents of starch. The minor component of amylose containing about 20–30% consists mainly of α-(1-4) linked glucose units and is thus often referred to as linear polymer. However, the presence of slight degree of branching has been reported to have α-(1-6) branch points. Hizukuri et al. (1981) reported a slight degree of branching (9–20 branch point per molecules) in amylose from starches of several plant sources. The side chains range in chain length from 4 to over 100 (Hizukuri et al. 1981; Takeda et al. 1987) and a decrease in β-amylolysis limit has been shown with an increase in molecular size of amylose (Greenwood and Thompson 1959). The degree of polymerization (DP) which is differed both between and within plant species is more frequently used to denote the size of the polymers than molecular weight (Perez and Bertoft 2010). The DP of this linear molecule is about 500–600 units (Jacobs and Delcour 1998). The molar fraction which varies from 0.1 to 0.7 in branched amylose generally has higher molecular weight than linear polymer (Takeda et al. 1987). Incomplete conversion of amylose to maltose by the exoenzyme β-amylase indicates the presence of branch points (Morrison and Karkalas 1990). Though slight branching is present, amylose essentially behaves like a linear polymer forming films and complexes with ligands (Biliaderis 1998). The confirmation of amylose has been subject to controversy and has been shown to vary from helical to an interrupted helix, to a random coil. Two important features of amylose in solution that determine the applicable functionality of this polymer in starch-based food products are its ability to form helical inclusion complexes when an appropriate ligand is present and the interchain associations mediated by local ordering of the polysaccharide chains (Biliaderis 1998). Single amylose chains have unique ability to complex with ligands such as monoacyl lipids and emulsifiers and smaller ligands such as alcohol and flavor. When amylose forms complexes with various ligands, a crystallographically distinct structure of starch, the V-polymorph, is formed (Biliaderis 1998). Depending on the size of the complexing agents, the amylose helices of six, seven, and eight glucose residues per turn can be seen (Perez and Bertoft 2010). When the ligand is bulkier than a hydrocarbon chain, helices of seven or eight glucose residues per turn are also formed (French and Murphy 1977). The typical chain conformation of V-amylose is a left-handed single helix with six residues per turn and the pitch height varies between 7.92 and 8.04 Å (Perez and Bertoft 2010). X-ray diffraction patterns of granular starches do not usually show the presence of V-structure with the exception of wrinkled pea starch, amylomaize, and some other maize genotypes (dull, au) (Zobel 1988, 1992; Gernat et al. 1993). The lack of V-type characteristics is not because of the absence of amylose-lipid complex but due to the absence of organized helices into well-defined three-dimensional structures (Biliaderis 1998). The development of the V-type polymorph can be induced by heat-moisture treatments of starch (Zobel 1988), extrusion cooking (Mercier et al. 1980), or simply gelatinization and cooling of starch dispersions. In the solution behavior, interchain association of amylose occurs forming double helices over chain segments less than 100 units (Biliaderis 1998). Concentration and chain length of amylose play important roles in the solubility, molecular association/crystallization of amylose (Gidley and Bulpin 1989; Pfannemuller et al. 1971). In the location and state of amylose within the starch granules, it has been reported that amylose is more concentrated at the periphery of the starch granules (Boyer et al. 1976). According to the crosslinking studies done on maize and potato starch by Jane et al. (1992), amylose was crosslinked with amylopectin and that there was no crosslinking between amylose molecules. After critically analyzing the current wisdom of the location and state of amylose in the starch granules, Perez and Bertoft (2010) explained that individual radial oriented amylose is randomly distributed among the radial amylopectin chains.

Amylopectin is the major extensively branched component of starch granule reported to have Mw 60×10^6 to 110×10^6 (Aberle et al. 1994). The polymer is composed of linear chains (glucose residues contented with α-(1,4) glycosidic bonds) contacted through α-(1,6) linkages (4–5%) forming a highly branched compact molecule (Biliaderis 1998). The molecular structure of the highly branched large molecule of amylopectin is more complex than that of amylose. The amylopectin chains are grouped into certain categories in order to distinguish the unit chain composition. According to Hizukuri (1996), amylopectin unit chains can be categorized into A-, B-, and C-chains. The unbranched A-chains are linked to B-chains and do not carry any other chains, the B-chains (B1–B4) carry one or more A-chains and/or B-chains, while the C-chain contains the reducing end group of the molecule. The shortest chains (A and B1) have chain length (CL) of 14–18 while the longest B2–B4 have CL of 45–55. For most amylopectin, the average chain length ranged 17–26 depending on the type of crystallinity (Perez and Bertoft 2010). The reported values for weight-average chain length of amylopectin of A-, B-, and C-type starches were in the ranges of 23–29, 30–44, and 26–29, respectively. In general,

A-crystalline starch contains shorter CL than B-types (Hizukuri 1985). The average CL of amylopectin poorly reflects the true structure since all samples (normal types of starch) of amylopectin possess at least two major groups, long and short chains (Perez and Bertoft 2010). The shortest reported CL for any amylopectin sample is six (Koizumi et al. 1991).

Though detailed information of structural properties of starch polymers has been extensively studied and characterized, the molecular order of the starch granules (arrangement of amylose and amylopectin within the granule), which governs the physicochemical properties of native starch, is still under investigation. In the semi-crystalline structure of starch granule, the crystalline lamellae exist in the granule alternatively with amorphous lamellae (Donald et al. 1997). The combined thickness of crystalline lamellae plus amorphous lamellae is 9 nm and 9.2 nm for A-type starches and B-type starches respectively (Jenkins et al. 1993; Jane et al. 1997). Clusters of amylopectin short chains occur within the crystalline domains of the granule (Yamaguchi et al. 1979). The amorphous region that has been shown to be very susceptible to chemical and enzymatic modification (Hood and Mercier 1978; Robyt 1984) accounts for 70% of the starch granule (Oostergetel and Van Bruggen 1993) and consists of free amylose, lipid-complexed amylose, and some branch points of amylopectin (Hizukuri 1996). Amylopectin molecules are involved in the granule crystallinity intertwining short linear chains of the branches forming double helices and clusters and α-(1,6) branch points are located in more amorphous region between the clusters of double helices (Robin et al. 1975; Perez and Bertoft 2010). The double helical clusters are packed together to form alternating crystalline and amorphous lamellae. Starch is classified according to the packing arrangement of the amylopectin double-stranded helices in the granules, namely A, B, and C type as determined by X-ray diffraction patterns. The A-type crystallinity is found mainly in cereal starches and is characterized by peaks at 15°, 17°, 18°, 20°, and 23° 2θ angles (Zobel 1988; Cheetham and Tao 1998). Typical B-Type X-ray pattern can be seen from most of the tuber and root starches (Zobel 1988) with peaks that are both broad and weak with two main reflections centered at 5.5 and 17 angles. The C-type starches are mixtures of A-type and B-type starches having special properties according to the distribution and proportion of A- and B-type polymorphs (He and Wei 2017). Bogracheva et al. (1998) showed that in C-type starches the B polymorphs are arranged centrally while the A polymorphs are located peripherally within the granules. The amylopectin of A-type starches has a close packing arrangement compared with that of B-type starches. The calculation of starch crystallinity and the correlation of double helical content with crystallinity have been developed through the combined use of X-ray diffraction and solid NMR (Lehman and Robin 2007). Contribution of different crystal polymorphs of starch to the total crystallinity can be calculated from this novel method.

4.3 Heat-Moisture Treatment

Heat-moisture treatment of starches is defined as a physical modification that involves incubation of starch granules at low moisture level (<35% water w/w) during a certain period of time, at a temperature above the glass transition temperature but below the gelatinization temperature (Hoover 2010). The conditions used for HMT of starches from various botanical origins are listed in Table 4.1.

Table 4.1 Heat-moisture treatment conditions used for different types of starches

Starch/flour	Temp (°C)	Time (h)	Water content (%)	References
Tuber and roots				
Potato	100, 120	2	30, 35	Brahma and Sit (2020)
	110	1	12, 15, 18, 21, 24	Bartz et al. (2017)
	100	10	30	Gunaratne and Hoover (2002)
	90–130	24	17–26	Vermeylen et al. (2006)
Cassava	120	0.33	19	Dudu et al. (2019)
	100	10	30	Gunaratne and Hoover (2002)
Sweet potato	110	8	30	Hung et al. (2021)
	105	1	26,30,34	Liao et al. (2019)
Yam (*Dioscore alata*)	100	10	30	Gunaratne and Hoover (2002)
Yam (*Dioscorea alata purpurea)*	105	24	20,25,30	Mustapha et al. (2019)
Canna	100	16	18,20,22,25	Watcharatewinkul et al. (2009)
Cereals				
Wheat	110	8	30	Hung et al. (2021)
White sorghum Sorghum	110 100	16 10	18, 21, 24, 27 20,25	Olayinka et al. (2008) Sun et al. (2014)
Waxy, low- and high-amylose rice	100	16	25	Sui et al. (2017)
Corn and waxy corn	100	12	25	Liu et al. (2020)
Legume				
Pea	120	1, 4, 8, 12	25	Han et al. (2021)
Lentil	100–120	2	30	Chung et al. (2009)
Faba bean, black bean	60, 100, 120	12	23	Ambigaipalan et al. (2014)
Palm				
Sago	120	1	10, 15, 20, 25, 30	Dewi et al. (2022)

Adapted from Hoover (2010)

4.3.1 Impact of HMT on Granule Morphology

Granule morphology, size and distribution, and surface characteristics play an important role in some food and non-food applications. Granule morphology of maize, wheat, potato, yam, lentil, and breadfruit starches has shown to remain unchanged after HMT (Kulp and Lorenz 1981; Stute 1992; Hoover and Vasanthan 1994; Franco et al. 1995; Hoover and Manuel 1996; Tan et al. 2017). However, HMT performed on rice starch with varying amylose content at 110 °C with 25% moisture has exhibited aggregated granules with irregular surfaces, and in the case of low amylose rice starch, signs of loss of physical integrity and characteristics of partial gelatinization have been observed by Zavareze et al. (2010). Jiranuntakul et al. (2011) have also reported a slight rough surface on some granules of normal rice starch and an obvious change in waxy-type granule. Formation of voids and the disappearance of birefringence at the granule center of potato starch (Kawabata et al. 1994; Vermeylen et al. 2006) and maize starch (Kawabata et al. 1994) have been observed at temperature exceeding 110 °C. However, granule periphery of both starches has remained highly birefringence even after HMT. The changed or unchanged granular morphology of HMT starches depends on the heating temperature and time, moisture content, and the type of starch used in HMT. High moisture contents in HMT can lead to partial gelatinization and agglomeration in the starch granules depending on the starch gelatinization temperature (Fonseca et al. 2021).

4.3.2 Impact of HMT on Starch Double Helical Arrangement and Crystallinity

X-ray diffractometry has been extensively used to reveal the impact of HMT on the characteristics of crystalline structures of starch. HMT has been shown to change the wide angle X-ray pattern from the B- to A-type (or A + B) for potato starch (Sair 1967; Hoover and Vasanthan 1994; Gunaratne and Hoover 2002; Vermeylen et al. 2006; Varatharajan et al. 2011; Ambigaipalan et al. 2014), yam starch (Hoover and Vasanthan 1994; Gunaratne and Hoover 2002), and breadfruit starch (Tan et al. 2017). A shift from C- to A-type on HMT has been observed for sweet potato starch by Shin et al. (2005). However, A-type of starch structure pattern was unaffected for cereal starches (Sair 1967; Hoover and Manuel 1996; Hoover and Vasanthan 1994) and cassava and yam starches (Gunaratne and Hoover 2002) after HMT. B-polymorphic high-amylose maize starch structure has been found to be more susceptible to HMT than the A-polymorphic regular maize starch, resulting in more structural changes (Wang et al. 2016). X-ray intensities have been found to reduce for potato (Hoover and Vasanthan 1994), cassava (Abraham 1993), and barley (Lorenz and Kulp 1982) after HMT. However, cereal starches generally exhibit either increased or unchanged intensities after HMT. Temperature and the moisture content of HMT appear to be affected on the changes of X-ray pattern and intensities (Hoover and Vasanthan 1994; Vermeylen et al. 2006). Bartz et al. (2017) applied different moisture contents (12, 15, 18, 21, and 24%) in HMT of potato starch and found that X-ray diffraction pattern was changed from B-type to A+B-type along with the moisture content used in the treatment and at the highest moisture content (24%) the B-type pattern of native potato starch was changed to A-type. Further they have observed a decrease in relative crystallinity with moisture contents of 12% and 15% but an increase in relative crystallinity with moisture contents of 21% and 24%. Several theories have been suggested to explain the change of X-ray pattern, intensities, and crystallinity of starch on HMT: (1) Destruction of crystallites (decrease X-ray intensities). Hoover and Vasanthan (1994) explained that in B-type starch rupture of hydrated water bridges leads adjacent double helices to move apart and form less ordered crystalline array. Double helical movement during heat-moisture treatment could disrupt starch crystallites and/or change the crystalline orientation (Gunaratne and Hoover 2002). The same authors found that in contrast to B-type (potato, true yam) starch, A-type tuber (new coco yam, taro, cassava) crystallinity was unaffected after HMT. However, HMT decreased relative crystallinity of A-type rice starch with differing amylose content (Zavareze et al. 2010; Kuhunae et al. 2007). (2) Growth of new crystallites. According to Hoover and Vasanthan (1994), double helices shifted due to heat and moisture during HMT could pack them in a more ordered crystalline array than in native starch result in increased X-ray intensities mainly in A-type starches. Vermeylen et al. (2006) observed increased crystallinity for potato starch heated at 130 °C whereas HMT between 90 and 120 °C decreased the crystallinity. These authors hypothesized based on debranching results of native and HMT amylopectin that at high temperature (130 °C) breaking of covalent linkages decouples the double helices from the amylopectin backbone and renders them sufficiently mobile to become organized in more perfect/larger crystallites. The interaction between amylose-amylose, amylose-amylopectin, and amylopectin-amylopectin chains increases X-ray intensities due to the formation of new crystallites (Hoover and Vasanthan 1994; Hoover and Manuel 1996). (3) Reorientation of the already existing crystallites may increase or decrease X-ray intensities (Lorenz and Kulp 1982; Hoover and Vasanthan 1994; Hoover and Manuel 1996). (4) Changes in the packing arrangement (B- to A-type crystallinity) of the double helices which results in a change in the X-ray pattern. HMT has been shown to trigger the transformation of B-type packing arrangement to either pure A-type or a mixed A + B (Sair 1967; Vermeylen et al. 2006; Gunaratne and Hoover 2002; Hoover and Vasanthan 1994; Kawabata et al. 1994). The change in X-ray pattern (B to A+B) on HMT can be attributed to dehydration of water molecules occupied in the central channel of the B-unit cell and movement of pair of double helices into the central channel that was originally occupied by the vaporized water molecules (Gunaratne and Hoover 2002). From WAXD patterns, Vermeylen et al. (2006) have confirmed the transition from B to A packing arrangement and at higher temperature (HMT 23–130 °C) a typical A-type diffraction is resulted (Fig. 4.1). (5) Formation of crystalline amylose-lipid complexes increases X-ray intensities. Hoover and Vasanthan (1994) and Hoover and Manuel (1996) observed a decreased apparent amylose content on HMT of

Fig. 4.1 Transformation of B-type starch structure of potato starch to A-type as the temperature of HMT is increased (Vermeylen et al. 2006). Native potato starch (a), HMT-23-90 (e), HMT-23-100 (f), HMT-23-110 (g), HMT-23-120 (h), HMT-23-130 (i). (Reused with permission from Elsevier Publications)

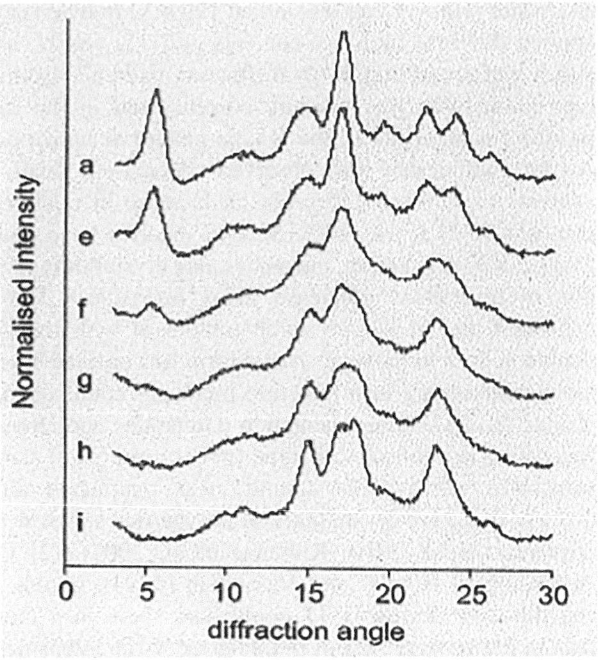

wheat and potato starches, indicating additional interaction between native starch lipid and amylose chains. A peak of crystalline V-amylose-lipid complexes has been clearly observed for HMT (100 °C, 6 h, 18–27% mc) rice starches, and the peak has been progressively increased with an increase in the moisture content as a result of an increase in the mobility of V-amylose-lipid complexed chains. This mobility facilitates the formation of more orderly large V-polymorph (Kuhunae et al. 2007).

4.3.3 Impact of HMT on Granular Swelling and Amylose Leaching

HMT has been shown to reduce the granular swelling and amylose leaching of cereal (Kulp and Lorenz 1981; Hoover and Vasanthan 1994; Olayinka et al. 2008), tuber and root (Hoover and Vasanthan 1994; Gunaratne and Hoover 2002), and legume starches (Hoover and Vasanthan 1994; Hoover et al. 1994). The decrease in amylose leaching on HMT could attribute to the additional interactions between amylose-amylose and amylose-amylopectin chains (Gunaratne and Hoover 2002) and formation of amylose-lipid complex (Hoover and Manuel 1996) during HMT. This type of mechanism may also be partly responsible for the reduction in granular swelling in HMT starches along with a decrease in granular stability due to unraveling of double helices and crystallite disruption, increase in crystallinity, and change in polymorphic starch structure from (B → A + B) (Hoover and Vasanthan 1994; Gunaratne and Hoover 2002; Shin et al. 2005). Intermolecular bond formation during HMT could result in a rigid granular structure and also reorientation of starch structure could reduce the available hydroxyl groups to bind with water molecules which in turn reduce the swelling ability of HMT starch (Ali et al. 2020).

4.3.4 Impact of HMT on Gelatinization Properties

Differential scanning calorimetry has been the most widely used technique to study the gelatinization characteristics of HMT starches. The impact of HMT on gelatinization properties is dependent on the source of starch, heating temperature, time, and level of moisture prevailing during the treatment. Generally, HMT increases the gelatinization transition temperatures, the onset (T_o), peak (T_p), and conclusion (T_c) and broaden the gelatinization temperature range ($T_c - T_o$) (Sair 1967; Donovan et al. 1983; Stute 1992; Hoover and Vasanthan 1994; Hoover and Manuel 1996; Gunaratne and Hoover 2002; Ambigaipalan et al. 2014; Sun et al. 2014). Figure 4.2 shows the increased gelatinization transition temperatures of HMT corn, pea, and

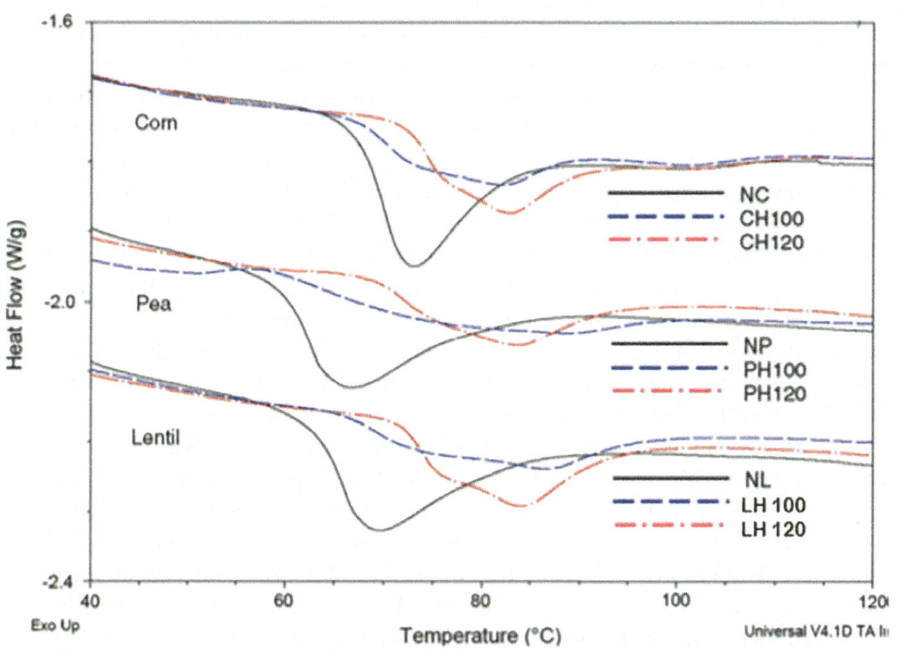

Fig. 4.2 DSC thermograms of native and HMT corn, pea, and lentil starches: *NC* native corn, *NP* native pea, *NL* native lentil, CH100, PH100, LH100 and CH120, PH120 and LH120 represent HMT corn, pea, and lentil starches heat-moisture treated at 30% moisture for 2 h at 100 °C and 120 °C respectively (Chung et al. 2009). (Reused with permission from Elsevier Publications)

lentil starches as measured by DSC. The melting temperatures of starch crystallites are controlled indirectly by the surrounding amorphous region. HMT triggers the interaction between amylose-amylose, amylose-amylopectin and amylose-lipids (Hoover and Vasanthan 1994; Gunaratne and Hoover 2002; Ambigaipalan et al. 2014). These interactions suppress the mobility of starch chains in the amorphous region. As a result, HMT starches require higher temperatures to achieve the swelling that imparts a destabilizing effect on the starch crystalline domain, leading to an increase in transition temperatures (Hoover 2010; Gunaratne and Hoover 2002; Hoover and Vasanthan 1994; Zavareze and Dias 2011). In the gelatinization process, hydration and swelling of the amorphous region of starch granules exert pressure on the crystalline domain, causing melting of the crystalline region and double helices (Adebowale et al. 2009). Some authors explained that the formation of more perfect crystallites upon HMT leads to melting them at high temperatures (Sun et al. 2014; Yadav et al. 2013). Tester (1997) has postulated that the extent of crystalline perfection is reflected in the gelatinization temperature whereas the ΔH reflects the overall crystallinity (quality and quantity). Ji et al. (2004) suggested that the onset of gelatinization would be a measure of the perfection of starch crystallites and thus more perfect crystallites have higher onset gelatinization temperatures. Pukkahuta et al. (2008) noticed biphasic broadening of T_p for HMT sago starch. They explained an inhomogeneous heat transfer during HMT could be the possible reason for observed biphasic T_p. Varatharajan et al. (2010) found double endotherms in HMT waxy potato starch in contrast to normal potato starch and suggested that more heterogeneous crystallites, in terms of thickness and stability, may have been formed in HMT process in waxy potato starch. Structural transformation occurred in the crystalline micelles during HMT could increase the thermodynamic stability resulting in a newly developed high-temperature endotherm (Jiranuntakul et al. 2011). The gelatinization enthalpy has been shown to decrease (Lorenz and Kulp 1981; Hoover and Vasanthan 1994; Gunaratne and Hoover 2002; Ambigaipalan et al. 2014; Adebowale et al. 2009), remain unchanged (Ambigaipalan et al. 2014; Hoover and Vasanthan 1994; Collado and Corke 1999), or increase (Sun et al. 2014) after HMT. The reduction of melting of enthalpy by HMT is much greater in tuber starches than that of cereal starches (Hoover and Vasanthan 1994; Donovan et al. 1983). Gunaratne and Hoover (2002) explained that the reduction of melting enthalpy of HMT starches is due to the disruption of some of the double helices present in crystalline and non-crystalline regions of the granules during HMT leaving fewer double helices to unravel and melt during the gelatinization of HMT starches. However, Hormdok and Noomhorm (2007) reported that partial gelatinization of less stable amylose and amylopectin molecules could reduce the melting enthalpy of HMT starches. In contrast, Ali et al. (2020) have observed the increased relative crystallinity and gelatinization enthalpy for A-type heat-moisture treated lotus seed starch attributing to the more closely packed and ordered crystallites in HMT starch than their native counterpart. Gunaratne and Hoover (2002) studied the effect of HMT on tuber and root starches and reported that the ΔH of B-type starches reduced to a greater extent than those of A-type starches. According to them, double helical starch chains in B-type starches are more mobile than A-type starches due to more water molecules occupied in B-type unit cell and hence more prone to disruption than those of A-type starches. Negatively charged phosphate groups on adjacent amylopectin chains of potato starch could have high impact on the disruption of double helices of potato starch because these negatively charged phosphate groups could hinder the strong interaction between double helices, making them more susceptible to disruption during HMT (Gunaratne and Hoover 2002).

4.3.5 Impact of HMT on Pasting and Gelling Properties

Starch heated in excess water undergoes various changes as a result of heat and moisture transfer. Gelatinization and pasting occurring in the same system can be used to describe many changes that occur. Gelatinization may be used to refer to early changes whereas pasting includes later changes. Pasting is defined as the phenomena following the dissolution of starch, involving granular swelling, exudation of molecular components from the granule, and eventually total disintegration of the granules (Atwell et al. 1988). Properties of starch paste are highly correlated with the textural characteristics and storage ability of many food applications. HMT significantly alters the pasting profile of starches. Investigations on pasting behavior and properties of HMT starches have shown that HMT increases the pasting temperature and thermal shear stability of starch paste but decreases the peak viscosity and granular breakdown (Lorenz and Kulp 1981; Hoover and Vasanthan 1994; Lawal and Adebowale 2005; Gunaratne and Corke 2007; Shih et al. 2007; Varatharajan et al. 2010; Yadav et al. 2013; Sun et al. 2014). The setback which reflects the extent of retrogradation has been reported to either increase or decrease (Sun et al. 2013; Shih et al. 2007; Varatharajan et al. 2010; Gunaratne and Corke 2007). The extent of these changes has been found to depend on the source of starch, HMT parameters (treatment conditions), and the instrument used to detect the pasting properties (Hoover 2010). Stable paste viscosity, limited swelling, and high gelatinization temperature of HMT starch are similar to those of chemically modified crosslinked starch (Watcharatewinkul et al. 2009) and thus HMT might be used to enhance the starch properties analogous to chemically modified crosslinked starch. Sun et al. (2014) found that pasting temperature of HMT sorghum starch and flour increased with the increase of moisture level of the HMT. Restricted swelling and amylose leaching, increased granular rigidity, interaction between starch chains, and changes in granular crystalline domain resulted in HMT are believed to be associated with the changes of pasting properties of HMT starches (Hoover and Vasanthan 1994; Jiranuntakul et al. 2011). HMT promotes the intermolecular association of starch chains reinforcing the granular bonding and thus high heat energy is required to disrupt the HMT granules in the formation of paste. In a typical pasting curve, rapid increase of viscosity is attributed to granular swelling and starch leaching. The restricted swelling and amylose leaching in HMT starch restrict the viscosity development. Varatharajan et al. (2010) studied the pasting characteristics of normal and waxy potato starches and suggested that enhanced interactions

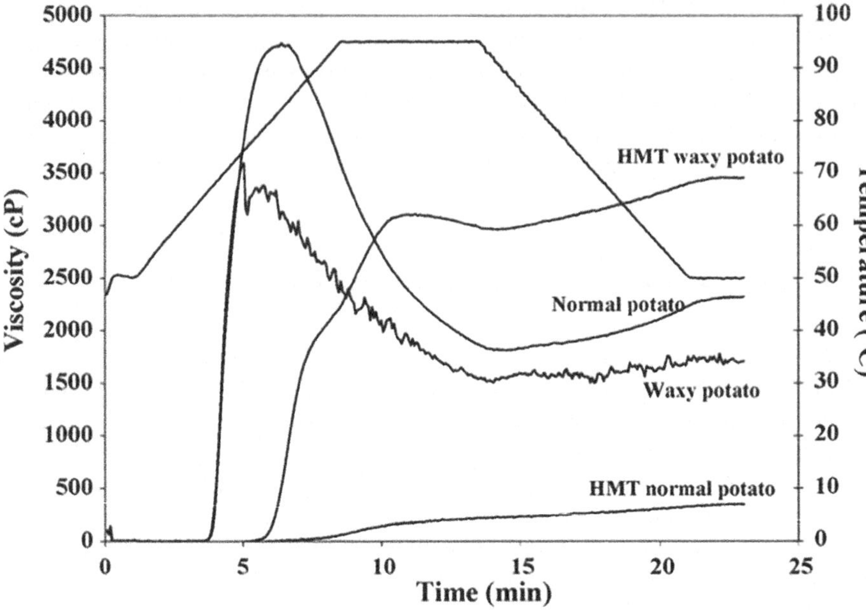

Fig. 4.3 RVA curves of native and HMT normal and waxy potato starches (Varatharajan et al. 2010). (Reused with permission from Elsevier Publications)

between starch chains (amylose-amylose, amylose-amylopectin, amylopectin-amylopectin) in HMT starches reduce the peak viscosity and granular breakdown while increasing the pasting temperature. This effect is lower in waxy starch as the contribution of amylose to the above interaction is negligible. Jiranuntakul et al. (2011) also reported that the degree of granular strengthening after HMT was more pronounced in normal starches than waxy starches. The same authors, by studying the pasting and gel morphology of HMT normal and waxy rice, corn, and potato starches, have concluded that HMT had more impact on pasting properties of the B-type starch than those of A-type starches attributing to the greater mobility of starch chains in B-type starches during the HMT process. In the final step (cooling step) of pasting profile, a decrease in energy in the system and subsequent hydrogen bond formation between starch chains mainly in amylose increases the viscosity (setback) (Hoseney 1994). Varatharajan et al. (2010) reported that the increased setback on HMT is attributed to the increase in the proportion of intact granule in the gel matrix. The extent of the setback (retrogradation) is governed by amylose gelation and the presence of rigid granules embedded in leached amylose network (Jayakody et al. 2007). Figure 4.3 shows the pasting curves (RVA) of HMT normal and waxy potato starch showing that HMT has decreased peak viscosity and granular breakdown while increasing the pasting temperature (Varatharajan et al. 2010).

Gel hardness of potato and wheat starch (Gunaratne and Corke 2007) and sorghum starch (Singh et al. 2011) and rice flour (Sun et al. 2013) has been reported to increase after HMT. Gel hardness of HMT sorghum starch increased with the severity of HMT (Singh et al. 2011). Starch gel can be regarded as a hydrated polymer composite where swollen amylopectin-rich granules are embedded in and reinforce continuous matrix of entangled amylose molecules (Ring 1985). Mechanical properties of a starch gel would depend on the rheological characteristics of the amylose matrix, the volume fraction and the rigidity (deformability) of the gelatinized granules, and the interactions between the dispersed and the continuous phases (Eliasson 1986). More intact and more orderly crystalline granules formed in HMT could form more rigid amylopectin-rich swollen granules embedded in the continuous phase of gel matrix resulting in harder gel. Hoover et al. (1994) reported that the restriction of swelling by HMT favors amylose aggregation in the formation of a gel matrix, leading to a firmer gel after HMT. During HMT, crosslinking between starch chains in the particular portion of amylose could form more junction zones in the continuous phase of gel matrix forming firmer gels (Hormdok and Noomhorm 2007). Gunaratne and Corke (2007) concluded that the interplay of the following factors, amylose concentration and conformational ordering and intermolecular association of amylose chains, rigidity of the filler components, and interaction between filler components and gel matrix, affect the gel hardness of HMT starch.

4.3.6 Impact of HMT on Retrogradation

Reassociation of starch polymers via hydrogen bonding in gelatinized starch on cooling is generally termed as retrogradation, and is time and temperature dependent. The association of starch molecules in the starch gel results in precipitation, gelation, and changes in consistency, followed by the gradual formation of crystallites, resulting in phase separation in the polymer and solvent (Karim et al. 2000). Retrogradation is of great interest for food scientists since it has a profound effect on quality, shelf-life, acceptability, and nutritional value of starch-based food products (Biliaderis 1991). Starch gel can be considered as composites containing gelatinized granules embedded in amylose matrix (Miles et al. 1985). Both amylose and amylopectin are involved in retrogradation. Miles et al. (1985) attributed the initial gel firmness to be dominated by irreversible gelation and crystallization within the amylose matrix and branched reversible amylopectin recrystallization is correlated with long-term development of starch gel. The ordered retrograded starch crystallites show a B-type X-ray diffraction pattern (Zobel 1988) containing both crystalline and amorphous region. Hoover and Vasanthan (1994) studied the retrogradation behavior of HMT (100 °C, 16 h, 30% moisture) wheat, potato, oat, and lentil starches by storing the starch gel at 25 °C for 20 days and postulated that the extent of retrogradation as measured by DSC is governed by the interplay of crystallinity changes and amylose-amylopectin (AM-AMP) interaction during HMT. Gunaratne and Hoover (2002) found that HMT decreased retrogradation in B-type starches but caused no significant effect on A-type starches. Crystallites are disrupted in B-type starches during HMT than those of A-type starches. The disruption of crystallites in B-type starch during HMT would enhance the degree of separation between the outer branches of adjacent amylopectin chains resulting in a much slower association of amylopectin chains and the formation of amylopectin double helices in starch gel.

4.3.7 The Impact of HMT on Enzyme Hydrolysis

The enzyme digestibility of starch has been shown to be affected by many factors such as starch source, granule size, amylose/amylopectin ratio, amylose-lipid complexes, crystalline structure, and granule surface organization (Zhang and Oates 1999; Ring et al. 1998; Holm et al. 1983; Snow and O'Dea 1981). The digestion of starch granule is a complex process that involves accessibility (diffusion and adsorption) and hydrolytic activity of enzyme toward substrate (Watcharatewinkul et al. 2010) and both morphological features and internal structural features affect the in vitro hydrolysis (Li et al. 2004). Depending on the botanical origin and HMT treatment conditions, increased (Kawabata et al. 1994; Hoover and Vasanthan 1994; Gunaratne and Hoover 2002) or decreased (Kweon et al. 2000; Franco et al. 1995) α-amylase hydrolysis have been observed. Gunaratne and Hoover (2002) studied the enzyme hydrolysis of HMT (100 °C, 16 h, and 30% moisture level) on root and tuber starches and postulated that crystallite disruption near granule surface on HMT facilitates the rapid entry of enzyme into the granule surface which increases the extent of enzyme hydrolysis. The temperature, moisture content, duration of the treatment, and starch sources used in HMT have been reported to influence on starch digestibility by α-amylase (Hoover 2010). The increased enzyme resistance of HMT starch could be attributed to the rearrangement of molecular chains and more compact granule structure (Tan et al. 2017). Studding the impact of different temperatures of HMT on normal and waxy potato starches, Varatharajan et al. (2011) suggested that the interplay of number of A-type crystallites, starch chain interactions, and changes to double helical conformation occurred in HMT is influenced on α-amylase hydrolysis of HMT starches. These authors further explained that the structural disorganization occurred near the granule surface as a result of HMT and transformation of B-type crystallite to A-type crystallites facilitate the α-amylase hydrolysis whereas starch chain interactions (AM-AM, AM-AMP) hinder the susceptibility to α-amylase hydrolysis. The occurrence of these factors and subsequent influence on starch digestibility of HMT starches depend on the temperature used in HMT. According to Jane et al. (1997) inferior crystalline structure containing α-(1-6) linked branch points and short double helices in A-type starch are more susceptible to enzyme hydrolysis compared to the B-type starch in which more branch points are clustered in the amorphous region and there are fewer short branch chains which make a superior less susceptible structure.

4.3.8 The Impact of HMT on Acid Hydrolysis

It is believed that acid hydrolysis occurred in a two-stage pattern in which acid molecules preferentially attack more amorphous parts of the granules which is a relatively fast hydrolysis rate whereas in the second stage crystalline material is slowly degraded (Hoover 2010). To account for the slower hydrolysis rate of the crystalline parts of the starch granule, two hypotheses have been proposed (Kainuma and French 1971). First, dense packing of starch chains within the starch crystallites does not readily allow the penetration of H_3O^+ into the regions. Second, the acid hydrolysis of a glycosidic bond may require a change in conformation (chair → half chair) of D-glycopyranosyl unit. The mode of the distribution of the α-(1,6) branch points differs between A-type starches and B-types starches. In B-type starches more branch points are clustered mainly in the amorphous region and thus more susceptible to acid hydrolysis than branch points located in the crystalline region of A-type starches (Jane et al. 1997). The influence of HMT on acid hydrolysis varied among the starch sources. The susceptibility to acid hydrolysis decreased after HMT of maize (Hoover and Manuel 1996), potato (Hoover and Vasanthan 1994), and wheat, lentil, oat, and yam (Hoover and Vasanthan 1994) whereas Gunaratne and Hoover (2002) found increased acid hydrolysis for HMT (100 °C, 16 h, and 30% moisture) taro, cassava, and potato starches up to 4th, 6th, and 5th day of storage, respectively. From the investigations carried out on the influence of HMT on acid hydrolysis four main influential factors have been proposed: crystallites disruption, starch chain interaction within the amorphous and crystalline domain, disruption of double helices in the amorphous region, and polymorphic transformation from B to (A + B). HMT increases the susceptibility toward acid hydrolysis through the crystallite disruption and the disruption of double helices in the amorphous region but interaction between starch chains occurred in HMT reduces the extent of acid hydrolysis (Gunaratne and Hoover 2002). Interactions between starch chains reduce the chain flexibility, and thereby hinder the conformational change (chair to half chair required for efficient protonation of glycosidic oxygen) (Gunaratne and Hoover 2002). Varatharajan et al. (2010) observed the reduction of acid hydrolysis for HMT normal and waxy potato starches and explained starch chain interaction (amylose and amylopectin) and the polymorphic transformation (B → A + B) could be the reasons for decreased acid hydrolysis of HMT potato starch. In A-type unit cell structure, amylopectin chains are closely packed in than those of B-type. Closely packed amylopectin chains hinder the accessibility of acid molecules.

4.3.9 Impact of HMT on the Formation of Slow Digestible and Resistant Starch

Based on the rate of starch hydrolysis by digestive enzymes, starch can be classified into rapidly digestible (RDS), slowly digestible (SDS), and resistant starch

(RS) (Englyst et al. 1992). RDS is the starch fraction that rapidly and completely digested in the small intestine causing a sudden increase of blood glucose level and SDS slowly but completely digested in the small intestine. Starch fraction not hydrolyzed in the small intestine in the healthy human is considered as resistant starch. Resistant starch can be subdivided into four groups: RS1 (type1) represents the physically inaccessible form of starch such as partly milled grains and seeds and in some very dense type of processed starch, RS2 (Type 2) represents the granular form of starch such as in raw potato and bananas, RS3, the most resistant type, is formed mainly from retrograded amylose in gelatinized starch, RS4 is considered to be formed in chemically modified starch (Lehman and Robin 2007). RS is believed to have health beneficial effects similar to those of some form of dietary fiber (Rabe 1999). RS has been reported to be beneficial for colon health and protection against collateral cancers, hypoglycemic effects, inhibition of fat metabolism, lowering cholesterol, and promoting the growth of probiotic microorganisms (Walter et al. 2005; Sajilata et al. 2006; Lehman and Robin 2007). With low to medium glycemic index SDS reduces the glycemic load (Lehman and Robin 2007) and thus RS and SDS can be considered as important nutritional components of foods. Many investigations have shown that HMT can modify the SDS and RS level without destroying the granular structure attributing that HMT can generate more highly ordered structure. The alteration of digestibility with increased levels of RS and SDS by HMT contributes to enhance the nutritional value of starch-based food products (Iuga and Mironeasa 2020). The amount of RS and SDS produced in HMT starches appeared to be dependent on the level of moisture content, temperature, incubation period, and the source of starch. Li et al. (2011) produced HMT mung bean starch with different moisture content (15%, 20%, 25%, and 35%) heating at 120 °C for 12 h and found that RS content significantly increased at 20% moisture level. Sweet potato which was subjected to HMT with various combinations of moisture and temperature produced the highest content of granular slowly digestible starch with 50% moisture at 55 °C (Shin et al. 2005). The structure of SDS had very rigid amorphous regions and partially disrupted crystalline region. Chung et al. (2009) reported that enhanced interaction between starch chains and the perfection of already existing starch crystallites resulted in increased RS and SDS content in HMT, and this is more pronounced in HMT than in annealing. Further, they mentioned that amylose-amylose interaction occurred during treatment is not disrupted in gelatinization, which reduces the accessibility of enzyme for digestion. In the strategies for the manufacture of resistant starch, partial acid hydrolysis prior to HMT can be used to enhance the RS yield (Brumovsky and Thompson 2001). The limited acid hydrolysis facilitates the mobility of the molecules, which allows the formation of highly ordered structure resistant to α-amylase hydrolysis (Brumovsky and Thompson 2001; Thompson 2000). Further, Brumovsky and Thompson (2001) have shown that partial acid hydrolysis (78 h) followed by HMT at 120 °C can produce high content of boiling stable granular RS, which should be stable to many subsequent thermal treatments. The amount as well as the quality of RS is important in terms of food applications. Jacobasch et al. (2006) reported that HMT of Novelose 330 produces high yield of RS3 with beneficial prebiotic properties and thus

HMT-Novelose with excellent prebiotic properties could be used to produce functional food ingredients. It is evident from literature that the formation of RS and SDS is strongly affected by botanical source and the treatment conditions and therefore process optimization through comprehensive investigations is required for the effective application of HMT starches with high levels of RS and SDS.

4.4 Food Applications of HMT Starch

Thermal, pasting, and gelling properties are the key functionalities useful in food applications. HMT improves the stability for mechanical agitation and acid in pasting and thus could be applied in food as an alternative for chemically modified starch. HMT starch appears to be a suitable material for the production of retort foods, dressings, noodles, baked foods, batter products, confections, dairy products, creams, fat mimetics, and resistant starches (Kurahashi and Yoshion 2000). Baking quality and freeze-thaw stability of potato starch have been shown to improve with HMT (Collado and Corke 1999). Sweet potato starch has been modified by HMT and used as a substrate and composite with maize starch to produce Bihon-type noodles. Acceptability values of raw starch noodles, plain boiled and sautéed noodles made from 100% HMT sweet potato starch and sweet potato composite with maze (50% HMT sweet potato with 50% maize starch) have been found to be similar to that of commercial Bihon (Collado et al. 2001). Incorporating 50% of HMT (20 g/100 g moisture at 110 °C for 1.5 h) rice starch with rice flour resulted in remarkable effect on the cooking and textural quality of cooked noodles indicating the possibility of utilizing the HMT rice starch within composite with poor quality rice flour (Hormdok and Noomhorm 2007). Noodles with firmer texture, along with augmented taste and distinct flavor, have been found for HMT (110 °C for 2.5 h) amaranth starch-based noodles compared to the native amaranth and corn starch (Chandla et al. 2017). Chung et al. (2014) found that heat-moisture treated germinated brown rice could partially or completely replace the wheat starch in the preparation of cookies with improved nutrients and high consumer acceptability. Chung et al. (2012) reported that when heat-moisture treated germinated brown rice is incorporated with wheat flour the overall cooking and nutritive quality of noodles can be improved. Lorenz and Kulp (1981) evaluated the bread and cake-baking potentials of HMT wheat and potato starches and found that bread quality and cake-making potential of HMT wheat starches were decreased with HMT whereas HMT increased the baking potentials, grain texture, and bread volume and the cake-making potentials of potato starch. However, compared to the native wheat starch, overall bread and cake-baking quality of HMT potato was lower. These authors believed that decreased baking potentials of HMT wheat starch could attribute to the starch damage occurred in HMT and improved baking potentials of HMT potato starch could be due to the changes of physicochemical properties as a result of structural changes that occurred in HMT. Effect of HMT maize starch on dough and bread making properties has been studied by incorporating 20% HMT maize or native maize starch with wheat starch (Miyazaki and Morita 2005). The results have shown that the HMT maize starch reduced the elasticity of dough compared to that of native maize starch or control (without any maize starch). However, by the addition of shortening, the specific volume of bread baked with HMT maize increased and grain structure became finer but bread crumb firmness baked with HMT was same as at the presence or absence of shortening. These authors believed that HMT reduces the ability of maize starch to bind with gluten protein, resulting in bread with lower quality, while shortening improves the binding efficiency of HMT maize starch with gluten. Hoover (2010) suggested that starch chain interaction and amylose-lipid extraction during HMT may be the main causative factors responsible for decreased interaction of HMT maize starch with gluten. There is a growing interest in the application of starch in the development of food packaging which has led to explore the production of biodegradable and/or edible packaging materials from starch, and HMT has potentials to enhance the properties of biodegradable films and/or edible food packages produced from starch (Fonseca et al. 2021). Zavareze et al. (2012) found improved mechanical properties such as higher tensile strength, young modulus, and elongation at break of biodegradable film produced from HMT potato starch but water vapor permeability has been increased which is not desirable for food packaging. The higher water permeability was also reported for the biodegradable films produced from HMT rice flour and starch (Majzoobi et al. 2015). They proposed that this type of films may be suitable for packaging of frigid products or product with less sensitivity to moisture content. However, Indrianti et al. (2018) reported that the biodegradable films produced from sweet potato starch modified by HMT with 25% moisture at 110 °C for 1, 2, and 3 h exhibited high thickness, tensile strength and elongation and lower water vapor permeability. Further they observed that the treatment time had no impact on the properties of edible films. Another emerging application of modified starch is the production of starch nanoparticles, which have novel food applications such as reinforcement materials, encapsulating agent, and emulsion stabilizers (Kumari et al. 2020). Since HMT can improve the thermal properties and crystallinity of starch nanoparticles it is widely used to modify the starch nanoparticles (Kumari et al. 2020).

4.5 Summary

Most of the studies on HMT have been focused on the impact of HMT on the structure and physicochemical properties of starch. From these studies it is evident that starch chain interactions and disruption of starch crystallites that occurred within the amorphous and crystalline region of starch in HMT alter the thermal, pasting, gelling, and digestible properties of starch without destroying the starch granules. The functional properties of native starches that can mainly improve by HMT are the acid stability and mechanical shear resistance in pasting. The extent of structural and subsequent changes of physicochemical properties has been shown to be affected by the amount of water, temperature, and time of heat treatment given in the treatment process. In case of type of starch structure, B-type starches are more susceptible to HMT. Though the structural and physicochemical changes occurred in HMT have been extensively studied, still little has been done to study the influence of HMT on lamellar organization and molecular characteristics of amylopectin and amylose within the starch granule. Extensive studies are also required to broaden the applications of HMT starch in food product development, with a view to optimizing the treatment process for the respective food applications of HMT starch.

References

Aberle T, Burchard W, Vorweg W, Radosta S (1994) Conformation, contributions of amylose and amylopectin to the properties of starches from various sources. Starch 46:329–335

Abraham TE (1993) Stabilization of paste viscosity of cassava starch by heat-moisture treatment. Starch 45:131–135

Adebowale KO, Henle T, Schwarzenbolz U, Doert T (2009) Modification and properties of African yam bean (*Sphenostylis stenocarpa Hochst. Ex A. Rich*) harms starch. I: heat-moisture treatments and annealing. Food Hydrocoll 23:1947–1957

Ali NA, Dash KK, Routray W (2020) Physicochemical characterization of modified lotus seed starch obtained through acid and heat-moisture treatment. Food Chem 319:126513

Ambigaipalan P, Hoover R, Donner E, Liu Q (2014) Starch chain interactions within the amorphous and crystalline domains of pulse starches during heat-moisture treatment at different temperature and their impact on physicochemical properties. Food Chem 143:175–184

Atwell WA, Hood LF, Lineback DR, Varriano-Marston E, Zobel HF (1988) The terminology and methodology associated with basic starch phenomena. Cereal Foods World 33:306–311

Bartz J, Zavareze ER, Dias ARG (2017) Study of heat-moisture treatment of potato starch by chemical surface gelatinization. J Sci Food Agric 97:3114–3123

Biliaderis CG (1991) The structure and interactions of starch with food constituents. Can J Physiol Pharmacol 69:60–78

Biliaderis CG (1998) Structures and phase transition of starch polymers. In: Walter RH (ed) Polysaccharide association structures in food. Marcel Dekker, Inc., New York, pp 57–168

Bogracheva TY, Morris VJ, Ring SG, Hedley CL (1998) The granular structure of C-type pea starch and its role in gelatinization. Biopolymers 45:323–332

Boyer CD, Shannon JC, Garwood DL, Creech RG (1976) Changes in starch granules size and amylose percentage during kernel development in several *Zea mays* L. genotypes. Cereal Chem 53:327–333

Brahma B, Sit N (2020) Physicochemical properties and digestibility of heat moisture–treated potato starches for different treatment conditions. Potato Res 63:367–383

Brumovsky JO, Thompson DB (2001) Production of boiling stable granular resistant starch by partial acid hydrolysis and hydrothermal treatments of high amylose maize starch. Cereal Chem 79:680–689

Chandla NK, Saxena DC, Singh S (2017) Processing and evaluation of heat-moisture treated (HMT) amaranth starch noodles; an inclusive comparison with corn starch noodles. J Cereal Sci 75:306–313

Cheetham NWH, Tao L (1998) Variation in crystalline type with amylose content in maize starch granules: and X-ray powder diffraction study. Carbohydr Polym 36:277–284

Chung HJ, Liu Q, Hoover R (2009) Impact of annealing and heat-moisture treatment on rapidly digestible, slowly digestible and resistant starch levels native and gelatinized corn, pea and lentil starches. Carbohydr Polym 75:436–447

Chung SY, Han SH, Lee SW, Rhee C (2012) Effect of Maillard reaction products prepared from glucose-glycine model systems on starch digestibility. Starch-Stärke 64(8):657–664
Chung HJ, Cho A, Lim ST (2014) Utilization of heat-moisture treated brown rices in sugar-snap cookies. LWT-Food Sci Technol 57:260–266
Collado LS, Corke H (1999) Heat-moisture treatment effects on sweet potato starches differing in amylose content. Food Chem 65:339–346
Collado LS, Mabesa LB, Oates CG, Corke H (2001) Bihon-type noodles from heat-moisture treated sweet potato starch. J Food Sci 66:604–609
Dewi AMP, Santoso U, Pranoto Y, Marseno DW (2022) Dual modification of sag starch via heat-moisture treatment and octenyl succinylation to improve starch hydrophobicity. Polymers 14: 1086
Donald AM, Waigh TA, Jenkins PJ, Gidley MJ, Debet M, Smith A (1997) Internal structure of starch granules revealed by scattering studies. In: Fraizer PJ, Richmond P, Donald AM (eds) Starch, structure and functionality. The Royal Society of Chemistry, Cambridge, pp 172–179
Donovan JW, Lorenz K, Kulp K (1983) Differential scanning calorimetry of heat-moisture treated wheat and potato starches. Cereal Chem 60:381–387
Dudu DE, Oyedegi AB, Oyebjinka SA, Ma Y (2019) Impact of steam- heat-moisture treatment on structural and functional properties of cassava flour and starch. Int J Biol Macromol 126:1056–1064
Eliasson AC (1986) Viscoelastic behavior during the gelatinization of starch. I. Comparison of wheat, maize, potato and waxy barely starches. J Text Stud 17:253–265
Englyst HN, Kingman SM, Cummings JH (1992) Classification and measurement of nutritionally important starch fractions. Eur J Clin Nutr 46:533–550
Fonseca LM, Halal SLME, Dias ARG, Zavareze EDR (2021) Physical modification of starch by heat-moisture treatment and annealing and their applications: a review. Carbohydr Polym 274: 118665
Franco CML, Ciacco CF, Tavares DQ (1995) Effect of heat-moisture treatment on enzymatic susceptibility of corn starch granules. Starch 47:223–228
French D, Murphy VG (1977) Computer modeling in the study of the starch. Cereal Foods World 22:61–70
Gernat C, Radosta S, Anger H, Damaschun G (1993) Crystalline parts of three different conformations detected in native and enzymatically degraded starches. Starch 45:309–314
Gidley MJ, Bulpin PV (1989) Aggregation of amylose in aqueous systems. The effect of chain length on phase behavior and aggregation kinetics. Macromolecules 22:341–346
Greenwood CT, Thompson JA (1959) Comparison of the starches from barely and malted barley. J Inst Brew 65:346–353
Gunaratne A, Corke H (2007) Effect of hydroxypropylation and alkaline treatment on some structural and physicochemical properties of heat-moisture treated wheat, potato and waxy maize starches. Carbohydr Polym 68:305–313
Gunaratne A, Hoover R (2002) Effect of heat-moisture treatment on the structure and physicochemical properties of tuber and root starches. Carbohydr Polym 49:425–437
Han L, Cao S, Yu Y, Xu X, Cao X, Chen W (2021) Modification of physicochemical, structural and digestive properties of pea starch during heat moisture process assisted by pre- and post-treatment of ultrasound. Food Chem 360:129929
He W, Wei C (2017) Progress in C-type starches from different plant sources. A review. Food Hydrocoll 73:162–175
Hizukuri S (1985) Relationship between the distribution of the chain length of amylopectin and the crystalline structure of starch granules. Carbohydr Res 141:295–306
Hizukuri S (1996) Starch analytical aspects. In: Eliason AC (ed) Carbohydrates in foods. Marcel Dekker Inc., New York, pp 347–429
Hizukuri S, Takeda Y, Yasuda M, Suzuki A (1981) Multi-branched nature of amylose and the action of debranching enzymes. Carbohydr Res 94:205–213

Holm J, Bjorck I, Ostrowska S, Eliasson AC, Asp NG, Larson K, Lungquist I (1983) Digestibility of amylose-lipid complexes in vitro and in vivo. Starch 35:294–297

Hood LF, Mercier C (1978) Molecular structure of unmodified and chemically modified manioc starches. Carbohydr Res 61:53–66

Hoover R (2010) The impact of heat-moisture treatment on molecular structures and properties of starches isolated from different botanical sources. Crit Rev Food Sci Nutr 50:835–847

Hoover R, Manuel H (1996) The effect of heat-moisture treatment on the structure and physicochemical properties of normal maize, waxy maize, dull waxy maize and amylomaize. V. starches. J Cereal Sci 23:153–162

Hoover R, Vasanthan T (1994) Effect of heat-moisture treatment on the structure and physicochemical properties of cereal, legume, and tuber starches. Carbohydr Res 252:33–53

Hoover R, Vasanthan T, Senanayake NJ, Martin AM (1994) The effect of defatting and heat-moisture treatment on the retrogradation of starch gels from wheat, oat, potato and lentil. Carbohydr Res 26:13–24

Hormdok R, Noomhorm A (2007) Hydrothermal treatment of rice starch for improvement of rice noodle quality. LWT-Food Sci Technol 40:1723–1731

Hoseney RC (1994) Principles of cereal science and technology, 2nd edn. American Association of Cereal Chemist, St. Paul, pp 29–34

Hung PV, Duyen TTM, Thanh HV, Widiastuti D, An NTH (2021) Starch digestibility and quality of cookies made from acid and heat-moisture treated sweet potato starch and wheat flour composites. J Food Measure Charact 15:3045–3051

Indrianti N, Pranoto Y, Abbas A (2018) Preparation and characterization of edible films made from modified sweet potato starch through heat-moisture treatment. Indian J Chem 18:679–688

Iuga M, Mironeasa S (2020) A review of hydrothermal treatments impact on starch based systems properties. Crit Rev Food Sci Nutr 60:3890–3915

Jacobasch G, Dongowski G, Schmiedl D, Muller-Schmehi K (2006) Hydrothermal treatment of Novelose 330 results in high yield of resistant starch type-3 with beneficial prebiotic properties and decreased secondary bile acid formation in rats. Br J Nutr 95:1063–1074

Jacobs M, Delcour JA (1998) Hydrothermal modifications of granular starch with retention of the granular structure. A review. J Agric Food Chem 46:2895–2905

Jane J, Xu A, Radosavljevic M, Seib A (1992) Location of amylose in normal starch granules. I. Susceptibility of amylose and amylopectin to cross-linking agents. Cereal Chem 69:405–409

Jane J, Kasemsuwan T, Leas S, Zobel HF, Robyt JF (1994) Anthology of starch granule morphology by scanning electron microscope. Starch-Starke 46:121–129

Jane JL, Wang KS, McPherson AE (1997) Branch structure difference in starches of A and B type X-ray pattern revealed by their Naegeli dextrins. Carbohydr Res 300:219–227

Jayakody L, Hoover R, Liu Q, Donner E (2007) Studies on tuber starches. II. Molecular structure, composition, and physicochemical properties of yam (*Dioscorea sp.*) grown in Sri Lanka. Carbohydr Polym 69:148–163

Jenkins PJ, Cameron RE, Donald AM (1993) A universal feature in the structure of starch structure of starch granules from different botanical sources. Starch 45:417–420

Ji Y, Ao Z, Han J-A, Jane J-L, BeMiller JN (2004) Waxy maize starch subpopulations with different gelatinization temperatures. Carbohydr Polym 57:177–190

Jiranuntakul W, Puttanlek C, Rungsardthong V, Puncha-arnon S, Uttapap D (2011) Microstructural and physicochemical properties of heat-moisture treated waxy and normal starches. J Food Eng 104:246–258

Kainuma K, French D (1971) Naegeli amylodextrin and its relationships to starch granule structure. I. Preparation and properties of amylodextrins from various starch types. Biopolymers 10:1673–1680

Karim AA, Norziah MH, Seow SS (2000) Methods for the study of starch retrogradation. Food Chem 71:9–36

Kawabata A, Takase N, Miyoshi E, Sawayama S, Kimura T, Kudp K (1994) Microscopic observation and X-ray diffractometry of heat-moisture treated starch granules. Starch 46:463–469
Koizumi K, Fukuda M, Hizukuri S (1991) Estimation of the distributions of chain length of amylopectins by high-performance liquid chromatography with pulsed amperometric detection. J Chromatogr 585:233–238
Kuhunae P, Tran T, Sirivongpaisl P (2007) Effect of heat-moisture treatment on structural and thermal properties of rice starches differing in amylose content. Starch 59:593–599
Kulp K, Lorenz K (1981) Heat-moisture treatment of starches. I. Physicochemical properties. Cereal Chem 58:46–48
Kumari S, Yadav BS, Yadav RB (2020) Synthesis and modification approaches o starch nanoparticles for their emerging food industrial applications. A review. Food Res Int 128: 108765
Kurahashi Y, Yoshion Z (2000) Heat-moisture treated starches; its production, properties and uses. J Appl Glycosci 47:125–132
Kweon M, Haynes L, Slade L, Levine H (2000) The effect of heat and moisture treatments on enzyme digestibility of AeWx, Aewx, and aeWx corn starches. J Therm Anal Calorim 59:571–586
Lawal OS, Adebowale KO (2005) An assessment of changes in thermal and physicochemical parameters of jack bean (Canavalia ensiformis) starch following hydrothermal modifications. Eur Food Res Technol 221:631–638
Lehman U, Robin F (2007) Slowly digestible starch-its structure and health benefits. A review. Trends Food Sci Technol 18:346–355
Li JH, Vasanthan T, Hoover R, Rossnagel BG (2004) Starch from hull-less barley: in-vitro susceptibility of waxy, normal, and high-amylose starches towards hydrolysis by alpha-amylases and amyloglucosidase. Food Chem 84:621–632
Li S, Rachelle W, Gao Q (2011) Effect of heat-moisture treatment on the formation and physico-chemical properties of resistant starch from Mung bean (Phaseolus radiates) starch. Food Hydrocoll 25:1702–1709
Liao L, Liu H, Gan Z, Wu W (2019) Structural properties of sweet potato and its vermicelli quality as affected by heat-moisture treatment. Int J Food Prop 22:1122–1133
Liu C, Song M, Liu L, Hong J, Guan E, Bian K, Zheng X (2020) Effect of heat-moisture treatment on the structure and physicochemical properties of ball mill damaged starches from different botanical sources. Int J Biol Macromol 156:403–410
Lorenz K, Kulp K (1981) Heat-moisture treatment of starches. II. Functional properties and baking potentials. Cereal Chem 58:49–52
Lorenz K, Kulp K (1982) Cereal and root starch modification by heat-moisture treatment. I. physicochemical properties. Starch 34:50–54
Majzoobi M, Pesaran Y, Mesbahi G, Golmakani MT, Farahanaki A (2015) Physical properties of biodegradable film from heat-moisture treated rice flour and ice starch. Starch 67:1053–1060
Mercier C, Charbonniere R, Grebaut J, de la Gueriviere JF (1980) Formation of amylose-lipid complexes by twin-screw extrusion cooking of manioc starch. Cereal Chem 57:4–9
Miles MJ, Morris VJ, Orford PD, Ring SG (1985) The roles of amylose and amylopectin in the gelation and retrogradation of starch. Carbohydr Res 135:271–281
Miyazaki M, Morita N (2005) Effect of heat-moisture treated maize starch on the properties of dough and bread. Food Res Int 38:369–376
Morrison WR, Karkalas J (1990) Methods in plant biochemistry. In: Starch, vol 2. Academic Press, New York, pp 323–352
Mustapha NA, Roslen SNH, Abd Gafar FS, Ibadullah WZW, Sukiri R (2019) Characterization of Dioscorea alata purpurea flour: Impact of moisture level. J Food Measure Charact 13:1636–1644
Olayinka OO, Adebowale KO, Olu-Owolabi BI (2008) Effect of heat-moisture treatment on physicochemical properties of white sorghum starch. Food Hydrocoll 22:225–230

Oostergetel GT, Van Bruggen EFJ (1993) The crystalline domain in potato starch granules are arranged in a helical fashion. Carbohydr Polym 21:7–12
Perez S, Bertoft E (2010) The molecular structures of starch components and their contribution to the architecture of starch granules. A comprehensive review. Starch 62:389–420
Pfannemuller B, Mayerhofer H, Schulz RC (1971) Conformation of amylose I aqueous solution: optical rotator dispersion and circular dichroism of amylose-iodine complexes and dependence on chain length of retrogradation of amylose. Biopolymers 10:243–261
Pukkahuta C, Suwannawat B, Shobsngob S, Varavinit S (2008) Comparative study of pasting and thermal transition characteristics of osmotic pressure and heat-moisture treated corn starch. Carbohydr Polym 72:527–536
Rabe E (1999) Complex carbohydrates in foods. Marcel Dekker, New York, pp 395–409
Ring SG (1985) Some studies on starch gelation. Starch 37:80–83
Ring SG, Gee MJ, Whittam M, Orford P, Johnson IT (1998) Resistant starch: its chemical form in food stuffs and effect of digestibility in vitro. Food Chem 28:97–109
Robin JP, Mercier C, Duprat F, Guilbot A (1975) Lintnerized starches. Chromatographic and enzymatic studies of insoluble residues from acid hydrolysis of various cereal starches, particularly waxy maize starch. Starch 27:36–45
Robyt JF (1984) Enzymes in the hydrolysis and synthesis of starch. In: Whistler RL, BeMiller JN (eds) Starch chemistry and technology, 2nd edn. Academic Press, New York, pp 87–124
Sair L (1967) Heat-moisture treatment of starch. Cereal Chem 44:8–26
Sajilata MG, Singhal RS, Kulkani PR (2006) Resistant starch. A review. Compr Rev Food Sci Food Saf 5:1–17
Shih F, King J, Daigle K, An HJ, Ali R (2007) Physicochemical properties of rice starch modified by hydrothermal treatment. Cereal Chem 84:527–553
Shin SI, Kim HJ, Ha HJ, Lee SH, Moon TW (2005) Effect of hydrothermal treatment on formation and structural characteristics of slowly digestible non-pasted granular sweet potato starch. Starch 57:421–430
Singh H, Chang YH, Lin JH, Singh N, Singh N (2011) Influence of heat-moisture treatment and annealing on functional properties of sorghum starch. Food Chem 44:2949–2954
Snow P, O'Dea K (1981) Factors affecting the rate of hydrolysis of starch in food. Am J Clin Nutr 34:2721–2727
Stute R (1992) Hydrothermal modification of starches. The difference between annealing and heat-moisture treatment. Starch 44:205–2014
Sui Z, Yao T, Ye X, Bao J, Kong X, Wu Y (2017) Physicochemical properties and starch digestibility of in-kernel heat-moisture treated waxy, low and high amylose rice starch. Starch 69:1600164
Sun Q, Wang T, Xiong L, Zhao Y (2013) The effect of heat-moisture treatment on physicochemical properties of early indica rice. Food Chem 141:853–857
Sun Q, Han Z, Wang L, Xiong L (2014) Physicochemical differences between sorghum starch and sorghum flour modified by heat-moisture treatment. Food Chem 145:756–764
Takeda Y, Hizukuri S, Takeda C, Suzuki A (1987) Structures of branched molecules of amyloses of various origins and the molar fractions of branched and unbranched molecules. Carbohydr Res 165:139–145
Tan X, Li X, Xie F, Li L, Huang J (2017) Effect of heat-moisture treatment on multi-scale structures and physicochemical properties of breadfruit starch. Food Chem 161:286–294
Tester RF (1997) Starch: the polysaccharide fractions. In: Frazier PJ, Richmond P, Donald AM (eds) Starch: structure and functionality. The Royal Society of Chemistry, Cambridge, pp 163–171
Thompson DB (2000) Strategies for the manufacture of resistant starch. Trends Food Sci Technol 11:245–253
Varatharajan V, Hoover R, Liu Q, Seetharaman K (2010) The impact of heat-moisture treatment on the molecular structure and physicochemical properties of normal and waxy potato starches. Carbohydr Polym 81:466–475

Varatharajan V, Hoover R, Li J, Vasanthan T, Nantanga KKM, Seetharaman K, Liu Q, Donner E, Jaiswal S, Chibbar RN (2011) Impact of structural changes due to heat-moisture treatment at different temperatures on the susceptibility of normal and waxy potato starches towards hydrolysis by porcine pancreatic alpha amylase. Food Res Int 44:2594–2606

Vermeylen R, Goderis B, Delcour JA (2006) An X-ray study of hydrothermally treated potato starch. Carbohydr Polym 64:364–375

Walter M, Silva LP, Denadrin CC (2005) Rice and resistant starch: Different content depending on chosen methodology. J Food Compos Anal 18:279–285

Wang H, Zhang B, Chen L, Li X (2016) Understanding the structure and digestibility of heat-moisture treated starch. Int J Biol Macromol 88:1–8

Watcharatewinkul Y, Puttanlek C, Rungsardthong V, Uttapap D (2009) Pasting properties of heat-moisture treated canna starch in relation to its structural characteristics. Carbohydr Polym 75: 505–511

Watcharatewinkul Y, Uttapap D, Puttanlek C, Rungsardthong V (2010) Enzyme digestibility and acid/shear stability of heat-moisture treated canna starch. Starch 62:205–216

Yadav BS, Guleria P, Yadav RB (2013) Hydrothermal modification of Indian water chestnut starch: influence of heat-moisture treatment and annealing on the physicochemical, gelatinization, and pasting characteristics. LWT-Food Sci Technol 53:211–217

Yamaguchi M, Kainuma K, French DJ (1979) Electron microscopy observation of waxy maize starch. J Ultrastruct Res 69:249–261

Zavareze EDR, Dias AR (2011) Impact of heat-moisture treatment and annealing in starches. A review. Carbohydr Polym 83:317–328

Zavareze RED, Storck RC, Castro LASD, Schirmer MA, Dias ARG (2010) Effect of heat-moisture treatment on rice starch of varying amylose content. Food Chem 121:358–365

Zavareze ER, Pinto VZ, Klein B, Halal SLME, Elias MC, Herandez CP, Dias ARG (2012) Development of oxidized and heat-moisture treated potato starch film. Food Chem 132:344–350

Zhang T, Oates CG (1999) Relationship between α-amylase degradation and physico-chemical properties of sweet potato starches. Food Chem 65:155–163

Zobel HF (1988) Starch crystal transformations and their industrial importance. Starch 40:1–7

Zobel HF (1992) Starch granule structure. In: Alexander RJ, Zobel HF (eds) Development in carbohydrate chemistry. American Association of Cereal Chemists, St. Paul, pp 1–36

Chapter 5
Annealing

Tianming Yao, Zhongquan Sui, and Srinivas Janaswamy

Abstract Annealing is a physical modification of starch through heat treatment in the presence of water with controlled time. The heating temperature is below the gelatinization temperature and above the glass transition temperature. It only uses water and heat energy without any chemical reagents; thus annealing is an eco-friendly and cost-effective process. Unlike other starch modification technologies, such as chemical and enzyme modification, annealing induces relatively mild property changes. Thus, annealed starches maintain intact granule architecture but with changes in physicochemical properties such as crystallinity, gelatinization, swelling factor, solubility, viscosity, and hydrolysis rate. In addition to heating time, temperature, and moisture, intrinsic starch properties (e.g., source, amylose content) influence the annealing properties. Overall, annealing is a cost-effective, simple, and viable technique to modify starch functionality to develop value-added food products.

Keywords Annealing · Starch · Functionality · Application · Functional foods

5.1 Introduction

Starch is a natural botanical biomaterial from diverse sources such as cereals, roots, fruits, and legumes. It is composed of two biopolymers, amylose and amylopectin, and their ratio and molecular arrangement predominantly account for a range of functional properties and widespread applications in food and non-food industries,

T. Yao
Department of Food Science, Whistler Center for Carbohydrate Research, Purdue University, West Lafayette, IN, USA

Z. Sui
Department of Food Science and Engineering, Shanghai Jiao Tong University, Shanghai, China

S. Janaswamy (✉)
Department of Dairy and Food Science, South Dakota State University, Brookings, SD, USA
e-mail: Srinivas.Janaswamy@sdstate.edu

Z. Sui, X. Kong (eds.), *Physical Modifications of Starch*,
https://doi.org/10.1007/978-981-99-5390-5_5

Table 5.1 Summary of starch utilization in various industrial applications[a]

Industry	Application	Key starch function
Adhesive	Glue, clean-label adhesive products	Thickening, low gelatinization, high viscosity
Agriculture	Pesticide delivery, herbicide encapsulation, seed coating	High thermal stability, low water binding
Cosmetics	Make-up powder, skincare emulsion, dry shampoo, bath cream	High temperature resistance, anti-thickening, light color, moderate viscosity
Detergent	Surfactants, builders, co-builders, bleaching agents and activators	Emulsion material (offering hydrophilic head groups), low molecular weight
Food	Major ingredients, viscosity modifier, glazing agents, fat replacer, emulsion agents, encapsulation material, bulking agents	Diverse needs of high/low gelatinization temperature, viscosity, thickening
Medical	Plasma extender, transplant organ preservation materials, delivery materials	Strong water binding, low molecular weight
Oil drilling	Viscosity modifier	Thickening, rheology stability
Paper	Binding, sizing, and coating	Bonding agent, low viscosity for surface sizing while high viscosity in paper mills
Pharmaceuticals	Diluent, binder, drug delivery	Thermal stability, water binding, emulsion function
Plastics	Biodegradable fillers, starch-based bags and boxes	Low gelatinization temperature, high viscosity
Textile	Sizing, finishing, printing, fire resistance	Strong water binding, high viscosity

[a] Adopted from Zhang et al. (2015), Li et al. (2021), Adewale et al. (2022), Zarski et al. (2021)

e.g., feed, foundry, medicine, textile, paper making, construction, and petroleum, to name a few. Native starch, however, cannot completely meet the criteria due to its undesirable properties, e.g., impaired thermal stability (low gelatinization temperature) and robust water-binding effect (high swelling factor). In addition, depending on the source of extraction, starch properties such as granular shape, size, molecular arrangement (A-, B-type polymorphs), and chemical composition (amylose to amylopectin ratio) vary, influencing the functional attributes. For instance, higher amylopectin content yields easier retrogradation and strong film formation, while elevated amylose results in soft gels. In this regard, modified starches are being engineered to gain the required gelatinization temperature, viscosity, transparency, stability, and film-forming properties, to name a few, and to meet the burgeoning demands (Table 5.1).

Generally, there are four starch modification approaches, namely, chemical, physical, enzymatic, and genetic. Among them, physical modification is nontoxic, cost-effective, energy saving, and stands out as a "clean label" technology (Jacobs and Delcour 1998; Aschemann-Witzel et al. 2019), with minimum processes, free from chemicals, and simple natural ingredients to produce ideal starch-based functional products. It further supports the mushrooming "green consumerism." Among the several physical modification processes, annealing is well studied that was originally applied in metal manufacturing to strengthen blades and swords. It greatly improves the strength of polymer crystals, induces morphological changes in elastomer, raises hot carrier ability in glassy materials, and even affects the dielectric properties of materials (Van Bogart et al. 1981; Kizilyalli et al. 1997; Chatterjee et al. 2008). It has been successfully applied on biopolymers too, e.g., starch, for modifying semi-crystalline structure to circumvent undesirable properties. In the case of starch, it modifies water solubility, thermal stability, pasting and viscosity properties, particle size, and digestibility (Iuga and Mironeasa 2020; Fonseca et al. 2021; Khatun et al. 2019) and even enhances the viscosity profile without gelatinization, starch granule damage, and loss of birefringence (Stute 1992). Treatment parameters such as amount of time, moisture, and temperature influence the modified starch functionality to varied extents. However, annealing could only induce mild physicochemical changes compared to chemical modifications that usually involve harsh reactions leading to substantial alterations in starch properties. More importantly, it does not involve toxic or harmful chemical reagents and thus is preferred highly for food applications. In recent years, there have been renewed interests on starch annealing and several articles have appeared in the literature. This chapter is aimed at (1) covering the basic principles and concepts of starch annealing along with the influence of experimental parameters and starch origins, (2) summarizing the newer findings and developments on annealing treatments coupled with the shift of starch properties, and (3) emphasizing the current and future applications of annealing-treated starch products.

5.2 The Basic Principle

5.2.1 Definition

Annealing is a process that combines water and heat with starch for a fixed period in the temperature range above the glass transition temperature and below the gelatinization temperature. This definition is somewhat vague to symbolize the actual annealing process. Findings about the heat and water effect on modulating starch properties generate hydrothermal treatment, and thus the term "annealing" along with "heat-moisture treatment" (HMT) was coined to describe the whole process. It certainly is debatable about the differentiation of annealing from the HMT, as both are similar in the experimental processes and are associated with controlling the moisture level, temperature, and heating time (Iuga and Mironeasa 2020). To some extent, there is no difference between these protocols and the gelatinization temperature and viscosity increase as well as the shape, size, and optical birefringence of starch granules, to name a few, are essentially the same. Products with higher temperature treatment are more altered than those at the low temperature treatment. Hence to distinguish these two processes the term annealing is used for the low

Table 5.2 A comparison between the annealing and heat-moisture treatment (HMT)

	Parameter	Annealing	HMT
Treatment	Moisture level (w/w)	Intermediate (40–80%)	Reduced (<35%)
	Temperature (°C)	Mild (25–80)	High (>80)
	Time period	Long (hours to days)	Short (1–24 h)
Property	Gelatinization temperature	Increase	Increase
	Retrogradation extent	High	Low
	Peak viscosity	Decrease	Decrease
	Polymorph structure	Remains as the original	Shifts from B to A

temperature hydrothermal treatment and HMT for the high temperature treatment (Table 5.2). Variations between them could be assessed by Brabender viscosity, Differential Scanning Calorimetry (DSC), and X-ray Diffraction (XRD) analyses, to name a few. For example, annealing the potato starch does not alter the crystal structure even after prolonged durations of treatment (Stute 1992). On the other hand, HMT changes the B-type native starch polymorph to the A- or C-type (Stute 1992; Boonna and Tongta 2018; Brahma and Sit 2020). Thus, annealing could be attributed as a lower grade hydrothermal modification compared to HMT. Based on these observations, annealing could be precisely defined as a process of physically treating starch by suspending in excess water (above 40% w/w) under mild temperature (<70 °C) conditions for longer durations.

5.2.2 *Mechanism of Annealing*

Native starch granules possess semi-crystalline organization made up of crystalline and amorphous regions (Bertoft 2017). Crystalline regions have tight and ordered helical structures that are predominantly composed of branched amylopectin chains, while amorphous regions possess loose and randomly packed amylose chains (Fig. 5.1). Due to the tightly packed chains organization, there is no space in the crystalline regions and hence free water cannot penetrate. The presence of heat, however, provides the required energy and melts the crystalline regions; thus water could seep through and fill the space between starch chains. This phenomenon is known as gelatinization (Biliaderis et al. 1980; Donmez et al. 2021). However, when the heat energy is not strong enough, i.e., lower than the melting enthalpy, crystalline regions cannot be thawed, and consequently, the initial reorganization of molecules will be retained. Thus, annealing could influence the amorphous regions only.

Generally, there are two states of amorphous regions, glass state and rubber state (Liu et al. 2009). The glass state has less mobility and the chains are highly stable. On the other hand, the rubber state contains higher energy and consequently it could be modified easily. The transition between the glass state to the rubber state could be modulated by temperature, known as the glass transition temperature. In the native

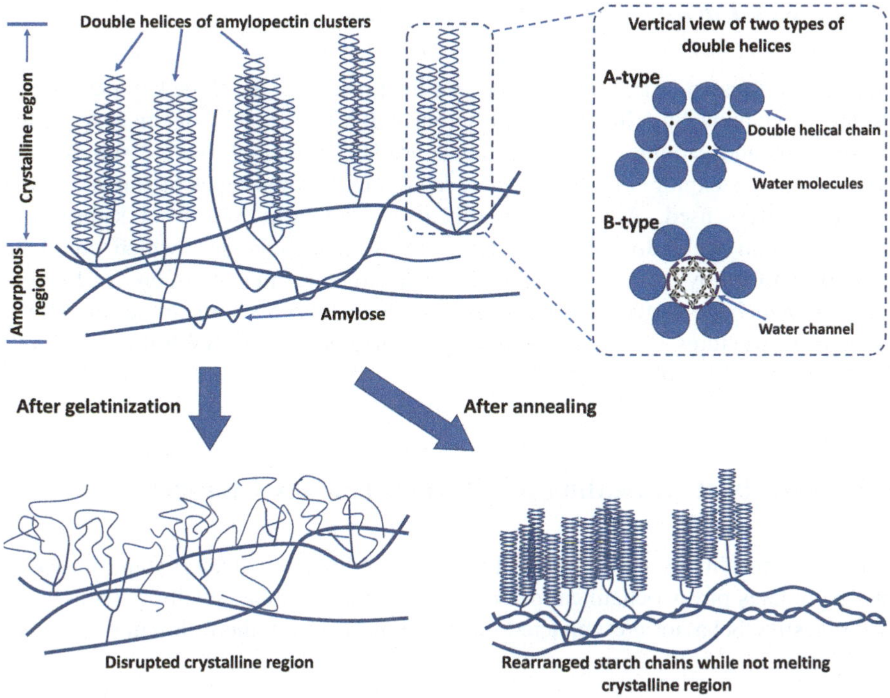

Fig. 5.1 Schematic illustration of molecular arrangements in semi-crystalline lamellae of starch granules. Melting of crystalline regions takes place when starch gelatinizes at high temperatures. In contrast, annealing induces subtle effect on crystalline regions but profound changes in the amorphous regions, e.g., increased interaction between amorphous starch chains (Rocha et al. 2011) and removal of amylose molecules (Wang et al. 2014a). (Reused with permission from Elsevier Publications)

starch, amorphous regions are in the glass state and thus they are less feasible for modification. During annealing, excess water penetrates the soft amorphous regions and hydrates them, thus providing the necessary chain mobility toward modification. Furthermore, annealing reduces the glass transition temperature, making the starch more plastic, and thus, the chains are susceptible to modification. The elevated temperature provides the required energy to transfer the glass state of amorphous starch to the rubber state (Waigh et al. 2000; Jeong et al. 2021). Consequently, starch chains gain freedom and the probability of their association leading to a steadier state increases significantly. After annealing, when the temperature reverts to ambient conditions, the chains form a thermally stable network arrangement and could eventually result in crystallization of amorphous chains. Although annealing does not melt the crystalline regions, Jayakody and Hoover (2008) proposed a sliding diffusion model of semi-crystalline polymers, in which the complete double helices clusters could move within a crystalline lattice, creating a fusion of crystals along with the reorganization of their amylose backbone. Such intricate changes alter starch properties and expand starch utility in food and non-food applications.

5.2.3 Key Factors in Starch Annealing

Heating temperature, moisture content, and treatment time are three key parameters during starch annealing. In addition, different starch sources with distinct crystalline polymorphs and variable amylose contents are also critical. Table 5.3 summarizes selected studies explored in this chapter, regarding different types of starch and the key parameters used. Annealing treatment on starches, even from same plant sources, could yield diverse and often controversial outcomes, which could be partially implied by the refined starch chemical structures that are often overlooked, e.g., internal chain length distribution and amylopectin branching structures, as fine molecular structures of starch could vary among phenotypes (Whitt et al. 2002; Sparla et al. 2014; Nakamura 2018).

5.3 Effects of Annealing on Starch Property Changes

Although annealing is a low degree of physical modification, it alters starch structure and modulates physicochemical properties. Thermal, pasting, swelling, hydrolysis, and digestive behavior are impacted most and discussed in the following sections.

5.3.1 Morphological Changes

Starch granule size, shape, and texture are important physical attributes in food and non-food applications. During annealing, starch granules retain their overall morphology (Falade and Ayetigbo 2015; Puelles-Román et al. 2021). For example, retention of starch granules is evidenced in several cultivars such as wheat, corn, oats, potato, lentil, and rice (Su et al. 2020; Yu et al. 2016; Vamadevan et al. 2013; Wang et al. 2017; Sun et al. 2021). Furthermore, granules could get aggregative with rough surfaces, for example, rice (da Rosa Zavareze et al. 2010). Porous surface structure as native in high amylose rice, cassava, waxy corn, and waxy barley starch could remain unperturbed (Nakazawa and Wang 2003; Dias et al. 2010; Wang et al. 2014b; Samarakoon et al. 2020). However, there could be a slight increase in the number and size of pores, resulting in a more fragile starch granular structure that is susceptible to erosion (Rocha et al. 2012). This phenomenon could be attributed to the expanding effect, as due to the presence of excess water in the system, granules gain the ability to absorb excessive amounts of water, increasing the distance between clusters in the granule and leading to pore-like channels on the granule surface.

Table 5.3 Summary of annealing parameters of various starches

Starch source	Temp (°C)	Moisture (%)	Time (h)	Significant outcome[a]	Reference
Wheat (common buck)	50	80.0	24	Increase in crystallinity and T (gel); reduction in swelling power, ΔH, peak viscosity, and digestibility	Liu et al. (2015)
Wheat (tartary buck)	50	80.0	24	Increase in crystallinity, T (gel), and ΔH; reduction in swelling power	Liu et al. (2016)
Wheat (normal)	30, 40, 50	16.6	24	Increase in T (gel) and pasting viscosity	Wang et al. (2017)
Wheat (normal)	50	75.0	12	Increase in amylose content and crystallinity; reduction in digestibility	Su et al. (2020)
Corn (normal)	45, 50	90.0	72	Increase in T (gel) and firmness; slight increase in crystallinity; reduction in ΔH and cohesiveness	Yu et al. (2016)
Corn (normal)	55	67.5	24	Increase in T (gel) and ΔH; reduction in swelling power	Ariyantoro et al. (2018)
Barely (normal)	50	80.0	3, 24	Increase in T (gel) and ΔH	Vamadevan et al. (2013)
Barely	50	75.0	72	Increase in crystallinity, reduction in swelling, shearing viscosity	Devi and Sit (2019)
Oats	45	80.0	3, 24	Increase in T (gel) and ΔH	Vamadevan et al. (2013)
Rice	58	80.0	3, 24	Increase in T (gel) and ΔH	Vamadevan et al. (2013)
Rice (normal)	50	83.3	24	Increase in swelling power and peak viscosity	Sun et al. (2021)
Rice (waxy)	50	83.3	24	Increase in swelling power; reduction in peak viscosity	Sun et al. (2021)
Rye	45	80.0	3, 24	Increase in T (gel) and ΔH	Vamadevan et al. (2013)
Sorghum	50	80.0	24	Increase in crystallinity, T (gel), and ΔH; reduction in swelling	Liu et al. (2016)
Pea	45	66.6	24, 72	Increase in crystallinity, T (gel), ΔH, pasting viscosity, and digestibility; reduction in swelling power	Wang et al. (2013)
Chickling pea	50	66.6	24	Increase in T (gel), ΔH, and digestibility; reduction in swelling power	Piecyk et al. (2018)
Cassava	55	80.0	3, 24	Increase in T(gel) and ΔH	Vamadevan et al. (2013)
Cassava	50	50.0	24	Increase in peak, final viscosity, and pasting temperature; reduction in swelling power	Falade et al. (2019)
Potato	30, 40, 50	16.6	24	Increase in T (gel) and ΔH; reduction in digestibility	Wang et al. (2017)
Sweet potato	55	90	72	Increase in T (gel) and digestibility; reduction in ΔH and crystallinity	Song et al. (2014)

(continued)

Table 5.3 (continued)

Starch source	Temp (°C)	Moisture (%)	Time (h)	Significant outcome[a]	Reference
Yam	30, 40, 50	16.6	24	Reduction in pasting viscosity	Wang et al. (2017)
Banana	65	70	24	Increase in crystallinity and digestibility; reduction in pasting viscosity	De la Rosa-Millan et al. (2014)

[a] T (gel) stands for gelatinization temperature and ΔH for melting enthalpy

5.3.2 *Influence on Starch Crystallinity*

The native starches are semi-crystalline materials. Their structural and network arrangements are assessed through X-ray diffraction patterns and the crystal polymorphic type dictates the overall crystallinity. For example, A-type starch networks with tightly packed double helices restrict water mobility while B-type starch networks possess water channels (Fig. 5.1). Consequently, A-type starches possess more crystallinity than B-types (Whittam et al. 1990; Martens et al. 2018). Upon gelatinization, however, starch granules lose their intrinsic structural integrity and the resulting patterns would be deprived of profiles with no crystallinity.

As the annealing treatment does not interfere with the granule shape and the heat energy is not enough to melt the crystalline regions, the initial semi-crystalline structural arrangement gets preserved. There are no reports on the shift of starch crystalline type through annealing modification in contrast to heat-moisture treatment (Table 5.2). For example, corn starch, a typical cereal starch, displays an A-type diffraction pattern for low amylose cultivators and B-type for high amylose cultivators. After annealing, diffraction patterns remain unaltered (Yu et al. 2016; Li et al. 2020), which supports the notion that annealing does not significantly alter the crystalline regions in the starch granules. However, diffraction patterns contain more defined sharper peaks and sometimes newer peaks, portraying increased molecular ordering and crystallinity. For example, barely starch, a typical A-starch like other cereal starches, displays an additional peak at 20° (2θ) after annealing (Waduge et al. 2006; Devi and Sit 2019), suggesting the presence of V-type crystallites likely due to complexation of amylose chains with lipid molecules.

After annealing, a consistent increase in relative crystallinity has been noticed among different botanical starches. Samarakoon et al. (2020) compared the relative crystallinity among waxy corn, waxy rice, and waxy barley starches. Although diffraction patterns are not affected by annealing, the relative crystallinity increases slightly for all the starches. The increase in relative crystallinity is also observed for wheat, corn, sorghum, pea, banana, and adlay starches (Liu et al. 2015, 2016; Yu et al. 2016; Wang et al. 2013; De la Rosa-Millan et al. 2014; Octavia Yusuf et al. 2022). On the other hand, the annealing effect on tuber starches is quite interesting. In the case of potato starch, substantial influence is not evidenced; however, for sweet potato starch, which is a C-type starch, a mixture of A- and B-type, the ratio of

A- to B-type changes with an increase in the overall crystallinity (Genkina et al. 2004). This interesting observation could presumably be due to the close proximity of the annealing temperature (2–3 °C) with the gelatinization temperature, which is below the A-type starch melting temperature and above the B-type melting. Normally, the melting point of A-starches is roughly 20 °C higher than B-starches (Whittam et al. 1990). Overall, annealing does not impart significant influence on the crystalline structure of starches.

5.3.3 Influence on Starch Gelatinization and Thermal Properties

Gelatinization is one of the important attributes that dictates starch quality. In the presence of a suitable amount of water coupled with high temperature (above the melting temperature), the semi-crystalline arrangement in the starch granule gets converted to an amorphous form. This unique property induces changes in the starch functional properties, mainly viscosity and gel-forming ability, a couple of important functional attributes for industrial food applications of starches. Since gelatinization is modulated by water and heat, annealing the starch granules will also induce measurable influence. Differential Scanning Calorimetry (DSC) analysis is used to study the impact of annealing on starch gelatinization and thermal properties. Annealing generally increases the onset temperature (T_o) and peak temperature (T_p), but decreases the temperature range of crystalline melting (T_c–T_o) (Liu et al. 2015; Wang et al. 2013, 2017; Yu et al. 2016; Ariyantoro et al. 2018; Devi and Sit 2019; Piecyk et al. 2018). As T_o represents the first melting of the starch crystalline regions, its increase suggests enhanced stability of the crystalline arrangement due to annealing, matching with the increase of relative crystallinity mentioned in the above section. These changes indeed specify the narrow melting range and homogenized crystalline packing structure (Samarakoon et al. 2020). More importantly, re-establishment of the amylose–amylose and/or amylose–amylopectin interactions in the amorphous regions is one of the reasons for an increase in the gelatinization temperature (Jacobs and Delcour 1998; Fonseca et al. 2021).

There is a debate that the gelatinization enthalpy (ΔH) changes due to annealing. For instance, evidence exists for a significant ΔH increase despite the narrow temperature range, indicating a high endothermal peak (Liu et al. 2016; Vamadevan et al. 2013; Wang et al. 2017; Piecyk et al. 2018), while a few studies suggest no change or decrease in the ΔH (Stute 1992; Liu et al. 2015; Song et al. 2014; Yu et al. 2016). The higher the ΔH, the more extra energy needs to be absorbed during starch gelatinization, suggesting a more perfect crystalline starch structure. As annealing deems to form a tighter starch network leading to enhanced crystallinity, it is reasonable to associate annealing with higher ΔH. Changes in ΔH could also be influenced by amylose content, amylose and amylopectin arrangement as well as the chain length distribution (Vamadevan and Bertoft 2020; Puelles-Román et al. 2021).

5.3.4 Influence on Swelling Power and Solubility

Reduction in starch swelling power and solubility upon annealing has been widely reported, especially for wheat, corn, barely, sorghum, potato, cassava, and pea starches (Liu et al. 2016; Wang et al. 2013, 2017; Ariyantoro et al. 2018; Devi and Sit 2019; Falade et al. 2019; Piecyk et al. 2018). Swelling power relates to the water holding ability of starch and very well reflects the mouth feeling of starch-based foods and the breakdown effect of medical tablets. Generally, greater swelling power is a better-quality attribute for food and non-food applications of starches. However, reduced swelling power could be useful in starch pasting and gel formation functions (Mathobo et al. 2021). The swelling power of native starch is high since it absorbs significant water. It is also related to starch gelatinization. The higher the gelatinization temperature, the lesser would be the swelling power of starch. Annealing strengthens the starch internal structure and hence more energy is required to expand starch chains and to trap water molecules. Furthermore, the elevated amount of crystalline material increases the gelatinization temperature. Thus, it is not surprising that annealing retards starch swelling. In addition, amylose content could also alter the extent of swelling power reduction. A significant decrease in the swelling power is observed in normal and high amylose corn starch after annealing, with little effect on waxy starch (Wang et al. 2014b). Overall, the intricate amylose and amylopectin interactions coupled with increased crystalline regions modulated by the amount of moisture and heat greatly influence starch swelling (Fonseca et al. 2021). The solubility is also reduced predominantly due to strong interactions between amylose and amylopectin chains and tight amylopectin–amylopectin organization that restrains amylose leaching.

5.3.5 Influence on Pasting and Viscosity Properties

Pasting is an important attribute that reflects the starch state in the presence of water and heat, which is an essential parameter for industrial applications. Interestingly, upon annealing, pasting property alters significantly and shows dependence on starch type and source. An increase in the pasting temperature and a decrease in peak and final viscosity have been observed for the potato starch (Song et al. 2014). On the other hand, wheat, pea, and adlay starches display higher peak and final viscosity (Wang et al. 2013, 2017; Octavia Yusuf et al. 2022), whereas oat starch viscosity remains unchanged (Werlang et al. 2021). Annealing starch renders better temperature resistance and stirring behavior mainly due to intricate changes in the crystalline regions. Crystallinity increase along with strong internal forces among starch chains results in elevated pasting temperature, and the reorganized chains contribute to reduced viscosity. Viscosity differences between starch sources could be attributed to amylose:amylopectin ratio, internal helix arrangement, and chain length distribution.

Annealing on starch pasting properties could be affected by correlated factors of amylose content, amylopectin branch chain length distribution, swelling power, gelatinization temperature, and relative crystallinity (Fonseca et al. 2021). As mentioned above, annealing reduces swelling power and increases relative crystallinity and gelatinization temperature. Such property shifts attenuate granule deformation by strengthening the intra-granular binding forces (i.e., the perfection of crystalline structure).

5.3.6 *Influence on Chemical Hydrolysis*

Starch granules could be hydrolyzed by mineral or organic acids to prepare products such as dextrin and resistant starch. Generally, amorphous regions in the starch network are susceptible to acid hydrolysis due to their soft network structure and allow acid to seep through the system breaking the glycosidic linkages. Later, acid attacks crystallites decoupled from amorphous chains, and thus more amorphous parts get exposed that further facilitates acid hydrolysis (Babu et al. 2019). This is very well supported by decreased amylose content during the initial stage of hydrolysis (Fabrice Fabien et al. 2020). The impact of annealing on starch hydrolysis behavior is quite interesting. Annealing the potato, cassava, high amylose maize, and millet starches makes them susceptible to hydrolysis (Nakazawa and Wang 2003; Babu et al. 2019) but with decreased hydrolysis rate in waxy barely and waxy potato starches (Samarakoon et al. 2020) and subtle changes on the waxy corn and rice starch (Samarakoon et al. 2020). These results clearly indicate that merely the perfection of crystalline regions cannot explain the acid hydrolysis. It could be influenced by several factors such as crystallinity, the helical structure of free amylose chains, void space increase in the amorphous areas, porous surface, the interaction between amylose-lipid complexes, and increase of branch point in the crystalline regions, to name a few (Jayakody and Hoover 2008). In addition, acid type and concentration, reaction conditions, and temperature influence starch hydrolysis.

5.3.7 *Influence on Enzymatic Digestion*

Enzymatic hydrolysis is an essential property of starch and aids in predicting the in vivo starch digestion for developing low glycemic index (GI) foods. Since annealing promotes changes in the amorphous areas of granules and perfection of crystallites, and the crystalline region is relatively resistant to digestion, annealed starch is less likely to have a fast digestion rate. The crystalline regions will be resistant to enzymes because of their highly tight packing arrangement. Depending on the digestion rate, starch could be categorized into rapidly digestive starch (RDS), slowly digestive starch (SDS), and resistant starch (RS). Reduction in RDS while increment in SDS and RS is key to reducing and smooth blood glucose level response. After annealing, decreased RDS but increased SDS and RS amounts have been observed for normal and waxy wheat starches (Su et al. 2020). In the case of Castanopsis starch a decrease in RDS and RS fractions with an increase in SDS has been observed (Shi et al. 2021). A similar trend has been reported in wheat, potato, and yam starches (Liu et al. 2015; Wang et al. 2017), suggesting that annealing modification could be one of the potential approaches to produce healthy functional food ingredients.

Interestingly, sweet potato, banana, and legumes starches could be easily hydrolyzed upon annealing, leading to increased starch digestion and elevated RDS amount (Song et al. 2014; Wang et al. 2013; Piecyk et al. 2018). The high digestive rate could be attributed to pores forming on starch granules after annealing. The porous structure could alter starch enzymatic hydrolysis but depends on the type of starch and enzyme (O'Brien et al. 2009). Further, this phenomenon is more significant in starches with intrinsic porous structures such as corn starch. These pores go through the whole granule and make it easy for enzymes to penetrate the internal soft regions leading to increased starch hydrolysis. On the other hand, a decrease in enzyme susceptibility is also noticed and has been attributed to the perfection of crystallites and enhanced amylose–amylose and/or amylose–amylopectin interactions. Overall, annealing induces two main structural effects: increasing porous structure on the granules and making perfect crystalline regions. It appears that annealing starches with porous structure is likely to increase enzyme hydrolysis, while starches with higher crystallinity are associated with a low digestion rate.

5.4 Applications of Annealed Starches

In general, annealing starches improves thermal stability, decreases viscosity and swelling, and modulates starch digestion. Hence, a variety of functional foods such as frozen foods, rice noodles, and tablets could be developed. The viscosity changes and low degree swelling of annealed starches benefit products such as bread, cakes, and noodles, to name a few. The reduced RDS and altered SDS amounts serve as new ingredients in the starch-based products to develop biocompatible functional foods and address diabetic and obesity concerns. In addition, novel non-food ingredients could be utilized in papermaking and cosmetic industries.

5.4.1 Applications in Breadmaking

Gelatinization and retrogradation are the important attributes to attain the bread quality. Annealed starches play a critical role in modifying the dough viscoelastic properties and texture (Miyazaki et al. 2006). Adding hydrothermally treated starch to wheat dough reduces dough viscosity, promotes a harder gel structure, and leads to gluten aggregates due to network strength decrease (Iuga and Mironeasa 2020). The adhesiveness and chewiness of rice dough improve with the presence of annealed rice starch (Hormdok and Noomhorm 2007). Although the texture of some annealing starch foods could be tougher or no change compared to native starch foods (Molavi and Razavi 2018), unique bread texture could be obtained, controlled and designed with suitable annealing parameters (e.g., temperature and time). The cost of annealed starch production is low and is quite economical. As a result, annealed starches have been gaining popularity in the breadmaking industry.

5.4.2 Applications in Frozen and Canned Foods

After annealing, starch granule structure is recrystallized and gets refined in obtaining a stable status that in turn reduces retrogradation and increases heat stability. Such improved thermal stability and shear resistance are helpful in food making. Frozen and canned foods are generally cooked foods that only need simple heating for consumption. However, cooling process is a major concern while retaining the food quality due to retrogradation at low temperatures. As the annealed starches have much lower retrogradation temperature than native starches, they are advantageous for applications in canned foods as well as frozen foods (Ariyantoro et al. 2018).

5.4.3 Applications for Improving Noodle Quality

Annealing decreases starch swelling power, solubility, and amylose leaching, but enhances shear stability and toughness, which are suitable in noodle processing. The overall noodle textural attributes, such as chewiness and adhesiveness could be improved extensively by using annealed treated noodle flours. For example, RVA profile and gel formation of rice flour could be promoted through annealing, leading to quality improvement (Hormdok and Noomhorm 2007). Overall, annealing results in a paradigm shift in utilizing fewer quality starches toward making quality products, e.g., noodles.

5.5 Summary

Annealing as a hydrothermal physical modification is an inexpensive and powerful tool to modify the physicochemical and digestive properties of starches. It is defined as treating starch at intermediate and/or excessive water content and heat above T_g while below T_m. The physicochemical changes are mainly due to structural alternations induced annealing. In the amorphous regions, free amylose chains could get re-oriented as well as the amylopectin chains and lipids, leading to more thermal stable areas in addition to enhancing the perfection among the crystalline regions. In the rearranged amorphous regions, some void spaces are created that in turn result in a porous structure through the whole granule. These microstructural changes induce alternations in starch macro properties such as reduced swelling power, peak viscosity, higher gelatinization temperature, easy hydrolysis, and modulated RS properties, to name a few. All these alterations are deemed to be helpful in developing novel food supplements, functional foods, and medicinal foods, in the near future.

Acknowledgments Support from the USDA National Institute for Food and Agriculture (SD00H648-18 and SD00H722-22) is greatly appreciated.

References

Adewale P, Yancheshmeh MS, Lam E (2022) Starch modification for non-food, industrial applications: market intelligence and critical review. Carbohydr Polym 291:119590. https://doi.org/10.1016/j.carbpol.2022.119590

Ariyantoro AR, Katsuno N, Nishizu T (2018) Effects of dual modification with succinylation and annealing on physicochemical, thermal and morphological properties of corn starch. Foods 7(9):9. https://doi.org/10.3390/foods7090133

Aschemann-Witzel J, Varela P, Peschel AO (2019) Consumers' categorization of food ingredients: do consumers perceive them as 'clean label' producers expect? An exploration with projective mapping. Food Qual Prefer 71:117–128

Babu AS, Mohan RJ, Parimalavalli R (2019) Effect of single and dual-modifications on stability and structural characteristics of foxtail millet starch. Food Chem 271:457–465. https://doi.org/10.1016/j.foodchem.2018.07.197

Bertoft E (2017) Understanding starch structure: recent progress. Agronomy 7(3):56

Biliaderis CG, Maurice TJ, Vose JR (1980) Starch gelatinization phenomena studied by differential scanning calorimetry. J Food Sci 45(6):1669–1674

Boonna S, Tongta S (2018) Structural transformation of crystallized debranched cassava starch during dual hydrothermal treatment in relation to enzyme digestibility. Carbohydr Polym 191:1–7. https://doi.org/10.1016/j.carbpol.2018.03.006

Brahma B, Sit N (2020) Physicochemical properties and digestibility of heat moisture–treated potato starches for different treatment conditions. Potato Res 63(3):367–383

Chatterjee S, Kuo Y, Lu J (2008) Thermal annealing effect on electrical properties of metal nitride gate electrodes with hafnium oxide gate dielectrics in nano-metric MOS devices. Microelectron Eng 85(1):202–209

da Rosa Zavareze E, Storck CR, de Castro LAS, Schirmer MA, Dias ARG (2010) Effect of heat-moisture treatment on rice starch of varying amylose content. Food Chem 121(2):358–365

De la Rosa-Millan J, Agama-Acevedo E, Osorio-Díaz P, Bello-Pérez LA (2014) Effect of cooking, annealing and storage on starch digestibility and physicochemical characteristics of unripe banana flour. Rev Mex Ing Quím 13(1):151–163

Devi R, Sit N (2019) Effect of single and dual steps annealing in combination with hydroxypropylation on physicochemical, functional and rheological properties of barley starch. Int J Biol Macromol 129:1006–1014. https://doi.org/10.1016/j.ijbiomac.2019.02.104

Dias ARG, da Rosa Zavareze E, Spier F, de Castro LAS, Gutkoski LC (2010) Effects of annealing on the physicochemical properties and enzymatic susceptibility of rice starches with different amylose contents. Food Chem 123(3):711–719

Donmez D, Pinho L, Patel B, Desam P, Campanella OH (2021) Characterization of starch–water interactions and their effects on two key functional properties: starch gelatinization and retrogradation. Curr Opin Food Sci 39:103–109. https://doi.org/10.1016/j.cofs.2020.12.018
Fabrice Fabien DD, Asongni W, Inocent G, Leng M (2020) Effect of annealing, acid hydrolysis and branching enzyme on Dioscorea Schimperiana starch technological and functional properties. J Sci Res Rep 26:25–37. https://doi.org/10.9734/JSRR/2020/v26i1030319
Falade KO, Ayetigbo OE (2015) Effects of annealing, acid hydrolysis and citric acid modifications on physical and functional properties of starches from four yam (Dioscorea spp.) cultivars. Food Hydrocoll 43:529–539. https://doi.org/10.1016/j.foodhyd.2014.07.008
Falade KO, Ibanga-Bamijoko B, Ayetigbo OE (2019) Comparing properties of starch and flour of yellow-flesh cassava cultivars and effects of modifications on properties of their starch. J Food Measure Charact 13(4):2581–2593. https://doi.org/10.1007/s11694-019-00178-5
Fonseca LM, Halal SLME, Dias ARG, da Zavareze ER (2021) Physical modification of starch by heat-moisture treatment and annealing and their applications: a review. Carbohydr Polym 274: 118665. https://doi.org/10.1016/j.carbpol.2021.118665
Genkina NK, Wasserman LA, Yuryev VP (2004) Annealing of starches from potato tubers grown at different environmental temperatures. Effect of heating duration. Carbohydr Polym 56(3): 367–370
Hormdok R, Noomhorm A (2007) Hydrothermal treatments of rice starch for improvement of rice noodle quality. LWT - Food Sci Technol 40(10):1723–1731
Iuga M, Mironeasa S (2020) A review of the hydrothermal treatments impact on starch based systems properties. Crit Rev Food Sci Nutr 60(22):3890–3915. https://doi.org/10.1080/10408398.2019.1664978
Jacobs H, Delcour JA (1998) Hydrothermal modifications of granular starch, with retention of the granular structure: a review. J Agric Food Chem 46(8):2895–2905
Jayakody L, Hoover R (2008) Effect of annealing on the molecular structure and physicochemical properties of starches from different botanical origins–a review. Carbohydr Polym 74(3): 691–703
Jeong D, Lee JH, Chung H-J (2021) Effect of molecular structure on phase transition behavior of rice starch with different amylose contents. Carbohydr Polym 259:117712. https://doi.org/10.1016/j.carbpol.2021.117712
Khatun A, Waters DL, Liu L (2019) A review of rice starch digestibility: effect of composition and heat-moisture processing. Starch 71(9-10):1900090
Kizilyalli IC, Lyding JW, Hess K (1997) Deuterium post-metal annealing of MOSFET's for improved hot carrier reliability. IEEE Electron Device Lett 18(3):81–83
Li H, Dhital S, Flanagan BM, Mata J, Gilbert EP, Gidley MJ (2020) High-amylose wheat and maize starches have distinctly different granule organization and annealing behaviour: a key role for chain mobility. Food Hydrocoll 105:105820. https://doi.org/10.1016/j.foodhyd.2020.105820
Li X, Jiang G, He Y, Chen G (2021) Novel starch composite fluid loss additives and their applications in environmentally friendly water-based drilling fluids. Energy Fuel 35(3): 2506–2513. https://doi.org/10.1021/acs.energyfuels.0c03258
Liu P, Yu L, Liu H, Chen L, Li L (2009) Glass transition temperature of starch studied by a high-speed DSC. Carbohydr Polym 77(2):250–253. https://doi.org/10.1016/j.carbpol.2008.12.027
Liu H, Guo X, Li W, Wang X, Peng Q, Wang M (2015) Changes in physicochemical properties and in vitro digestibility of common buckwheat starch by heat-moisture treatment and annealing. Carbohydr Polym 132:237–244
Liu H, Lv M, Wang L, Li Y, Fan H, Wang M (2016) Comparative study: how annealing and heat-moisture treatment affect the digestibility, textural, and physicochemical properties of maize starch. Starch 68(11–12):1158–1168
Martens BM, Gerrits WJ, Bruininx EM, Schols HA (2018) Amylopectin structure and crystallinity explains variation in digestion kinetics of starches across botanic sources in an in vitro pig model. J Anim Sci Biotechnol 9(1):1–13

Mathobo VM, Silungwe H, Ramashia SE, Anyasi TA (2021) Effects of heat-moisture treatment on the thermal, functional properties and composition of cereal, legume and tuber starches—a review. J Food Sci Technol 58(2):412–426

Miyazaki M, Van Hung P, Maeda T, Morita N (2006) Recent advances in application of modified starches for breadmaking. Trends Food Sci Technol 17(11):591–599

Molavi H, Razavi SM (2018) Dynamic rheological and textural properties of acorn (Quercus brantii Lindle.) starch: effect of single and dual hydrothermal modifications. Starch 70(11-12):1800086

Nakamura Y (2018) Rice starch biotechnology: rice endosperm as a model of cereal endosperms. Starch 70(1-2):1600375. https://doi.org/10.1002/star.201600375

Nakazawa Y, Wang Y-J (2003) Acid hydrolysis of native and annealed starches and branch-structure of their Naegeli dextrins. Carbohydr Res 338(24):2871–2882

O'Brien S, Wang Y-J, Vervaet C, Remon JP (2009) Starch phosphates prepared by reactive extrusion as a sustained release agent. Carbohydr Polym 76(4):557–566

Octavia Yusuf MT, Dwi Masahid A, Ratnawati L, Indrianti N, Ekafitri R, Sholichah E, Afifah N, Sarifudin A, Hikal DM, Sami R, Khojah E, Aljahani AH, Al-Moalem MH, Fikry M (2022) Impact of heating temperature on the crystallization, structural, pasting, and hydration properties of pre-gelatinized adlay flour and its implementation in instant porridge product. Crystals 12(5): 5. https://doi.org/10.3390/cryst12050689

Piecyk M, Drużyńska B, Ołtarzewska A, Wołosiak R, Worobiej E, Ostrowska-Ligęza E (2018) Effect of hydrothermal modifications on properties and digestibility of grass pea starch. Int J Biol Macromol 118:2113–2120. https://doi.org/10.1016/j.ijbiomac.2018.07.063

Puelles-Román J, Barroso NG, Cruz-Tirado JP, Tapia-Blácido DR, Angelats-Silva L, Barraza-Jáuregui G, Siche R (2021) Annealing process improves the physical, functional, thermal, and rheological properties of Andean oca (Oxalis tuberosa) starch. J Food Process Eng 44(6): e13702. https://doi.org/10.1111/jfpe.13702

Rocha TS, Cunha VAG, Jane J, Franco CML (2011) Structural characterization of peruvian carrot (Arracacia xanthorrhiza) starch and the effect of annealing on its semicrystalline structure. J Agric Food Chem 59(8):4208–4216. https://doi.org/10.1021/jf104923m

Rocha TS, Felizardo SG, Jane J, Franco C, M. L. (2012) Effect of annealing on the semicrystalline structure of normal and waxy corn starches. Food Hydrocoll 29(1):93–99. https://doi.org/10.1016/j.foodhyd.2012.02.003

Samarakoon ERJ, Waduge R, Liu Q, Shahidi F, Banoub JH (2020) Impact of annealing on the hierarchical structure and physicochemical properties of waxy starches of different botanical origins. Food Chem 303:125344. https://doi.org/10.1016/j.foodchem.2019.125344

Shi X, Ding Y, Wan J, Liu C, Prakash S, Xia X (2021) Effect of annealing on structural, physicochemical, and in vitro digestive properties of starch from Castanopsis sclerophylla. Starch 73(7-8):2100005. https://doi.org/10.1002/star.202100005

Song HY, Lee SY, Choi SJ, Kim KM, Kim JS, Han GJ, Moon TW (2014) Digestibility and physicochemical properties of granular sweet potato starch as affected by annealing. Food Sci Biotechnol 23(1):23–31

Sparla F, Falini G, Botticella E, Pirone C, Talamè V, Bovina R, Salvi S, Tuberosa R, Sestili F, Trost P (2014) New starch phenotypes produced by tilling in barley. PLoS ONE 9(10):e107779. https://doi.org/10.1371/journal.pone.0107779

Stute R (1992) Hydrothermal modification of starches: the difference between annealing and heat/moisture-treatment. Starch 44(6):205–214

Su C, Saleh ASM, Zhang B, Zhao K, Ge X, Zhang Q, Li W (2020) Changes in structural, physicochemical, and digestive properties of normal and waxy wheat starch during repeated and continuous annealing. Carbohydr Polym 247:116675. https://doi.org/10.1016/j.carbpol.2020.116675

Sun L, Xu Z, Song L, Ma M, Zhang C, Chen X, Xu X, Sui Z, Corke H (2021) Removal of starch granule associated proteins alters the physicochemical properties of annealed rice starches. Int J Biol Macromol 185:412–418. https://doi.org/10.1016/j.ijbiomac.2021.06.082

Vamadevan V, Bertoft E (2020) Observations on the impact of amylopectin and amylose structure on the swelling of starch granules. Food Hydrocoll 103:105663. https://doi.org/10.1016/j.foodhyd.2020.105663

Vamadevan V, Bertoft E, Soldatov DV, Seetharaman K (2013) Impact on molecular organization of amylopectin in starch granules upon annealing. Carbohydr Polym 98(1):1045–1055. https://doi.org/10.1016/j.carbpol.2013.07.006

Van Bogart JWC, Bluemke DA, Cooper SL (1981) Annealing-induced morphological changes in segmented elastomers. Polymer 22(10):1428–1438

Waduge RN, Hoover R, Vasanthan T, Gao J, Li J (2006) Effect of annealing on the structure and physicochemical properties of barley starches of varying amylose content. Food Res Int 39(1): 59–77

Waigh TA, Gidley MJ, Komanshek BU, Donald AM (2000) The phase transformations in starch during gelatinisation: a liquid crystalline approach. Carbohydr Res 328(2):165–176

Wang S, Jin F, Yu J (2013) Pea starch annealing: new insights. Food Bioprocess Technol 6(12): 3564–3575. https://doi.org/10.1007/s11947-012-1010-7

Wang S, Luo H, Zhang J, Zhang Y, He Z, Wang S (2014a) Alkali-induced changes in functional properties and in vitro digestibility of wheat starch: the role of surface proteins and lipids. J Agric Food Chem 62(16):3636–3643

Wang S, Wang J, Yu J, Wang S (2014b) A comparative study of annealing of waxy, normal and high-amylose maize starches: the role of amylose molecules. Food Chem 164:332–338. https://doi.org/10.1016/j.foodchem.2014.05.055

Wang S, Wang J, Wang S, Wang S (2017) Annealing improves paste viscosity and stability of starch. Food Hydrocoll 62:203–211. https://doi.org/10.1016/j.foodhyd.2016.08.006

Werlang S, Bonfante C, Oro T, Biduski B, Bertolin TE, Gutkoski LC (2021) Native and annealed oat starches as a fat replacer in mayonnaise. J Food Process Preserv 45(3):e15211

Whitt SR, Wilson LM, Tenaillon MI, Gaut BS, Buckler ES (2002) Genetic diversity and selection in the maize starch pathway. Proc Natl Acad Sci 99(20):12959–12962. https://doi.org/10.1073/pnas.202476999

Whittam MA, Noel TR, Ring SG (1990) Melting behaviour of A-and B-type crystalline starch. Int J Biol Macromol 12(6):359–362

Yu K, Wang Y, Wang Y, Guo L, Du X (2016) Effects of annealing and additives on the gelatinization, structure, and textural characteristics of corn starch. Int J Food Prop 19(6): 1272–1281. https://doi.org/10.1080/10942912.2015.1071842

Zarski A, Bajer K, Kapuśniak J (2021) Review of the most important methods of improving the processing properties of starch toward non-food applications. Polymers 13(5):5. https://doi.org/10.3390/polym13050832

Zhang Y, Ding L, Gu J, Tan H, Zhu L (2015) Preparation and properties of a starch-based wood adhesive with high bonding strength and water resistance. Carbohydr Polym 115:32–37. https://doi.org/10.1016/j.carbpol.2014.08.063

Chapter 6
Pre-gelatinized Modification of Starch

Yan Hong and Xingxun Liu

Abstract Pre-gelatinized starch (PGS) is a type of physical modification of starch that is accomplished by heating and mechanical shearing. Pre-gelatinized modification starch is produced by sufficient heat, followed with drying and grinding. The objective is to generate starch ingredients with instantaneous cold water solubility and thickening/gelling capabilities. PGS, sometimes called "instant" starches and α-starch, can be dissolved in water at temperatures below gelatinization of the native starches. PGS particles exhibit a complete lack of birefringence and generally retain very little original native granule structure. But there are some ungelatinized granules left during commercial preparation made on a hot roll, i.e., via drum drying and extrusion.

Keywords Starch · Pre-gelatinization · Characterization · Application · Properties

6.1 Introduction

Pre-gelatinized starch (PGS) is a type of physical modification of starch that is accomplished by heating and by mechanical shearing. Pre-gelatinized modification starch is produced by sufficient heat, followed by drying and grinding. The objective is to generate starch ingredients with instantaneous cold water solubility and thickening/gelling capabilities. PGS, sometimes called "instant" starches and α-starch, can be dissolved in water at temperatures below gelatinization of the native starches (BeMiller 2016). PGS particles exhibit a complete lack of birefringence and generally retain very little original native granule structure, although there are some ungelatinized granules left in any commercial preparation made on a hot roll, i.e.,

Y. Hong (✉)
School of Food Science and Technology, Jiangnan University, Wuxi, Jiangsu, People's Republic of China
e-mail: hongyan@jiangnan.edu.cn

X. Liu
College of Food Science and Engineering, Nanjing University of Finance and Economics, Nanjing, Jiangsu, China

Z. Sui, X. Kong (eds.), *Physical Modifications of Starch*,
https://doi.org/10.1007/978-981-99-5390-5_6

via drum drying (Fritze 1973). PGS can be easily dispersed in cold water due to its properties, i.e., solubility and good adhesion in cold water. It is widely used in food products, animal products, pharmaceutical products, and other products to improve product quality (Maria et al. 2020).

6.2 Application of Pre-gelatinized Starch in Industries

6.2.1 *Application of PGS in Feed Industry*

The feed industry consists primarily of livestock, poultry, and aquaculture subsectors. There are five major feed markets including pig, chicken, duck, fish, and shrimp. Among the commercially important livestock, all are ruminants except for pig. However, the pig industry is responsible for the bulk of the feeds consumed for livestock. Cattle and goats though with large populations are raised primarily on roughages. Poultry feed production consists mainly of chicken and duck feeds. Chickens are raised for meat (broiler) or for eggs (layer), while ducks are farmed for their eggs. Recently aquaculture feeds have been considered a minor subsector of the feed milling industry. However, today the production of fish feeds is the fastest growing feed market (Cruz 1997).

PGS possesses high adhesion, is easily dissolved in cold water, easy to digest, and has other characteristics (He et al. 2020). It can be used in the feed industry to increase the adhesive properties and improve the taste of plant-based meals. For the feed with powdered form, PGS could act as the adhesive for the feed to produce good viscoelasticity, maintain stability in water, and reduce the loss, especially for turtle and eel feed. On the other hand, PGS could increase the digestion rate compared with native starch, which may help the livestock grow. Actually, PGS has ideal adhesion effect on various nutrients in the feed, ensuring the nutrition for livestock. The addition of PGS will make the products not easy to be broken in the process of transportation with less dust (Kamarudin et al. 2018).

6.2.2 *PGS Products in Food Industry*

Native starches have unique properties. However, the limited versatility of starches restricts their application in the full range of food industries. Therefore, starches are modified in order to enrich their functions in food manufacturing. Pre-gelatinization is one of the physical methods to modify starches. PGS has a wide range of application, including low-fat salad dressings, high-solid fillings, bakery fillings, and dry mixes. In processing, hot process is preferred against cold process where problems may arise, if flour, fruit, spices, or other ingredients contain amylases. In a hot process, the enzymes will be inactivated before the starch swells, making it vulnerable to enzymes. In real practice, PGS should be mixed firstly with other dry ingredients before addition to the aqueous phase, to slow hydration and reduce lumping. Regarding formulations with little sugar or other dry ingredients, starches are agglomerated to facilitate dispersion. One way to help desperation is a light coating of oil.

PGS has good water absorption and can improve the gas production capacity of the product system, and it can be used in baked goods to get more soft food. It also has a porous structure, which can improve the crispy structure of snack food such as puffed beans and crispy peanuts, giving them a crisp and fluffy structure. Furthermore, PGS is also used in frozen food, meat food, condiments, soft drinks, and other food industries (Li et al. 2020).

6.2.3 Pre-gelatinized (Instant) Starch Products in Pharmaceutical Industry

Recently, PGS is also used in the pharmaceutical industry as a versatile and multifunctional pharmaceutical excipient and acts as the tablet binder, disintegrant, and filler in pharmaceutical excipients (Elgorashi 2016). Pharmaceutical excipients have substantial impact on manufacturability, quality, safety, efficacy, and stability of drug substances in a dosage form. Therefore, it is desired to employ the minimum number of excipients in efficient formulation design. As a result, the use of multifunctional excipients has increased.

Starches from different sources such as corn, potato, and wheat and other botanical source starches were used in pharmaceutical excipients. Compared with native starch, PGS features excellent wet stability, easy dispersion in cold water, high viscosity, and moisture sorption and swelling. They can be used as a hydrophilic excipient to formulate sustained release in solid dosage forms. Pre-gelatinized maize starch has been used as a diluent in capsules and as disintegrants and binders to produce tablets. PGS is an excellent dry binder in direct compression tablets. It provides uniform filling of dies to ensure correct dosage and has excellent disintegration/dissolution properties. In addition, these starches have provided some promising results as hydrophilic matrices for extended release (Petra et al. 2021).

Further, low elastic recovery values supported the good binding properties with less tendency for capping and lamination. The drug was released very rapidly from all the formulations, supporting robust excipient characteristics of these starches (Elgorashi 2016).

Compared to native starches, partially pre-gelatinized starch can be produced with improved flow and compression characteristics that allow its use as a tablet binder in dry compression or direct compression processes. In such processes, PGS is self-lubricating. However, when used with other excipients, the addition of some lubricant to the formulation may be necessary. Typically, 0.25% w/w magnesium stearate is used for this purpose. Higher concentrations may cause adverse effects on tablet strength and dissolution (Kibbe 2012).

6.2.4 *Pre-gelatinized (Instant) Starch Products in Other Industries*

PGS is widely used in putty powder due to its solubility in cold water, good adhesion, nontoxicity, and environmental protection (Prasenjit and Borgohain 2015). Putty powder in the domestic construction industry is widely used in the interior wall leveling base. Among them dry mix mortar putty is an auxiliary material of building decoration coating made of organic polymer or inorganic binder and stone powder and other additives. PGS is one of the most commonly used organic polymer binder. Another important application of PGS in building materials is as a ceramic body strengthening agent. PGS can effectively improve the plasticity and wet strength of clay as an additive. When the strength increases, the rupture of the products reduces. The products after high temperature firing, starch heated gas volatilization, will not bring any adverse effects on the product.

In recent years, PGS (~12.26%) was used in the preparation of peach gum tablets. The product has the advantages of good elasticity, fast forming, and moderate viscosity, while the dosage is less than 12.26%. When the elasticity and gelation of the finished product are poor, and when it is higher than 12.26%, the hardness of the finished product is high. PGS (30–60%) is used in building cold water instant rubber powder with environmental protection, cold water solubility, strong adhesion, good smoothness, easy storage, and low cost. It can be mixed or used alone. PGS is applied in oil well drilling mud to increase consistency and decrease the filter loss of drilling fluid; in non-destructive drilling, PGS serves as an effective substitute for fluid loss control, which provides stable mud control. However, there is a high degree of degradation of PGS, so the drilling rate is required as soon as possible quickly.

6.3 Methods to Prepare PGS

PGSs are cooked and then dried in one of several ways, including drum drying, spray drying, cooking in aqueous ethanol, and extrusion cooking. This section will discuss the commercial ways to prepare the PGS.

6.3.1 *Drum Drying*

Drum drying is the traditional method to prepare PGS by pouring a starch slurry onto a hot drum and scraping off the cooked sheet with a knife. Chemical modification, drum operation, and grinding of the dried sheets are the three methods to control the rate of rehydration and texture of the finished phase. The feed starch can be a chemically modified product to add more ultimate properties. To make different products, factors would be changed, such as solids concentration, drying time, and temperature. The size of particles can be controlled. Drum-dried products have one shortcoming. In comparison with cook-up starches (as non-pre-gelatinized products are called), drum-dried products display slightly less viscosity and produce less glossy and less smooth pastes when dispersed in water. This deficiency is caused by the destruction of starch granule integrity in drum drying. Newer technologies have paid attention to this problem. Despite the shortness, drum drying is recommended for economy and for its ability to enable the usage of large particles in texturizing starches. Texturizing starches are typically obtained from highly cross-linked starches that are drum-dried. They are slow to hydrate and are applied in cookie fillings and toppings to control boil-out when subjected to high oven temperatures and offer a pulpy fruit texture.

Recently, the production of dry PGS from maize starch and maize flour using four different drum drying processes is described (Fritze 1973). A comparison is made among the following items: the twin-drum sump dryer, the single-drum dryer with top applicator rolls and starch slurry feed ahead of the first roll only, the single-drum dryer with top applicator rolls and starch slurry feed ahead of each roll, and the single-drum dryer as described under the preceding item but preceded by a starch cooker. In particular the dryer capacity, water absorptivity, viscosity, cold water solubility, bulk weight, and degree of gelatinization were investigated as functions of the feed concentration of the starch slurry and partly also of the drum speed. It could be demonstrated that advantages are offered in particular by pre-gelatinization before drying and that as a result thereof, the physical properties could be varied (Fritze 1973).

6.3.2 Extrusion

While extrusion gelatinized (cooked) starch damages granules more than drum drying, it reduces viscosity. Extrusion may even fragment amylopectin molecules. Drum-dried and extruded wheat starch both produced a continuous phase of melted starch. Drum-dried starch had higher average molecular weights in the amylose and amylopectin fractions than extruded starch. It has been demonstrated that both temperature and sheer stresses were responsible for this molecular fragmentation. By comparing the rheology of drum-dried and extruded starches, Loisel et al. (2006) found that extruded starches had lower cold water swelling power and greater solubility. Starches extruded at lower temperatures and at higher moisture contents were more likely the drum-dried starches in these properties. Extrusion has been favored to reduce molecular weight and increase solubility for improved emulsification by octenyl succinylated starches. Recently, the effects of pre-gelatinization, mild and severe parboiling processes on paddy rice, and the utilization of the corresponding flours (PGF, MPF, and SPF) for gluten-free (GF) pasta-making were investigated. Two pasta-making processes (extrusion cooking and conventional extrusion) were carried out. Compared to pre-gelatinization, both parboiling processes induced lower pasting viscosity at any temperature, enzymatic susceptibility, and hydration. The magnitude of these changes significantly increased with the severity of the parboiling treatment. The lowest value for cooking loss was detected for samples prepared by 100% SPF (extrusion cooking) or by mixture of SPF and PGF (50:50) (conventional extrusion). Nevertheless, the extrusion cooking process promoted an extremely firm texture of cooked pasta when applied to parboiled flours (Marti et al. 2013).

6.3.3 Spray Drying

Starch slurry is injected through an atomization aperture in a nozzle assembly to form a fine spray. Steam is injected through another aperture in the nozzle assembly into the atomized starch spray to gelatinize the starch; the entire operation should take place in an enclosed chamber. The time for passage of the material through the chamber, i.e., from the atomization aperture through the vent aperture, defines the cooking or gelatinization time. The gelatinized starch is recovered essentially as granules. The technology is broadly applied to both native and chemically modified starches. For example, using this process, Katcher and Schara developed a pre-gelatinized, modified normal maize starch that is essentially flavor-free and which has a viscosity-building capacity equivalent to spray-dried and pre-gelatinized tapioca starch (Katcher 1994). In 1992, an improved process was reported that starch is uniformly and simultaneously atomized and cooked in an aqueous medium by means of a single atomization step in an apparatus comprising a two-fluid, internal mix, spray-drying nozzle, coupled to a strategy for drying the cooked, atomized starch to produce a uniformly pre-gelatinized cold water swelling (CWS) starch with desirable textural, visual, and organoleptic properties.

6.3.4 Other Methods

Except for those three commercial methods, there are also new innovative and emerging technologies to prepare the PGS, such as mechanical activation (Zhang et al. 2013), microwave heating (González Parada and Pérez Sira 2003), and vacuum freeze drying.

Mechanical activation (MA) using a customized stirring ball mill could be used to prepare partially pre-gelatinized cassava starch (PPCS) which has been studied recently. MA could significantly destroy the crystal structure of starch granules, contributing to the increase of cold water solubility, flowability, and flood ability. It also leads to variations in particle size and surface morphology, and a decrease in pasting temperature, viscosity of cooked PPCS paste, breakdown, and setback. PPCS with different degrees of gelatinization can be prepared by controlling the milling time. The unique properties induced by MA and the potential use of PPCS as excipient in pharmaceutical formulations were discussed (Zhang et al. 2013).

Microwave heating is different from traditional heating methods, microwave has a strong medium penetrating ability, which can increase the temperature of internal materials and lead to the gelatinization of starch, but also causes the change of starch structure and degradation of macromolecular chain, which is more conducive to the action of enzyme (Yuan et al. 2020). In general, the higher the water content of the material, the greater the dielectric loss, the more conducive to increasing the microwave heating efficiency. After microwave radiation, cassava starch with varying water content forms burst holes of different pore sizes, providing channels for enzyme action and easy hydrolysis (Fan et al. 2015).

6.4 The Structure Change in Pre-gelatinization Processing

Native starch granules are widely regarded to be a multi-scale structure polymer which is from nanometer to micrometer scales, i.e., amylose and amylopectin chains (~nm) (Wu and Gilbert 2013; Xie et al. 2012; Pérez and Bertoft 2010; Liao et al. 2014), crystalline and amorphous lamellae structure (9–10 nm), alternating amorphous and semicrystalline growth rings (100–400 nm), and starch granules (<1–100 μm). The two basic components in starch are near-linear amylose and highly multiple-branched amylopectin, which may affect higher-level structure of starch (aggregation structure) (Witt et al. 2012; Witt and Gilbert 2014; Chen et al. 2016). Generally, it is widely regarded that the amorphous zone is amylose and the branching points of amylopectin, while the short-branched chains in the amylopectin are the main crystalline component in granular starch (Liu et al. 2009). Recently, the techniques including fluorophore-assisted carbohydrate electrophoresis, high-performance anion-exchange chromatography, multiple-detector size-exclusion chromatography, and various scattering techniques (light, X-ray, and neutron), spectroscopy technology (FTIR and NMR), and microscopy technologies such as light and polarized microscopy, scanning electron microscopy, and confocal laser scanning microscopy were used to study starch structure and its relationship between structure and property.

The changes in starch molecular, aggregation structure, and granular structures, i.e., disruption of double helical crystallites and damage to starch granules, respectively, as a result of the grinding of cereal grains, will be discussed separately. Slurries of canna starch were heated at 60 °C to investigate the change of aggregation structure by Lan et al. (2016). For this purpose, solid-state 13C CP/MAS NMR, X-ray diffraction (XRD), and small-angle X-ray scattering (SAXS) were used to estimate the short- and long-range orders within the structure. Crystal stabilization was observed, and the critical time for the formation of double helical crystals was 3 h. With longer times, imperfect sub-crystal started to prevail (Lan et al. 2016).

DSC, XRD, and SEM were used to study the pre-gelatinized Dioscorea dumetorum (bitter yam) and Dioscorea oppositifolia (Chinese yam) starches

(Okunlola et al. 2015). The PGS had significantly larger structures that were gel-like, aggregated, and irregular. The DSC endotherms revealed that pre-gelatinized bitter and Chinese starches had higher gelatinization temperatures than the native forms, and XRD spectra showed that characteristic semicrystalline structures of the native starches (2θ between 13 and 23 °C) were disrupted by pre-gelatinization, forming amorphous structures.

Three rice starches with different amylose contents (glutinous, 1.4%; jasmine, 15.0%; and Chiang, 20.2%) were pre-gelatinized in a double drum dryer at 110, 117, and 123 °C. Starch granule and crystallinity were determined by SEM and XRD and FTIR, respectively (Nakorn et al. 2009). The disintegration of rice starch granules and the formation of holes were showed after pre-gelatinization. Pre-gelatinized glutinous-rich starch (PGS) and pre-gelatinized jasmine rice starch (PJS) had higher degree of disintegration of the granular and macromolecular structure. The increased release of soluble components during thermal treatment explained this phenomenon, which was proved by the completely amorphous X-ray diffractograms of PGS and PJS indicating a complete destruction of the granular order. On the contrary, the granular structure of high-amylose Chiang rice starch was less ruined. A V-type crystalline amylose-lipid complex structure was formed in the pre-gelatinized Chiang rice starch (PCS), as confirmed by XRD. FTIR spectroscopy was used to investigate the crystallinity change in the short-range order. This IR band appeared as a small peak for PCS, whereas the IR band at 1046 cm^{-1} disappeared for PGS and PJS, which was in accordance with XRD results (Nakorn et al. 2009).

6.5 The Physical-Chemical Properties During Pre-gelatinization

Pre-gelatinized (PG) starches are always processed at low temperature to increase viscosity and offer a desirable texture. The functional properties of PG starch can be influenced by other constituents used in food matrices. The functional properties such as gelatinization, pasting, solubility, swelling, and digestibility as well as other physical and chemical properties were discussed.

The functional properties of those three rice starches with different amylose contents (glutinous, 1.4%; jasmine, 15.0%; and Chiang, 20.2%) which were pre-gelatinized in a double drum dryer at 110, 117, and 123 °C (Nakorn et al. 2009) were also studied by Nakorn et al. The water absorption index (WAI) and water solubility index (WSI) indicate the mass of water absorbed by dry starch and the amount of soluble components that have leached out from the granules. The macromolecular disorganization and the degradation of starch during thermal treatment can result in higher WAI and WSI values of pre-gelatinized rich starches. PGS made from low-amylose rice exhibited a higher WAI and WSI. The rise in amylose content of the rice starches led to a less disrupted granule structure and accordingly a fall in the WAI and WSI values. Increased drum temperatures led to increased rupture of the granular structure and thus a decrease in WAI. Compared to PGS and PJS that displayed RVA cold peak viscosity and an amorphous structure, pre-gelatinized starch from high-amylose rice (Chiang) showed RVA hot peak viscosity and an amylose-lipid complex structure. This is because pre-gelatinization destroyed the granular structure of rice starch with lower amylose content more severely than that of rich starch with higher amylose content. Both PGS and PJS showed shear- thinning behavior at low temperature. The cold peak viscosity for the PGS and PJS decreased when increasing drum temperature. Nevertheless, apparent viscosity was only slightly affected by the rise of drum temperature.

Hedayati et al. (2016a) have studied the effects of different levels of sucrose and glucose (0%, 10%, 20%, 30%, and 40% of dry starch weight basis) as two common sweeteners on drum-dried pre-gelatinized maize starch. With the increase of sugar level, water absorption was increased due to the equatorial hydroxyl groups of sugars, formation of intermolecular hydrogen bonds, and cross-linking. The increased water uptake of PGS resulted in higher viscosity and mechanical properties. Samples with sucrose have significantly higher water absorption and viscosity than those with glucose. Sucrose had more obvious effects on gel clarity and is more appropriate for products in which clarity is important. Freeze-thaw stability of starch pastes was also improved in the presence of sugars owing to their cryoprotective effects. Thus, PGS has high potential to be used in frozen products, which contain sugars (Hedayati et al. 2016a).

The effects of pre-gelatinization on the rheological properties of selected botanical starches as pharmaceutical excipients were evaluated. Pre-gelatinization of cassava starch and sweet potato was prepared.

Microscopic analysis indicated definitive alteration in the granular character of the modified starches. The native starches were susceptible to retrogradation, which was alleviated by pre-gelatinization. Pre-gelatinization modification increased resistance to retrogradation, solubility, swelling power, and water absorption capacity of the starches. The increases in solubility, swelling power, and water absorption capacity were directly related to a decrease in amylose content (Lawal et al. 2015).

Rheological properties of native potato starch (NPS) and pre-gelatinized potato starch (PPS) were investigated under steady and dynamic shear conditions at 20 °C and 60 °C, respectively. The consistency index, apparent viscosity, and yield stress values of PPS were much lower than those of NPS. The flow behavior of PPS is more temperature-dependent. Dynamic moduli (G' and G'') and complex viscosity of PPS were much lower than those of NPS, while tan δ values of PPS were in the range of 2.30–3.50 (tan $\delta > 1$), indicating that PPS is more viscous than elastic. Rheological properties of NPS are strongly influenced by pre-gelatinization (Elgorashi 2016).

The influence of pH changes (3, 5, 7, and 9) on physical properties of pre-gelatinized (PG) and granular cold water swelling (GCWS) maize starches was also studied (Hedayati et al. 2016b). PG starches were fragmented in acidic pH. However, GCWS starches mainly reserved their granular integrity but were shriveled. For both modified starches, the water absorption, cold water viscosity, textural parameters, turbidity, and freeze-thaw stability of the samples decreased, whereas water solubility increased at pH 3 and 5. On the other hand, alkaline pH did not bring about evident changes on the morphology of PG starch, but the surface of

GCWS starch became smoother. Water absorption, solubility, rheological and mechanical properties, freeze-thaw stability, and turbidity of the starch pastes increased at high pH values. Overall, both starches were more stable at alkaline pH compared to acidic pH values, and GCWS starch was more resistant to pH changes than PG starch (Hedayati et al. 2016b).

The compression and mechanical properties of directly compressible pre-gelatinized sago starches at 65 °C with different pre-gelatinization times of 15, 30, 45, and 60 min were labeled as PS1, PS2, PS3, and PS4, respectively (Hedayati et al. 2016b). The compressibility of sago starch is found to be lower than that of its pre-gelatinized forms, and the compressibility increases with an increase in the pre-gelatinization time. The result of mechanical properties showed the same trends. This study confirms previous reports showing that pre-gelatinization is able to improve the compressibility and compatibility of a starch and therefore showing the potential of pre-gelatinized sago starch as a new directly compressible excipient in tablet formulations (Widodo and Hassan 2015).

Rheological properties of native potato starch (NPS) and pre-gelatinized potato starch (PPS) were investigated under steady and dynamic shear conditions at 20 °C and 60 °C. The consistency index, apparent viscosity, and yield stress values of PPS were much lower than those of NPS. The flow behavior of PPS is more temperature-dependent. Dynamic moduli (G' and G'') and complex viscosity of PPS were much lower than those of NPS, while $tan\delta$ values of PPS were in the range of 2.30–3.50 ($tan\delta > 1$), indicating that PPS is more viscous than elastic. Rheological properties of NPS are strongly influenced by pre-gelatinization (Elgorashi 2016).

The changes of gelatinization, powder characteristics, and pasting properties of customized stirring ball mill treated partially pre-gelatinized cassava starch (PPCS) were studied. MA significantly destroys the crystal structure of starch granules, contributing to the increase of cold water solubility, flowability, and flood ability. It also leads to variations in particle size and surface morphology, and a decrease in pasting temperature, viscosity of cooked PPCS paste, breakdown, and setback. With these unique properties induced by MA, PPCS was expected to be used as an excipient in pharmaceutical formulations as it could exhibit dual functionality as both binder and disintegrant and improve the production efficiency and qualities of solid oral drugs (Zhang et al. 2013).

The effect of physical aging on the physicochemical properties of pre-gelatinized tapioca starch was investigated by Manchun et al. (2012). The tapioca starch was pre-gelatinized by either heating at 80 °C or using high-power (400 W) ultrasonic treatment. After pre-gelatinization, dextrose equivalent (DE), viscosity, turbidity, swelling power, and solubility were determined and compared with native tapioca starch. Compared to fresh tapioca starch, the aged starch exhibited an increase in DE, turbidity, and solubility. The viscosity and swelling power were decreased after storage. Similar results were found for both tapioca starches pre-gelatinized by heat and ultrasonic treatments. The results of the physicochemical properties of pre-gelatinized starches were obtained from ultrasonic treatment related to the formation of low molecular weight components that aging starch is easily changed by disruption of the molecular structure within the starch granules (Widodo and Hassan 2015).

6.6 Future Perspectives

Over the last decade, important industry application and fundamental science study have been made in the production of PGS. As the most important physical modification starch, PGS has been widely used in feed, food, and pharmaceutical industry and also has a good most promising application in instant food and other high value industry fields.

Despite extremely interesting properties, such as solubility, thickening/gelling capabilities, and high digestion rate for PGS, the applications of PGS in the high value production area remain limited, mainly due to the extreme difficulty to control the physical modification extent and the starch molecular assembly. PGS has broad application prospects in the instant food field, but further research and development are necessary to achieve significant outcomes, such as precise control of starch molecular assembly, understanding interactions between starch and other food components, and creating more cost-effective PGS preparation methods. In summary, as the most important physical modification method for starch, further scientific research, new preparation methods and related equipment development, and new field applications are essential for the future.

References

BeMiller JN (2016) Reference module in food science. Elsevier, Amsterdam

Chen P, Wang K, Kuang Q, Zhou S et al (2016) Understanding how the aggregation structure of starch affects its gastrointestinal digestion rate and extent. Int J Biol Macromol 87:28–33

Cruz PS (1997) Quaculture feed and fertilizer resource atlas of the Philippines. Food and Agriculture Organization of the United Nations, Rome

Elgorashi DB (2016) Assessment of pregelatinized sorghum and maize starches as superior multifunctional excipients. J Pharm Innov 11:143–155

Fan DM, Shen HJ, Huang LL et al (2015) Microwave-absorbing properties of rice starch. Polymers 7:1895–1904

Fritze H (1973) Dry gelatinized starch produced on different types of drum dryers. Ind Eng Chem Process Des Develop 12:142–148

González Parada ZM, Pérez Sira EE (2003) Physicochemical and functional evaluation of pregelatinized and microwaved cassava (Manihot esculenta Crantz) starches. Acta Cient Venez 54:127–137

He XH, Xia W, Chen RY et al (2020) A new PGS preparing by gelatinization and spray drying of rice starch with hydrocolloids. Carbohydr Polym 229:115485

Hedayati S, Shahidi F, Koocheki A, Farahnaky A, Majzoobi M (2016a) Comparing the effects of sucrose and glucose on functional properties of pregelatinized maize starch. Int J Biol Macromol 88:499–504

Hedayati S, Shahidi F, Koocheki A, Farahnaky A, Majzoobi M (2016b) Physical properties of pregelatinized and granular cold water swelling maize starches at different pH values. Int J Biol Macromol 91:730–735

Kamarudin MS, Gruz GR, Saad GR et al (2018) Effects of extruder die head temperature and pre-gelatinized taro and broken rice flour level on physical properties of floating fish pellets. Anim Feed Sci Technol 236:122–130

Katcher RS (1994) Process for preparing modified, pregelatinized dent cornstarch and product thereof

Kibbe AH (2012) Handbook of pharmaceutical excipients, 7th edn. Pharmaceutical Press, London, pp 794–797
Lan X, Xie S, Wu J, Xie F et al (2016) Thermal and enzymatic degradation induced ultrastructure changes in canna starch: further insights into short-range and long-range structural orders. Food Hydrocoll 58:335–342
Lawal MV, Odeniyi MA, Itiola OA (2015) Material and rheological properties of native, acetylated, and pregelatinized forms of corn, cassava, and sweet potato starches. Starch 67:964–975
Li QQ, Liu SY, Obadi M et al (2020) The impact of starch degradation induced by pre-gelatinization treatment on the quality of noodles. Food Chem 302:125267
Liao L-S, Liu H-S, Liu X-X, Chen L et al (2014) Development of microstructures and phase transitions of starch. Acta Polym Sin 2014:761–773
Liu HS, Xie FW, Yu L, Chen L, Li L (2009) Thermal processing of starch-based polymers. Prog Polym Sci 34:1348–1368
Loisel C, Maache-Rezzoug Z, Esneault C, Doublier JL (2006) Effect of hydrothermal treatment on the physical and rheological properties of maize starches. J Food Eng 73:45–54
Manchun S, Piriyaprasarth S, Patomchaiviwat V, Limmatvapirat S, Sriamornsak P (2012) Adv Mater Res 506:35–38
Maria AVTG, Cleverson FG, André AGF (2020) Pharmaceutical and biomedical applications of native and modified starch: a review. Starch 72:1900270
Marti A, Caramanico R, Bottega G, Pagani MA (2013) Cooking behavior of rice pasta: effect of thermal treatments and extrusion conditions. LWT Food Sci Technol 54:229–235
Nakorn KN, Tongdang T, Sirivongpaisal P (2009) Crystallinity and rheological properties of pregelatinized rice starches differing in amylose content. Starch 61:101–108
Okunlola A, Adebayo SA, Adeyeye MC (2015) Solid state characterization of two tropical starches modified by pregelatinization and acetylation: potential as excipients in pharmaceutical formulations. Br J Pharm Res 5:58–71
Pérez S, Bertoft E (2010) The molecular structures of starch components and their contribution to the architecture of starch granules: a comprehensive review. Starch 62:389–420
Petra S, Jitka M, Pavel O (2021) Comparison of compressibility, compactability, and lubricant sensitivity of two partially pregelatinized starches. Starch 73:2000166
Prasenjit T, Borgohain GS (2015) A study on the role of pre-gelatinized starch (PGS) in the non damaging drilling fluid (NDDF) for the Tipam sand of Geleki oilfield of upper Assam basin. Int J Appl Sci Biotechnol 3:12552
Widodo RT, Hassan A (2015) Compression and mechanical properties of directly compressible pregelatinized sago starches. Powder Technol 269:15–21
Witt T, Gilbert RG (2014) Causal relations between structural features of amylopectin, a semicrystalline hyperbranched polymer. Biomacromolecules 15:2501–2511
Witt T, Doutch J, Gilbert EP, Gilbert RG (2012) Relations between molecular, crystalline, and lamellar structures of amylopectin. Biomacromolecules 13:4273–4282
Wu AC, Gilbert RG (2013) Characterization methods for starch-based materials: state of the art and perspectives. Aust J Chem 66:1550
Xie F, Halley PJ, Avérous L (2012) Rheology to understand and optimize processibility, structures and properties of starch polymeric materials. Prog Polym Sci 37:595–623
Yuan L, Hu JL, Yin JY (2020) Progress on the effect of microwave irradiation on structural of starch and its application in starch derived food processing. Sci Technol Food Ind 18:330–337
Zhang Y, Huang Z, Yang C, Huang A et al (2013) Material properties of partially pregelatinized cassava starch prepared by mechanical activation. Starch 65:461–468

Chapter 7
Gamma Irradiation of Starch

Xiangli Kong

Abstract As a physical treatment, gamma irradiation has many advantages, such as high efficiency, low cost, eco-friendship, and no remarkable rise in temperature during the treatment process. The degradation changes of starch by radiation are reproducible and quantitative, without needing special appliances to control temperature and the environment. The effects of irradiation on starch have attracted much more attention in the last decade. Gamma radiation may generate free radicals which can contribute to molecular changes and fragmentation of starch. Starch depolymerization leads to progressive reduction in the molecular size of amylose and amylopectin by random cleavage of glycosidic chains, and consequently changes the structural and physicochemical characteristics of starch. Visible or significant changes on starch granules upon irradiation were not found; however, some researchers found that Maltese cross became unclear and fissures or cracks developed on starch granules. To date, all researchers found that the C-type starches showed a decrease in amylose content upon gamma irradiation, and most researchers reported that gamma irradiation treatment decreased amylose content in A-type starches but increased amylose content in B-type starches. The X-ray diffraction pattern of starches remained unchanged after gamma irradiation, but the relative crystallinity decreased upon irradiation in most cases; however, some A-type starches exhibited an increase in relative crystallinity at low-dose irradiation and then decreased at higher doses. The Fourier transform-infrared spectral pattern of irradiated starch was not altered; however, most intensities increased indicating decreases in the ordered structure of starch upon irradiation. Gelatinization temperatures and enthalpy of gelatinization of A-type starches decreased upon irradiation in most cases. With regard to the B-type starches, the gelatinization parameters showed an increase at lower dosage but decreased with irradiation dose increasing in some cases. For C-type starches, the gelatinization parameters were observed to increase by most researchers. The swelling power was observed to decrease under gamma irradiation treatment in most cases, water solubility index of irradiated starches

X. Kong (✉)
College of Agriculture and Biotechnology, Zhejiang University, Hangzhou, Zhejiang, China
e-mail: xlkong@zju.edu.cn

Z. Sui, X. Kong (eds.), *Physical Modifications of Starch*,
https://doi.org/10.1007/978-981-99-5390-5_7

increased with irradiation dosage elevated in all reports, and the syneresis of gels formed with irradiated treated starches decreased. Most pasting and rheological parameters of irradiated starches decreased continuously with irradiation dosage increasing. Gamma irradiation can be employed to elevate the resistant starch fraction, whereas some researchers observed that irradiation treatment could decrease resistant starch content. Radiation processing can promote cross-linking in the starch matrix under oxygen. Graft copolymerization of chemicals onto starch by a simultaneous irradiation technique can also be achieved to produce biodegradable film or superabsorbent hydrogel.

Keywords Gamma irradiation · Depolymerization · Structure · Physicochemical properties

7.1 Introduction

Most of the techniques currently employed for starch modification are usually complex, expensive, and time-consuming; however, radiation processing has been highlighted to provide a low-cost and environment-friendly alternative to alter the structural, physicochemical, and functional characteristics of starch. During the radiation processing, no pollutant agents are needed to be used, the toxic substances are not allowed to be penetrated in the treated products, and catalysts and laborious preparation of sample are not required (Bhat and Karim 2009; Wani et al. 2014). The degradation changes of starch by radiation are reproducible and quantitative, without needing special appliances to control temperature and the environment. Based on the advantages mentioned above, irradiation was becoming an alternative technique to modify starch to replace common techniques of chemical and physical modifications. The modification of starch by irradiation technique is an impending development and the commonly used one is gamma irradiation (Reddy et al. 2015a; Atrous et al. 2017), which is an ionizing and nonthermal physical method. Gamma irradiation processing involves the use of a radioactive isotope, either in the form of cobalt-60 or cesium-137, which emits high-energy gamma rays or photons capable of intruding in depth into the target product, up to several meters (Ocloo et al. 2014). The Unites States Food and Drug Administration (FDA) and international standards for food irradiation permit gamma rays from cobalt-60 or cesium-137; however, all industrial gamma irradiation facilities employ cobalt-60 rather than cesium-137 as the radiation source because of practical difficulties in handling the latter one. The radiation dose unit is a function of the energy of the radiation source and the time of exposure. Irradiation dosage, previously referred to as the rad, is currently expressed in kilograys (kGy). One kilogray is equal to 1 kJ of energy absorption per kilogram of a material and is equivalent to 100,000 rads. The unit gray can be used for any type of radiation, but it does not describe the biological effects of the different radiations (Mahapatra et al. 2005). According to FDA, gamma irradiation under 2 kGy can be employed to improve the food quality, gamma irradiation between 2 kGy and 10 kGy can be used to increase the food products shelf-life, and the sterilization of food products can be achieved by gamma irradiation with more than 10 kGy (Komolprasert and Morehouse 2004).

Gamma radiation may generate free radicals that can contribute to molecular changes and fragmentation of starch. Starch depolymerization leads to progressive reduction in the molecular size of amylose and amylopectin by random cleavage of the glycosidic chains (Sujka et al. 2015), and consequently changes the structural and physicochemical characteristics of starch. Considering that the intensity of free radicals was observed to be dependent on starch water content, irradiation dose, temperature, and time of storage (Raffi and Agnel 1983), alterations of starch materials are dependent on irradiation conditions (e.g., dosage, dose rate, and temperature), starch moisture, and starch types to a large degree. When water is irradiated, the following highly reactive entities are formed:

$$H_2O \rightarrow \cdot OH, e^-{}_{aq}, H, H_2, H_2O_2, H_3O^+$$

where ·OH represents hydroxyl radical; $e^-{}_{aq}$ represents hydrated electron; H and H_2 represent hydrogen atom and hydrogen molecule, respectively; H_2O_2 and H_3O^+ are hydrogen peroxide and hydrated proton, respectively. Among these formed products, hydroxyl radicals and hydrated electrons are the powerful oxidizing agents and the strong reducing agents, respectively; and hydrogen atoms are somewhat weaker reducing agents. Hydrogen and hydrogen peroxide are the stable products, which are produced in a low yield even with high irradiation dosage. The presence of oxygen has effects on the course of radiolysis; oxygen can be reduced by hydrogen atoms into hydroperoxyl radical (·OOH), which is a mild oxidizing agent. Oxygen can also react with the solvated electron to form the superoxide radical ($\cdot O_2$), which is another oxidizing agent. These two oxidizing agents contribute to hydrogen peroxide. Furthermore, oxygen can also be added to other radicals formed while water is irradiated, giving rise to peroxy radicals ($\cdot RO_2$). During irradiation, the oxygen can form ozone (O_3), a powerful oxidant (Stewart 2001).

The general scheme of free radicals generation in starch molecules after exposure to radiation treatments is illustrated in Fig. 7.1. It is well known that radiation processing leads to the formation of radiolytic products or radiation degradation products (RDPs). Some of the major radiolytic products of carbohydrate molecules are formic acid, acetaldehyde, and formaldehyde (Bhat and Karim 2009; Zhu 2016). The major practical aim of high-level irradiation of starch is preparation of water soluble starch for food application, animal feed production, tablet preparation, or even as corrosion inhibitors for steel (Tomasik and Zaranyika 1995). Radiation processing can also promote cross-linking in the starch matrix under oxygen. This unique phenomenon is quite important as presently various methods are employed for cross-linking purposes. Graft copolymerization of chemicals onto starch by a simultaneous irradiation technique can also be achieved to produce biodegradable film or superabsorbent hydrogel. In summary, gamma radiation can be employed as a convenient technique for modification of starch material by means of degradation, grafting, and cross-linking.

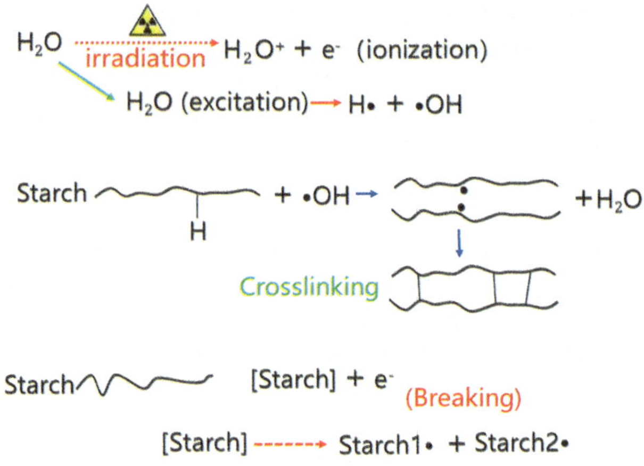

Fig. 7.1 Possible degradation and cross-linking of starch chains influenced by gamma irradiation

7.2 Starch Change Under Gamma Irradiation

7.2.1 Granule Morphology

The morphological properties of native and irradiated starch granules can be observed by regular or polarized light microscope and scanning or transmission electron microscope to analyze damages at the granular level. Easier breakdown and gelatinization of the irradiated starch were attributed to the changes in the morphological features, which were also observed to help in starch digestibility (Sunder et al. 2022). Most researchers did not find visible or significant changes on starch granules upon irradiation, whereas some researchers found that Maltese cross became unclear and fissures or cracks developed on starch granules; furthermore, granule size decreased in some studies (Table 7.1). The appearance of cracks on the granule surface, caused by highly energetic and penetrating radiations, is dependent of the starch source, irradiation dosage, dose rate, and water content. Essential changes in the potato granular appearance were not found after irradiation even with 600 kGy; however, deformed shape and rough surface observed on granules were attributed to the mechanical treatment (Cieśla and Eliasson 2002). Chung and Liu (2010) observed that B-type starch granules were more resistant than C-type under gamma irradiation and concluded that B-type starches might have more resistance than A-type starch to gamma irradiation treatment based on the results of thermal treatment. Low-dosage irradiation promoted aggregation of potato starch granules, which was attributed to starch-starch interactions; actually gamma irradiation could cause degradation or cross-linking of starch. Low-dosage irradiation might mainly cause cross-linking instead of degradation, which can be verified by the phenomenon that low dosage caused an increase in the final viscosity (Singh et al. 2011). With increasing irradiation dose, increases in surface cracking and the amount of small-sized granules were observed by some researchers. These changes were caused by free radicals generated by gamma irradiation cleaving large starch

Table 7.1 Granular change of starches under gamma irradiation

Starch samples		Irradiation treatment conditions		Results	
Source/type	Amylose content	Material status Dosage/dose rate	Water content	Granular change	Literatures
Rice[A]	~20%	Starch powder 1, 2, 5 kGy at 0.4 kGy/h	~9%	No change in granule morphology, the larger granules increased or decreased	Polesi et al. (2016)
Rice[A]	~20%	Starch powder 2, 5, 10 kGy at 0.4 kGy/h	~12%	No visible changes in granule morphology; however, a few granules were deformed at a dosage of 10 kGy, particle diameter of starch granules decreased upon irradiation	Gul et al. (2016)
Rice[A]	~26%	Starch powder 5, 10, 20 kGy at 2 kGy/h	~12%	Surface cracking of starch granules and the amount of small- sized granules were observed to increase with increasing irradiation dose	Ashwar et al. (2014)
Rice[A]	~19%	Grain 2, 5, 8, 10 kGy at 1 kGy/h	~30%	The amount of small-sized granules increased	Yu and Wang (2007)
Rice[A]	~1 to 26%	Grain 0.2, 0.4, 0.6, 0.8, 1 kGy at 1 kGy/h	ND	The shape of starch granules was deformed	Wu et al. (2002)
Wheat[A]	~34%	Starch powder 3, 5, 10, 20, 35, 50 kGy at 0.83 kGy/h	~13%	Unaffected by irradiation	Atrous et al. (2017)
Wheat[A]	~35%	Starch powder 3, 5, 10, 20, 35, 50 kGy at 0.83 kGy/h	~13%	Maltese cross in some starch granules irradiated at 50 kGy has become somewhat unclear, but the granules surfaces were unaffected by irradiation	Atrous et al. (2015)
Oat[A]	~22 to 27%	Grain 5, 10, 15, 20 kGy at 2 kGy/min	~12%	Ridges on the surface of starch were formed, and some granules fractured and cleaved as a result of irradiation	Mukhtar et al. (2017)
Arrowhead[A]	~31%	Starch powder 5, 10, 15 kGy at 5 kGy/h	~8%	Unaffected by irradiation	Wani et al. (2015)

(continued)

Table 7.1 (continued)

Starch samples		Irradiation treatment conditions		Results	
Source/type	Amylose content	Material status Dosage/dose rate	Water content	Granular change	Literatures
Arrowroot[A]	ND	Starch powder 5, 10, 15 kGy at 1 kGy/h	~13%	The size of starch granules showed a slight decrease	Barroso and Mastro (2019)
Indian horse chestnut[A]	~27%	Starch powder 5, 10, 15 kGy at 5 kGy/h	~12%	No significant differences between native and irradiated starches	Wani et al. (2014)
Lotus seed[A]	~20%	Starch powder 5, 10, 15 kGy	ND	No significant changes in shape and size were observed at the doses of 5 and 10 kGy, the roughness of granules increased at the dose of 15 kGy, dents and cavities on the surface of starch granules were formed, and a few granules ruptured on treatment of 20 kGy	Punia et al. (2020)
Kithul[A]	~39%	Starch powder 0.5, 1, 2.5, 5, and 10 kGy at 2 kGy/h	~11%	The granules split and reduced its smoothness significantly; large fissures and cracks on the surface of starch granules after irradiation were observed	Sudheesh et al. (2019)
Sorghum[A]	ND	Flour powder 10, 10 + 25 kGy	ND	External structure of sorghum starch granules was unaffected by irradiation	Mukisa et al. (2012)
Maize[A]	ND	Starch powder 1, 2, 5, 10, 20, 50, 100, 200, 500 kGy at 5 kGy/h	ND	Unaffected by irradiation	Liu et al. (2012)
Corn[A]	ND	Starch powder 3, 5, 10, 20, 50 kGy at 1.14 Gy/h	ND	No notable changes on the shapes and sizes of starch granules under irradiation	Ben Bettaïeb et al. (2014)
Corn[A]	~29%	Starch powder 2, 10, 50 kGy at 2 kGy/h and 0.40, 0.67, 2 kGy/h to a total dose of 10 kGy	~9%	Granule structure and birefringence remain unchanged up to 10 kGy; however, some starch granules became fractured at 50 kGy.	Chung and Liu (2009)

(continued)

Table 7.1 (continued)

Starch samples		Irradiation treatment conditions		Results	
Source/type	Amylose content	Material status Dosage/dose rate	Water content	Granular change	Literatures
				Polarization cross in some starch granules has become unclear at 50 kGy	
Corn[A or B]	2–72%	Starch powder 5, 10, 25, and 50 kGy at 1 kGy/h	~12%	Only marginal pin holes were observed on the surface of few granules; no significant change in starch granule surfaces	Lee et al. (2013)
Corn[A or B]	4–70%	Starch powder 1, 5, 10, 25, 50 kGy at 1 kGy/h	ND	Intact and visually unchanged	Chung et al. (2015)
Maize[B]	~58%	Starch powder 30, 60kGy at 1.2 kG/h	~10%	Unaffected by irradiation	Ocloo et al. (2014)
Potato[B]	~30%	Starch powder 5, 10, 20, 30 kGy at 3.6 kGy/h	~12%	Granules were covered with a layer of probably products of starch depolymerization above 10 kGy	Sujka et al. (2015)
Potato[B]	~30%	Starch powder 3, 5, 10, 20, 35, 50 kGy at 0.83 kGy/h	~18%	Unaffected by irradiation	Atrous et al. (2017)
Potato[B]	ND	Starch powder 5, 10, 20, 30 kGy at 3.6 kGy/h	~14%	Defective structures are present in large granules and the number of small granules increased at a dosage of 30 kGy	Cieśla et al. (2015)
Potato[B]	~32%	Starch powder 5, 10, 20 kGy at 2 kGy/h	~10%	The number of fractured granules increased with increasing irradiation dosage	Gani et al. (2014)
Potato[B]	~15%	Starch powder 0.01, 0.05, 0.1, 0.5kGy	ND	Presence of aggregated starch granules	Singh et al. (2011)
Potato[B]	~32%	Starch powder 10, 50 kGy at 2 kGy/h	~10%	Granule surface unaffected by irradiation; Maltese cross in some starch granules of irradiated starches became unclear	Chung and Liu (2010)
Potato[B]	~17%	Tuber 0.1, 0.5 kGy	NA		

(continued)

Table 7.1 (continued)

Starch samples		Irradiation treatment conditions		Results	
Source/type	Amylose content	Material status Dosage/dose rate	Water content	Granular change	Literatures
				The proportion of small- sized granules was increased	Ezekiel et al. (2007)
Potato[B]	ND	Starch powder 20 kGy at 1.66 kGy/h, 600 kGy at 2.02 kGy/h, 446 kGy at 3.1 kGy/h	~20%	Essential changes in the granular appearance were not found after irradiation even with 600 kGy. The granules with a deformed shape and rough surface were present in the amorphized sample due to the mechanical treatment. A large fraction of the needle-shaped long granules was noticed	Cieśla and Eliasson (2002)
Cassava[B]	~22%	Starch powder 1.0, 2.5, 5.0, 7.5, 10.0 kGy	ND	Granular shapes were somewhat deformed by irradiation at the dose higher than 5 kGy; the number of small- sized granules increased with the increase of radiation dose	Tran et al. (2022)
Elephant foot yam[B]	~28%	Starch powder 5, 10, 15, 20, 25 kGy at 2 kGy/h	~11%	No visual changes in granule morphology	Reddy et al. (2015a)
Chickpea[C]	~30%	Flour powder 0.5, 1, 2.5, 5, 10 kGy at 0.5 kGy/h	ND	Slight surface fissures in irradiated starch granules treated with 5 and 10 kGy	Bashir and Aggarwal (2017)
Chickpea[C]	~33%	Starch powder 4, 8, 12 kGy at 2 kGy/h	~12%	Fissures developed on surfaces, granule size reduced significant irradiated at 12 kGy	Bashir and Haripriya (2016)
Chickpea[C]	ND	Grain 2, 5, 10, 20, 30, 50kGy at 3.6 kGy/h	ND	No significant changes in size distribution and structure of starch granules	Graham et al. (2002)
Cowpea[C]	ND	Flour powder and paste 2, 10, 50 kGy	ND	Unaffected by irradiation	Abu et al. (2006)
	~52%		~8%		

(continued)

Table 7.1 (continued)

Starch samples		Irradiation treatment conditions		Results	
Source/type	Amylose content	Material status Dosage/dose rate	Water content	Granular change	Literatures
Broad bean[C]		Starch powder 5, 10, 15 kGy at 5 kGy/h		Fissures developed on starch granules, but mean granule length, mean granule width, and width range of native and modified starches did not vary significantly	Sofi et al. (2013)
Kidney beans[C]	36–41%	Starch powder 5, 10, 20 kGy at 2 kGy/h	~13%	Surface fracturing of granules increased with increasing irradiation dose	Gani et al. (2012)
Bean[C]	~36%	Starch powder 10, 50 kGy at 2 kGy/h	~10%	Cracks observed after irradiation with 10 kGy and increased with 50 kGy	Chung and Liu (2010)
Bean[C]	ND	Starch powder 2.5, 5, 10, 20 kGy	~12%	Gamma irradiation increased the suscepti-bility to central fissures	Duarte and Rupnow (1993)
Sago[C]	~27%	Starch powder 6, 10, 25 kGy at 8.3 kGy/h	~13%	No changes in granule size, shape, distribution, and surface	Othman et al. (2015)
Mung bean[C]	ND	Starch powder 0.5, 1, 3, 5 kGy	~10%	Slight imperfections as pores were observed through SEM	Castanha et al. (2019)

molecules, and some starch granules were fractured along the cleaved molecules, and this kind of breakage resulted in an increased number of small-sized starch granules (Ashwar et al. 2014). Some researchers also observed that granules seem to be the starch granules which were covered with a layer of probably products of starch depolymerization, making granules "fused." Gamma irradiation damage to some starch granules might exist only in the form of changes to the structure of the starch molecules when starch granules remained intact and visually unchanged. Considering that the researchers performed their studies under various conditions, it is necessary keep other factors unchanged in order to explore the effects of one factor.

7.2.2 Chemical Change

Chemical changes of starch under gamma irradiation included reducing sugar content, carboxyl content, acidity, pH value, amylose content, and starch molecules degradation. Gamma irradiation increased the reducing power of amylose and amylopectin continuously with increasing dose (Kertesz et al. 1959; Samec 1960; El Saadany et al. 1974, 1976; Nene et al. 1975; Sokhey and Hanna 1993; Ezekiel et al. 2007). Glucose, maltose, and small dextrins were found in the irradiated starch specimen by chromatographical analysis (Mishina and Nikuni 1959). The carboxyl content and acidity were observed to increase, and pH value decreased by all the researchers who measured these indices (Table 7.2); formic, acetic, pyruvic, and glucuronic acids were identified in irradiated starch (Ghali et al. 1979). So the main radiolytic products of starch were carboxylic acids, which resulted in an increase in carboxyl content and a decrease in pH value in all studies. Increased carboxyl content of irradiated starch was attributed to the breakdown of starch molecules by the action of free radicals; the slower irradiation dose rates produced much lower carboxyl content than a high-dose rate (Chung and Liu 2009, 2010).

Greenwood and Mackenzie (1963) reported gross molecular degradation of amylose and amylopectin isolated from irradiated starch. Chaudhry and Glew (1973) observed that the iodine binding capacity, the β-amylolysis limit, and the average chain length of the starch amyloses were decreased by radiation treatment, which suggested that degradation of the molecule had taken place; however, the iodine binding capacity and the β-amylolysis limit of the starch amylopectins were increased as a result of radiation treatment. A significant decrease of the amylopectin peak and a concomitant increase of the amylose-like peak area were observed by using gel permeation chromatography when starch was irradiated (Rayas-Duarte and Rupnow 1993; Polesi et al. 2016). Ionizing radiation caused a greater fragmentation of the amylopectin fraction of starch; the fragments became smaller in molecular weight and merged with the amylose in gel permeation chromatography (Sokhey and Hanna 1993). The increased proportion of β(1–3)- and β(1–4)-bonded starch as a result of transglucosidation was observed with increasing irradiation dose, which could in part explain the reduction in digestibility of some irradiated starches (Rombo et al. 2004). The number-average degree of polymerization (DP) decreased with the increase in gamma irradiation intensity; the degradation of amylopectin leading to increases in low molecular weight molecules was also observed (Lee et al. 2006). The proportion of short A chains increased; meanwhile, the longer chains decreased, resulting in a significant decrease in average chain length of amylopectin (Chung and Liu 2009). When amylopectin contained very long branch chains, this fraction could be degraded into long chains, resulting in an increase in the proportions of short chains and long chains (Chung and Liu 2010). However, Polesi et al. (2016) reported that increasing doses of gamma radiation promoted an increase or decrease in the number of short chains and long chains for different starch samples upon irradiation; the authors attributed this to cleavage or cross-linking under irradiation. Gamma irradiation reduced the molar mass, size, and

Table 7.2 Chemical change of starches under gamma irradiation

Starch samples		Irradiation treatment conditions		Results		
Starch/Type	Amylose content	Material status Dosage/dose rate	Water content	Carboxyl content, acidity, and pH	Amylose content and molecular changes	Literatures
Rice[A]	~27%	Grain 1, 2, 5 kGy at 0.4 kGy/h	~10%	Carboxyl: ND Acidity: ND, pH: ND	Amylose: ↓	Polesi et al. (2017)
Rice[A]	~19%	Grain 2, 5, 8, 20 kGy at 1 kGy/h	~30%	Carboxyl: ND Acidity: ND, pH: ND	Amylose: ↓	Yu and Wang (2007)
Rice[A]	~20%	Starch powder 1, 2, 5 kGy at 0.4 kGy/h	~9%	Carboxyl: ↑ Acidity: ↑, pH: ↓	Amylose: ↑ Amylose/amylopectin degraded; however, debranched short chains varied because of cleavage or cross-linking	Polesi et al. (2016)
Rice[A]	~20%	Starch powder 1, 5, 10 kGy at 0.4 kGy/h	~13%	Carboxyl: ↑ Acidity: ND, pH: ↓	Amylose: ↓	Gul et al. (2016)
Rice[A]	~26%	Starch powder 5, 10, 20 kGy at 2 kGy/h	~12%	Carboxyl: ↑ Acidity: ND, pH: ↓	Amylose: ↓	Ashwar et al. (2014)
Rice[A]	~26%	Starch powder 5, 10, 20, 30, 40, 50 kGy at 2 kGy/h	~5%	Carboxyl: ND, Acidity: ND, pH: ↓	The apparent amylose was reduced at the dose of 5 kGy and remained similar as in the range of 5–20 kGy; further drops were seen at doses of 30–50 kGy	Lam et al. (2021)
Rice[A]	27–40%	Flour powder 0.5, 1, 2, 3, 4, 5 kGy at 1.0 kGy/h	NA	Carboxyl: ND, Acidity: ND, pH: ND	Amylose: ↓ Maximum wavelength and blue value of starch–iodine complex decreased which indicated that the degree of polymerization and average chain	Shu et al. (2013)

(continued)

Table 7.2 (continued)

Starch samples		Irradiation treatment conditions		Results		
Starch/Type	Amylose content	Material status Dosage/dose rate	Water content	Carboxyl content, acidity, and pH	Amylose content and molecular changes	Literatures
					length of amylose and amylopectin reduced	
Rice[A]	ND	Grain 0.5, 1, 3, 5, 7, 9 kGy at 0.5 kGy/h	~12%	Carboxyl: ND, Acidity: ND, pH: ND	Amylose: ND Amylopectin molecules degraded, molecular weight of amylopectin and amylose decreased with the increase in irradiation dosage, gamma irradiation had no significant effect on amylopectin chain length profiles	Bao et al. (2005)
Rice[A]	~1%-26%	Grain 0.2, 0.4, 0.6, 0.8, 1 kGy at 1 kGy/h	ND	Carboxyl: ND Acidity: ND, pH: ND	Amylose: ↓ more for low amylose samples	Wu et al. 2002
Wheat[A]	~25%	Flour powder 0.5, 1, 2.5, 5, 10 kGy	~11%	Carboxyl: ND Acidity: ND, pH: ND	Amylose: ↑	Bashir and Aggarwal (2017)
Wheat[A]	~34%	Starch powder 3, 5, 10, 20, 35, 50 kGy at 0.83 kGy/h	~13%	Carboxyl: ND Acidity: ND, pH: ND	Amylose: ↓	Atrous et al. (2017)
Wheat[A]	~35%	Starch powder 3, 5, 10, 20, 35, 50 kGy at 0.83 kGy/h	~13%	Carboxyl: ND Acidity: ND, pH: ND	Amylose: ↓ Free radicals were detected by Electron Paramagnetic Resonance (EPR) spectrometry and increased with dosage	Atrous et al. (2015)
Oat[A]	~22%-27%	Grain 5, 10, 15, 20 kGy at 2 kGy/min	~12%	Carboxyl: ND Acidity: ND, pH: ND	Amylose: ↓	Mukhtar et al. (2017)
	~27%		~12%		Amylose: ↓	

Indian horse chestnut[A]		Starch powder 5, 10, 15 kGy at 5 kGy/h		Carboxyl: ↑ Acidity: ND, pH: ↓		Wani et al. (2014)
Arrowhead[A]	~31%	Starch powder 5, 10, 15 kGy at 5 kGy/h	~8%	Carboxyl: ND Acidity: ND, pH: ↓	Amylose: ↓	Wani et al. (2015)
Lotus seed[A]	~20%	Starch powder 5, 10, 15 kGy	ND	Carboxyl: ND Acidity: ND, pH: ND	Amylose content decreased as radiation dose increased	Punia et al. (2020)
Tapioca[A]	~35%	Starch powder 5, 10, and 20 kGy at 1.8 kGy/h	~10%	Carboxyl: ↑ Acidity: ND, pH: ↓	Amylose: ↓	Kanatt (2020)
Kithul[A]	~39%	Starch powder 0.5, 1, 2.5, 5, and 10 kGy at 2 kGy/h	~11%	Carboxyl: ↑ Acidity: ↑, pH: ↓	Amylose: ↓	Sudheesh et al. (2019)
Corn[A]	Waxy and normal	Starch powder at various pH and salt concentrations 5, 10, 20 kGy at 10 kGy/h	6%	Carboxyl: ND Acidity: ND, pH: ND	Amylose: ND Reduction in the molar mass, size, and specific volume was observed; adjustments of pH had little influence on the average molar mass and size of irradiated starch, whereas incorporation of salt greatly reduced the molar mass and size of irradiated waxy and normal maize starches	Baik et al. (2010)
Corn[A]	Waxy and normal	Starch powder 5, 10, 20 kGy at 10 kGy/h	5%, 12%	Carboxyl: ND Acidity: ND, pH: ND	Amylose: ND Molecular weight of starch decreased, starch molecules became smaller and denser, structural changes were more significant when the moisture content of the irradiated starch was 5% than the changes at 12%. The chain degradation appeared more significant for waxy corn starch than for normal corn starch	Yoon et al. (2010)

(continued)

Table 7.2 (continued)

Starch samples		Irradiation treatment conditions		Results		
Starch/Type	Amylose content	Material status Dosage/dose rate	Water content	Carboxyl content, acidity, and pH	Amylose content and molecular changes	Literatures
Corn[A]	~29%	Starch powder 2, 10, 50 kGy at 2 kGy/h and 0.40, 0.67, 2 kGy/h to a total dose of 10 kGy	~9%	Carboxyl: ↑ Acidity: ND, pH: ND	Amylose: ↓ Short A chains (DP 6-12) increased, while the longer chains (DP ≥ 37) decreased, resulting in a significant decrease in average chain length. The extent of this change was more pronounced at higher dosage	Chung and Liu (2009)
Corn[A]	ND	Starch powder 5, 10, 20, 30, 40 at 1 kGy/h	~12%	Carboxyl: ND Acidity: ND, pH: ND	Amylose: ND Number-average degree of polymerization decreased with dosage increasing, amylopectin degradation resulted in increase of low molecular weight molecules	Lee et al. (2006)
Maize[A]	~25%	Maize flour 5, 10, 20, 40 kGy at 0.5 kGy/h	~14%	Carboxyl: ND Acidity: ND, pH: ND	Amylose: ND The proportion of β(1–3)- and β(1–4)-bonded starch increased; higher irradiation doses led to a reduction in the molecular size of amylopectin involved debranching and an increase in the production of short, straight-chain molecules	Rombo et al. (2004)
Corn[A or B]	4–70%	Starch powder 1, 5, 10, 25, 50 kGy at 1 kGy/h	ND	Carboxyl: ND Acidity: ND, pH: ND	Amylose: ↑ for waxy at 10 kGy, slight decrease for all samples <10 kGy and ↓ significantly for all samples under 25, 50 kGy treatment Starch–iodine complex absorption spectrum analysis indicated that the	Chung et al. (2015)

					reduction of the molecular size of the samples has taken place	
Corn[A or B]	2–72%	Starch powder 5, 10, 25, and 50 kGy at 1 kGy/h	~12%	Carboxyl: ND Acidity: ND, pH: ND	Amylose: ↑ with lower doses then ↓ with higher doses	Lee et al. (2013)
Maize[B]	~58%	30, 60 kGy at 1.2 kGy/h	~10%	Carboxyl: ND Acidity: ND, pH: ND	Amylose: ↑ Weight-average MW values of amylose and amylopectin decreased with irradiation dosage increase	Ocloo et al. (2014)
Potato[B]	~30%	Starch powder 3, 5, 10, 20, 35, 50 kGy at 0.83 kGy/h	~18%	Carboxyl: ND Acidity: ND, pH: ND	Amylose: ↓	Atrous et al. (2017)
Potato[B]	~30%	5, 10, 20, and 30 kGy at 3.6 kGy/h	~12%	Carboxyl: ND, Acidity: ND, pH: ND	Amylose: ↓ Long linear chains degraded and showed lower iodine binding ability	Sujka et al. (2015)
Potato[B]	~32%	Starch powder 5, 10, 20 kGy at 2 kGy/h	~10%	Carboxyl: ↑ Acidity: ND, pH: ↓	Amylose: ↓	Gani et al. (2014)
Potato[B]	~31%	Tuber 0.1 kGy at 32.4 kGy/h	~75%	Carboxyl: ND Acidity: ND, pH: ND	Amylose: ↑ No significant difference was observed in chain length profile of amylopectin upon irradiation	Lu et al. (2012a, b)
Potato[B]	~15%	Starch powder 0.01, 0.05, 0.1, 0.5 kGy	ND	Carboxyl: remain constant, ↑ at 0.5 kGy. Acidity: ND, pH: ↓	Amylose: ↑	Singh et al. (2011)
Potato[B]	~32%	Starch powder 10, 50 kGy at 2 kGy/h	~10%	Carboxyl: ↑ Acidity: ND, pH: ↓	Amylose: ↓ The proportions of short chain (DP 6-12) and long chain (DP≥37) increased, whereas the proportion of DP 13–24 decreased with increasing irradiation dose	Chung and Liu (2010)
Potato[B]	~18%	Tuber 0.1 kGy, 0.5 kGy	ND	Carboxyl: ND Acidity: ND, pH: ND	Amylose: ↑ Reducing sugars increased	

(continued)

Table 7.2 (continued)

Starch samples		Irradiation treatment conditions		Results		
Starch/Type	Amylose content	Material status Dosage/dose rate	Water content	Carboxyl content, acidity, and pH	Amylose content and molecular changes	Literatures
						Ezekiel et al. (2007)
Cassava[B]	ND	Starch powder 1 kGy at 8.1 kGy/h	~10%	Carboxyl: ND, Acidity: ND, pH: ND	Free radicals formed	Bertolini et al. (2001)
Cassava[B]	~22%	Starch powder 1.0, 2.5, 5.0, 7.5, 10.0 kGy	ND	Carboxyl: ND Acidity: ND, pH: ND	Amylose: ↓ Viscosity average molecular weight was significantly decreased by radiation degradation	Tran et al. (2022)
Elephant foot yam[B]	~28%	Starch powder 5, 10, 15, 20, 25 kGy at 2 kGy/h	~11%	Carboxyl: ↑ Acidity: ND, pH: ↓	Amylose: ↓	Reddy et al. (2015a)
Banana[B]	~25%	5, 10, 15, 20, 25 kGy at 2 kGy/h	ND	Carboxyl: ↑ Acidity: ND, pH: ↓	Amylose: ↓	Reddy et al. (2015b)
Chickpea[C]	~33%	Starch powder 4, 8, 12 kGy at 2 kGy/h	~12%	Carboxyl: ND Acidity: ND, pH: ↓	Amylose: ↓	Bashir and Haripriya (2016)
Bean[C]	~36%	Starch powder 10, 50 kGy at 2 kGy/h	~10%	Carboxyl: ↑ Acidity: ND, pH: ↓	Amylose: ↓ The proportions of short chain (DP 6–12) and long chain (DP $\geq$ 37) increased, whereas the proportion of DP 13–24 decreased with increasing irradiation dose	Chung and Liu (2010)
Bean[C]	~35%	Maize flour 5, 10, 20, 40 kGy at 0.5 kGy/h	~9.8%	Carboxyl: ND Acidity: ND, pH: ND	The proportion of β(1–3)- and β(1–4)-bonded starch increased; higher	

					irradiation doses led to a reduction in the molecular size of amylopectin involved debranching and an increase in the production of short, straight-chain molecules	Rombo et al. (2004)
Bean[C]	ND	Starch powder 2.5, 5, 10, 20 kGy	~12%	Carboxyl: ND Acidity: ↑, pH: ↓	Amylose: ND Reducing value increased with increasing irradiation dosage; radicals were observed by ESR	Duarte and Rupnow (1994)
Bean[C]	ND	Starch powder 2.5, 5, 10, 20 kGy	~12%	Carboxyl: ND Acidity: ND, pH: ND	Amylose: ND Irradiation changed molecular weight distribution of amylose and amylo-pectin, produced more amylose-like fraction	Duarte and Rupnow (1993)
Mung bean[C]	ND	Starch powder 0.5, 1, 3, 5 kGy	~10%	Carboxyl: ND Acidity: ND, pH: ↓	The gradual size decreasing of the starch molecules was observed by chromatographic column	Castanha et al. (2019)
Broad bean[C]	~52%	Starch powder 5, 10, 15 kGy at 5 kGy/h	~8%	Carboxyl: ↑ Acidity: ND, pH: ↓	Amylose: ↓	Sofi et al. (2013)
Kidney beans[C]	36–41%	Starch powder 5, 10, 20 kGy at 2 kGy/h	~13%	Carboxyl: ↑ Acidity: ND, pH: ↓	Amylose: ↓	Gani et al. (2012)
Sago[C]	~27%	Starch powder 6, 10, 25 kGy at 8.3 kGy/h	~13%	Carboxyl: ND Acidity: ND, pH: ND	Amylose: ↓ at lower doses, ↑ at higher doses, reducing sugars content increased	Othman et al. (2015)
Ariá[C]	ND	Starch powder 1 kGy at 226.43 Gy/h, 5 kGy at 1131.49 Gy/h, 20 kGy at 4527.10 Gy/h, 50 kGy at 11,614.44 Gy/h	~8%	Carboxyl: ↑ Carbonyl: ↑ Acidity: ND, pH: ↓	The number of reducing groups levels increased	Pinto et al. (2021)

specific volume considerably and consistently (Baik et al. 2010), since starch depolymerization by gamma irradiation caused a decrease in amylopectin molecular weight.

To date, all researchers found that the C-type starches showed a decrease in amylose content upon gamma irradiation, and most researchers reported that gamma irradiation treatment decreased amylose content in A-type starches but increased amylose content in B-type starches (Table 7.2). Chung and Liu (2010) found that B-type starch granules had more resistance under gamma irradiation than C-type base on morphological observations. The increase in amylose content might be attributed to the degradation of the amylopectin branches (Bashir and Aggarwal 2017). The degradation of amylopectin upon irradiation could release more linear chains of amylose, which were capable of complexing with iodine and increased the apparent amylose content (Othman et al. 2015). So waxy starch samples showed an increase in apparent amylose content under irradiation (Chung et al. 2015). However, irradiation could cause a breakdown of both amylopectin and amylose chains, and produced short chains, which might not complex with iodine and diminish its iodine binding capacity; hence the amylose content of some irradiated starches became lower. It is proposed that low irradiation dose produced short and amylose-like straight-chain molecules from amylopectin degradation (Rombo et al. 2004; Singh et al. 2011), which can better bind iodine. In contrast, high irradiation doses degraded amylose and amylopectin into fragments that are too short to bind iodine effectively, so the apparent amylose content of the irradiated samples reduced. When the amylose content was determined by separation of amylopectin precipitates based on their high molecular weight, as irradiation breaks the starch chains, the amylopectin fragments generated by irradiation were not eliminated during the step of precipitation of amylopectin, and they were still quantified as amylose (Polesi et al. 2016). In summary, the change in amylose content of irradiated starches was dependent on starch type, amylose content, irradiation treatment conditions, and methods employed for amylose determination.

7.2.3 Crystalline Structure and Thermal Properties

Two major crystal types are identified by using wide angle X-ray diffraction for native starches. Most cereal starches have A-type crystals, while tuber and amylose-rich starches have B-type crystals, there is also C-type that is a combination of A-type and B-type. Reddy et al. (2015a) reported the pattern of elephant foot yam starch changed from B- to C-type, but the result can be challenged since the native starch showed a strange X-ray diffraction curve compared to those of irradiated starches; except this case, gamma irradiation did not change the X-ray diffraction pattern of starches in all studies reported, but the relative crystallinity decreased upon irradiation in most cases; however, some A-type starches exhibited an increase in relative crystallinity at low-dose irradiation and then decreased at higher doses (Table 7.3). Gamma irradiation treatments degraded both crystalline and amorphous

Table 7.3 Crystalline structure and thermal properties change of starches under gamma irradiation

Starch samples		Irradiation treatment conditions		Results		
Starch/Type	Amylose content	Material status Dosage/dose rate	Water content	Crystalline structure	Thermal properties	Literatures
Rice[A]	~20%	Starch powder 1, 2, 5 kGy at 0.4 kGy/h	~9%	No change in XRD pattern, relative crystallinity increased with 1 kGy, then decreased with higher doses	T_o, T_p, T_c, ΔH remained unaffected; however, gelatinization temperature increased at 5 kGy for some sample	Polesi et al. (2016)
Rice[A]	~20%	Starch powder 2, 5, 10 kGy at 0.4 kGy/h	~12%	No change in XRD pattern, relative crystallinity decreased with increasing doses. The spectral patterns did not change except for an increase in intensities at 995 cm^{-1} indicating a decrease in the ordered structure	T_o, T_p, T_c, ΔH showed a slight decrease initially and significant decrease at 10 kGy. The retrogradation enthalpy decreased with an increase in irradiation dose	Gul et al. (2016)
Rice[A]	~26%	Starch powder 5, 10, 20 kGy at 2 kGy/h	~12%	No significant change in XRD pattern, relative crystallinity decreased significantly	ND	Ashwar et al. (2014)
Rice[A]	~26%	Starch powder 5, 10, 20, 30, 40, 50 kGy at 2 kGy/h	~5%	Gamma irradiation has no significant effect on the crystal pattern of starches tested	ND	Lam et al. (2021)
Rice[A]	27–40%	Flour powder 0.5, 1, 2, 3, 4, 5 kGy at 1.0 kGy/h	NA	No significant change in XRD pattern and relative crystallinity	T_o, T_p, T_c, ΔH decreased more significantly for high amylose starches	Shu et al. (2013)
Rice[A]	ND	Grain 0.5, 1, 3, 5, 7, 9 kGy at 0.5 kGy/h	~12%	Relative crystallinity increased at 1 kGy, then decreased with more irradiation dosages	T_o remained unchanged, T_p, T_c, ΔH decreased slightly	Bao et al. (2005)
Wheat[A]	~25%	Flour powder 0.5, 1, 2.5, 5, 10 kGy	~11%	FTIR spectra pattern did not change with irradiation; intensities in the bands at 1001.84 cm^{-1}	ND	Bashir and Aggarwal (2017)

(continued)

Table 7.3 (continued)

Starch samples		Irradiation treatment conditions		Results		
Starch/Type	Amylose content	Material status Dosage/dose rate	Water content	Crystalline structure	Thermal properties	Literatures
				increased with increasing irradiation dosage		
Wheat[A]	ND	Starch powder 1, 3, 5, 7, 9 kGy at 1 kGy/h	~12%	The relative crystallinity showed a slight increase with irradiation dose increasing up to 7 kGy and a decrease at 9 kGy. No significant differences on FTIR spectra patterns were observed; gamma irradiation treatment increased the absorbance of wheat starch at band of 1018 cm^{-1}	T_o, T_p, T_c, ΔH showed a slight decrease but not statistically significant. Transition temperatures of retrograded wheat starches increased slightly with irradiation dose up to 5 kGy, then decreased with higher irradiation dose	Kong et al. (2016)
Wheat[A]	~35%	Starch powder 3, 5, 10, 20, 35, 50 kGy at 0.83 kGy/h	~13%	No change in XRD pattern and relative crystallinity, intensity decreased dramatically with increasing irradiation doses by FTIR	T_o, T_p, T_c, ΔH remained unaffected until 50 kGy	Atrous et al. (2015)
Oat[A]	~22 to 27%	Grain 5, 10, 15, 20 kGy at 2 kGy/min	~12%	Ratio of intensity of 1047 cm^{-1}/1022 cm^{-1} decreased and absorption peak at 3400 cm^{-1} became narrower reflecting more hydroxyl groups exposure upon irradiation	T_o, T_p, T_c increased; however, ΔH decreased significantly after irradiation	Mukhtar et al. (2017)
Amaranth[A]	6–10%	2, 4, 6, 8, and 10 kGy at 1 kGy/h	~10%	Slight decrease in relative crystallinity	T_o, T_p, T_c, ΔH showed a slight decrease, ΔH exhibited a certain level of fluctuation	Kong et al. (2009)
Indian horse chestnut[A]	~27%	Starch powder 5, 10, 15 kGy at 5 kGy/h	~12%	No change in XRD pattern; however, the intensity of	ND	Wani et al. (2014)

				characteristic peaks decreased. FTIR spectra pattern did not change; there was an increase in intensities of the band at 1018 cm^{-1} at high irradiation dose, indicating a decrease in the ordered structure		
Arrowhead[A]	~31%	Starch powder 5, 10, 15 kGy at 5 kGy/h	~8%	No change in XRD pattern; intensity of the characteristic peaks reduced. The intensity of absorption band at 1047 cm^{-1} gets reduced, referring to the decrease in crystalline order due to irradiation	T_o, T_p, T_c, ΔH showed a slight decrease upon irradiation, and a steep decrease at 15 kGy	Wani et al. (2015)
Arrowroot[A]	ND	Starch powder 5, 10, 15 kGy at 1 kGy/h	~13%	The FTIR spectra did not change substantially upon irradiation	Gelatinization temperatures and enthalpy increased with irradiation dose increasing	Barroso and del Mastro (2019)
Lotus seed[A]	~20%	Starch powder 5, 10, 15 kGy	ND	No change in XRD pattern; intensity, angular position, and shape of the different peaks did not change significantly. Relative crystallinity decreased with radiation dose increasing. No change in the FTIR spectra, but intensities of absorbance increased after irradiation	Gelatinization temperatures decreased upon radiation; however, enthalpy changes were not obvious	Punia et al. (2020)
Tapioca[A]	~35%	Starch powder 5, 10, and 20 kGy at 1.8 kGy/h	~10%	ND	T_o, peak T_p, and T_c increased significantly upon irradiation, while ΔH showed a dose -dependent decrease	Kanatt (2020)

(continued)

Table 7.3 (continued)

Starch samples		Irradiation treatment conditions		Results		
Starch/Type	Amylose content	Material status Dosage/dose rate	Water content	Crystalline structure	Thermal properties	Literatures
Kithul[A]	~39%	Starch powder 0.5, 1, 2.5, 5, and 10 kGy at 2 kGy/h	~11%	FTIR showed a marked reduction in the hydroxyl groups; the crystalline pattern was not affected but the relative crystallinity decreased after irradiation	The gelatinization parameters were reduced at irradiation doses up to 2.5 kGy; however, the parameters increased when the irradiation doses raised to 5 and 10 kGy	Sudheesh et al. (2019)
Corn[A]	ND	Starch powder 3, 5, 10, 20, 50 kGy at 1.14 kGy/h	ND	No change in XRD pattern; intensity decreased dramatically	T_o, T_p, T_c showed a slight decrease upon irradiation	Ben Bettaïeb et al. (2014)
Maize[A]	ND	Starch powder 1, 2, 5, 10, 20, 50, 100, 200, 500 kGy at 5 kGy/h	ND	No change in XRD pattern, relative crystallinity decreased, the ordered structure decreased as indicated by the increase in intensities of the band at 1018 cm^{-1} by FTIR	Decreases in T_o, T_p, T_c, ΔH were not statistically significant from up to 20 kGy; however, significant decreases at higher than 50 kGy	Liu et al. (2012)
Corn[A]	~29%	Starch powder 2, 10, 50 kGy at 2 kGy/h and 0.40, 0.67, 2 kGy/h to a total dose of 10 kGy	~9%	No change in XRD pattern, the relative crystallinity decreased, the degree of order analyzed by FTIR at starch granule surface decreased. High- dose rate disrupted the crystalline structure more significantly	T_o, T_p, T_c decreased slightly up to 10 kGy, and significantly at 50 kGy, but ΔH remained unaffected	Chung and Liu (2009)
Corn[A]	ND	Starch powder 5, 10, 20, 30, 40 at 1 kGy/h	~12%	No significant change in XRD pattern and relative crystallinity; the basic structure of anhydrous glucose was affected by FTIR	T_o, T_p, T_c, ΔH showed a continuous decrease with irradiation dosage increasing	Lee et al. (2006)

Maize[A]	~25%	Maize flour 5, 10, 20, 40 kGy at 0.5 kGy/h	~14%	ND	Remained unaffected	Rombo et al. (2004)
Corn[A or B]	4–70%	Starch powder 1, 5, 10, 25, 50 kGy at 1 kGy/h	ND	No change in XRD pattern, relative crystallinity decreased and more significantly with low amylose content samples	T_o, T_p, T_c decreased significantly; however, ΔH fluctuated	Chung et al. (2015)
Maize[B]	~58%	30, 60 kGy at 1.2 kGy/h	~10%	ND	T_o, T_p, T_c, ΔH decreased significantly	Ocloo et al. (2014)
Banana[B]	~25%	5, 10, 15, 20, 25 kGy at 2 kGy/h	ND	No change in XRD pattern, relative crystallinity decreased slightly	T_o, T_p, T_c, ΔH showed a slight decrease upon irradiation	Reddy et al. (2015b)
Elephant foot yam[B]	~28%	Starch powder 5, 10, 15, 20, 25 kGy at 2 kGy/h	~11%	XRD pattern changed from B to C type upon irradiation, relative crystallinity decreased slightly, intensities of the band at 1155 cm^{-1} increased by FTIR, crystalline and amorphous regions were destructed by irradiation	T_o, T_p, T_c, ΔH showed an increase at lower dosage, and a decrease with dose increasing	Reddy et al. (2015a)
Cassava[B]	~22%	Starch powder 1.0, 2.5, 5.0, 7.5, 10.0 kGy	ND	*Intensities of the characteristic* X-ray diffraction peaks and degree of crystallinity were slightly reduced after gamma irradiation. No significant differences among FTIR patterns of the initial and irradiated starches were observed. The absorbance peak related to the glycosidic bonds was decreased significantly by gamma irradiation,	Thermogravimetric analysis showed that gamma irradiation reduced thermal stability of cassava starch	Tran et al. (2022)

(continued)

Table 7.3 (continued)

Starch samples		Irradiation treatment conditions		Results		Literatures
Starch/Type	Amylose content	Material status Dosage/dose rate	Water content	Crystalline structure	Thermal properties	
				implying starch molecules were broken or depolymerized into small fragments during irradiation		
Potato[B]	~30%	Starch powder 5, 10, 20, 30 kGy at 3.6 kGy/h	~12%	ND	T_o, T_p, T_c, ΔH showed a slight decrease upon irradiation	Sujka et al. (2015)
Potato[B]	~31%	Tuber 0.1 kGy at 32.4 kGy/h	~75%	Decrease in relative crystallinity was observed	T_o, T_p, T_c, ΔH showed a slight decrease upon irradiation	Lu et al. (2012a, b)
Potato[B]	~15%	Starch powder 0.01, 0.05, 0.1, 0.5 kGy	ND	The intensity of peaks decreased. Starches irradiated at 0.5 kGy did not show the presence of any peaks	T_o, T_p, T_c, ΔH showed a slight increase	Singh et al. (2011)
Potato[B]	~32%	Starch powder 10, 50 kGy at 2 kGy/h	~10%	No change in XRD pattern, the relative crystallinity decreased, the degree of order analyzed by FTIR at starch granule surface decreased	T_o, T_p, T_c increased at 10 kGy but decreased at 50 kGy, ΔH decreased as irradiation dose increased	Chung and Liu (2010)
Potato[B]	~32%	Starch powder 5, 10, 20 kGy at 2 kGy/h	~10%	No change in XRD pattern, relative crystallinity decreased	ND	Gani et al. (2014)
Chickpea[C]	~30%	Flour powder 0.5, 1, 2.5, 5, 10 kGy at 0.5 kGy/h	ND	No change in FTIR spectra pattern, the ordered structure of starch decreased. The intensities of the peaks and relative crystallinity by XRD decreased	T_o, T_p, T_c, ΔH: decreased with dose	Bashir and Aggarwal (2017)

Chickpea[C]	~33%	Starch powder 4, 8, 12 kGy at 2 kGy/h	~12%	Relative crystallinity showed a decrease at 4 kGy and 8 kGy doses, but a steep increase at 12 kGy dose. Decrease in absorbance intensity of the band was observed by FTIR at higher doses of irradiation	ND	Bashir and Haripriya (2016)
Cowpea[C]	ND	Flour powder and paste 2, 10, 50 kGy	ND	No change in FTIR spectra pattern and the ordered structure of starch granule surface	Peak temperature of gelatinization increased; however, enthalpy of gelatinization fluctuated	Abu et al. (2006)
Broad bean[C]	~52%	Starch powder 5, 10, 15 kGy at 5 kGy/h	~8%	No change in XRD pattern, relative crystallinity decreased. FTIR spectral patterns did not change but there was an increase in intensities of the band at 1018 cm^{-1} at high irradiation dose, indicated a decrease in the ordered structure	ND	Sofi et al. (2013)
Kidney beans[C]	36–41%	Starch powder 5, 10, 20 kGy at 2 kGy/h	~13%	No significant change in XRD pattern, relative crystallinity decreased	ND	Gani et al. (2012)
Bean[C]	~36%	Starch powder 10, 50 kGy at 2 kGy/h	~10%	No change in XRD pattern, the relative crystallinity decreased, the degree of order at starch granule surface decreased	T_o, T_p, T_c increased at 10 kGy but decreased at 50 kGy, ΔH decreased as irradiation dose increased	Chung and Liu (2010)
Bean[C]	~35%	Maize flour 5, 10, 20, 40 kGy at 0.5 kGy/h	~9.8%	ND	T_o, T_p, ΔH increased	Rombo et al. (2004)
Bean[C]	ND	Starch powder 2.5, 5, 10, 20 kGy	~12%	No notable change in XRD pattern	Enthalpy and temperature of gelatinization increased	Duarte and Rupnow (1993, 1994)

(continued)

Table 7.3 (continued)

Starch samples		Irradiation treatment conditions		Results		
Starch/Type	Amylose content	Material status Dosage/dose rate	Water content	Crystalline structure	Thermal properties	Literatures
Mung bean^C	ND	Starch powder 0.5, 1, 3, 5 kGy	~10%	The crystalline pattern of the amylose and the amylopectin molecules was not changed by irradiation	ND	Castanha et al. (2019)
Sago^C	~27%	Starch powder 6, 10, 25 kGy at 8.3 kGy/h	~13%	No change in crystal type by XRD, but irradiation induced a decrease in degree of granule crystallinity	T_o and T_p showed a slight increase, T_c and ΔH were not altered significantly	Othman et al. (2015)
Ariá^C	ND	Starch powder 1 kGy at 226.43 Gy/h, 5 kGy at 1131.49 Gy/h, 20 kGy at 4527.10 Gy/h, 50 kGy at 11614.44 Gy/h	~8%	The amorphous phase increases slightly, the mean crystallite size and crystallinity decreased continuously, and the microstrains increased with doses of radiation increasing	ND	Pinto et al. (2021)

regions in starch granules, which could increase or decrease relative crystallinity, and this depended on which region was most affected by various irradiation doses (Chung and Liu 2010). The reduction in crystallinity with increasing irradiation is attributed to the breakage of the crystalline regions of starch granules. However, the packing of double helices within the A-type crystalline structure is relatively compact and less sensitive to irradiation than B-type and C-type polymorphs (Gul et al. 2016), resulting in a slight increase in relative crystallinity at lower dosages possibly.

The short-range order structure of starch can be probed by Fourier transform-infrared spectroscopy (FTIR) to analyze the breakdown of various bonds and the decrease of double helices. The FTIR spectral pattern of irradiated starch did not change; however, most intensities showed increases indicating decreases in the ordered structure of starch upon irradiation (Table 7.3). Because the bands at 1047 and 1022 cm^{-1} are sensitive to changes in the crystalline and amorphous regions, respectively, the ratio of 1047/1022 was employed by some researchers to analyze the degree of order at the surface of starch granules and observed that the degree of starch granule surface order decreased with an increase in irradiation dose (Chung and Liu 2009, 2010; Kong et al. 2016).

Gelatinization temperatures (onset, T_o; peak, T_p; conclusion, T_c) and enthalpy of gelatinization (ΔH) of A-type starches decreased upon irradiation in most cases. With regard to the B-type starches, the gelatinization parameters showed an increase at lower dosage but decreased with irradiation dose increasing in some cases. For C-type starches, the gelatinization parameters were observed to increase by most researchers (Table 7.3). The effects of gamma irradiation on gelatinization parameters were complicated. Gamma irradiation could cause defective crystalline structure and amylopectin short chains increasing, which resulted in a decrease in gelatinization temperatures. However, irradiation could also destroy weaker crystal structure making the remaining structures more stable, which would result in an increase in gelatinization temperatures. The differences in gelatinization parameters induced by gamma irradiation at the same dose might also be attributed to dose rate of treatment, ratio of amylose/amylopectin, and native starch structure (Kong et al. 2016).

7.2.4 Physical and Digestion Properties

Physical properties discussed here included swelling power, water solubility, syneresis, freeze-thaw stability, transmittance, oil and water absorption, pasting properties, and starch gel textural properties. Starch swelling power represents its capacity to trap and hold water in starch granules and primarily resulted from amylopectin fraction, which can be measured at various temperatures, and normally from 50 to 95 °C. The swelling power was observed to decrease under gamma irradiation treatment in most reports (Table 7.4); the decrease in swelling power could be attributed to fragmentation of amylopectin, and this is beneficial to improve the textural quality upon cooking as the bursting of starch could be prevented (Gani

Table 7.4 Physical and digestion properties change of starches under gamma irradiation

Starch samples		Irradiation treatment conditions		Results	
Starch/type	Amylose content	Material status Dosage/dose rate	Water content	Physical and digestion properties	Literatures
Rice[A]	~27%	Grain 1, 2, 5 kGy at 0.4 kGy/h	~10%	Values of viscosity parameters decreased with exception of breakdown and pasting temperature. Resistant starch of cooked samples reached the highest at 1 kGy	Polesi et al. (2017)
Rice[A]	~20%	Starch powder 2, 5, 10 kGy at 0.4 kGy/h	~12%	Values of pasting parameters and swelling power decreased significantly, the solubility increased	Gul et al. (2016)
Rice[A]	~26%	Starch powder 5, 10, 20 kGy at 2 kGy/h	~12%	Swelling power, syneresis, and pasting properties decreased, whereas water absorption capacity and transmittance increased	Ashwar et al. (2014)
Rice[A]	~26%	Starch powder 5, 10, 20, 30, 40, 50 kGy at 2 kGy/h	~5%	All measured pasting parameters decreased continuously with increasing irradiation dose. Only doses of 20–50 kGy cause an increase in solubility of starch. Content of indigestible fraction of starch increased with irradiation dose increasing	Lam et al. (2021)
Rice[A]	27–40%	Flour powder 0.5, 1, 2, 3, 4, 5 kGy at 1.0 kGy/h	NA	Starch digestibility decreased upon irradiation	Shu et al. (2013)
Rice[A]	~19%	Grain 2, 5, 8, 10 kGy at 1 kGy/h	~30%	Pasting properties decreased significantly	Yu and Wang (2007)
Rice[A]	ND	Grain 0.5, 1, 3, 5, 7, 9 kGy at 0.5 kGy/h	~12%	Pasting properties decreased	Bao et al. (2005)
Rice[A]	~1%-26%	Grain 0.2, 0.4, 0.6, 0.8, 1 kGy at 1 kGy/h	ND	Peak, hot pasting, cool pasting, setback viscosities decreased considerably	Wu et al. (2002)

(continued)

Table 7.4 (continued)

Starch samples		Irradiation treatment conditions		Results	
Starch/type	Amylose content	Material status Dosage/dose rate	Water content	Physical and digestion properties	Literatures
Wheat[A]	~25%	Flour powder 0.5, 1, 2.5, 5, 10 kGy	~11%	Values of viscosity parameters decreased, swelling, solubility, syneresis, and freeze- thaw stability were improved with dosage	Bashir and Aggarwal (2017)
Wheat[A]	ND	Starch powder 1, 3, 5, 7, 9 kGy at 1 kGy/h	~12%	Values of viscosity parameters decreased with exception of breakdown; gel hardness and gumminess decreased	Kong et al. (2016)
Wheat[A]	~34%	Starch powder 3, 5, 10, 20, 35, 50 kGy at 0.83 kGy/h	~13%	Viscosity, rheological and textural parameters decreased, water solubility increased, and swelling power showed maximum at 20 kGy	Atrous et al. (2017)
Wheat[A]	~35%	Starch powder 3, 5, 10, 20, 35, 50 kGy at 0.83 kGy/h	~13%	Solubility increased, swelling power increased to the maximum at 20 kGy, then decreased significantly. Rheological properties decreased	Atrous et al. (2015)
Oat[A]	~22%-27%	Grain 5, 10, 15, 20 kGy at 2 kGy/min	~12%	Significant decreasing trend in syneresis value and increasing trend in transmittance value were observed upon irradiation, pasting viscosities and temperature were reduced continuously with irradiation dose increasing, swelling power decreased, and water solubility index increased upon irradiation	Mukhtar et al. (2017)

(continued)

Table 7.4 (continued)

Starch samples		Irradiation treatment conditions		Results	
Starch/type	Amylose content	Material status Dosage/dose rate	Water content	Physical and digestion properties	Literatures
Amaranth[A]	6%-10%	2, 4, 6, 8, and 10 kGy at 1 kGy/h	~10%	Pasting and rheological properties decreased	Kong et al. (2009)
Arrowhead[A]	~31%	Starch powder 5, 10, 15 kGy at 5 kGy/h	~8%	Peak, trough, final, and setback viscosities were significantly reduced, whereas the breakdown viscosity increased. Solubility, water absorption capacity, and transmittance increased, whereas swelling power, bulk density, and syneresis decreased	Wani et al. (2015)
Arrowroot[A]	ND	Starch powder 5, 10, 15 kGy at 1 kGy/h	~13%	The pasting viscosities decreased with the increase of radiation dose; however, pasting temperature showed a slight increase	Barroso and del Mastro (2019)
Indian horse chestnut[A]	~27%	Starch powder 5, 10, 15 kGy at 5 kGy/h	~12%	Water absorption capacity, oil absorption capacity, solubility, freeze- thaw stability increased, whereas syneresis, swelling power, pasting properties decreased	Wani et al. (2014)
Lotus seed[A]	~20%	Starch powder 5, 10, 15 kGy	ND	Swelling power decreases whereas solubility increases significantly with irradiation dose increasing. Pasting properties showed drastic reductions after gamma irradiation as compared to native sample. The irradiated lotus seed starch gels showed a	Punia et al. (2020)

(continued)

Table 7.4 (continued)

Starch samples		Irradiation treatment conditions		Results	
Starch/type	Amylose content	Material status Dosage/dose rate	Water content	Physical and digestion properties	Literatures
				reduction in light transmittance	
Tapioca[A]	~35%	Starch powder 5, 10 and 20 kGy at 1.8 kGy/h	~10%	Increase in solubility and light transmittance was observed to be dose dependent while there was a decrease in swelling capacity, syneresis was reduced by radiation processing	Kanatt (2020)
Kithul[A]	~39%	Starch powder 0.5 kGy, 1 kGy, 2.5 kGy, 5 kGy and 10 kGy at 2 kGy/h	~11%	Irradiation increased the light transmittance and freeze-thaw stability, but reduced the syneresis and pasting viscosities; in vitro digestibility was improved after irradiation	Sudheesh et al. (2019)
Corn[A]	ND	Starch powder 3, 5, 10, 20, 50 kGy at 1.14 kGy/h	ND	Maximal consistency of the starch paste decreased determined by brabender viscoamylograph test	Ben Bettaïeb et al. (2014)
Maize[A]	ND	Starch powder 1, 2, 5, 10, 20, 50, 100, 200, 500 kGy at 5 kGy/h	ND	Peak, trough, final, and setback viscosities decreased, whereas solubility increased	Liu et al. (2012)
Corn[A]	Waxy and normal	Starch powder 5, 10, 20 kGy at 10 kGy/h	5%, 12%	Resistant starch content increased; slowly digestible starch content decreased	Yoon et al. (2010)
Corn[A]	Waxy and normal	Starch powder at various pH and salt concentrations 5, 10, 20 kGy at 10 kGy/h	6%	Pasting viscosities and rheological properties decreased	Baik et al. (2010)
Corn[A]	~29%	Starch powder 2, 10, 50 kGy at 2 kGy/h and 0.40, 0.67, 2 kGy/h to a total dose of 10 kGy	~9%	Pasting viscosities decreased, rapidly digestible starch (RDS) content slightly decreased up to 10 kGy, but	Chung and Liu (2009)

(continued)

Table 7.4 (continued)

Starch samples		Irradiation treatment conditions		Results	
Starch/type	Amylose content	Material status Dosage/dose rate	Water content	Physical and digestion properties	Literatures
				increased at 50 kGy, resistant starch (RS) content slightly decreased at 2 kGy and then increased up to 50 kGy. The slowly digestible starch (SDS) content showed the opposite trend to RS content	
Corn[A or B]	4–70%	Starch powder 1, 5, 10, 25, 50 kGy at 1 kGy/h	ND	Peak, final, and set-back viscosities decreased; however, peak and final viscosities increased for high amylose samples	Chung et al. (2015)
Corn[A or B]	2–72%	Starch powder 5, 10, 25, and 50 kGy at 1 kGy/h	~12%	Resistant starch content increased	Lee et al. (2013)
Maize[B]	~62%	Starch powder 30, 60 kGy at 1.2 kG/h	ND	Pasting viscosity reduced	Ocloo et al. (2016)
Maize[B]	~58%	Starch powder 30, 60 kGy at 1.2 kGy/h	~10%	Swelling power decreased, solubility increased. Oil and water absorption capacities increased at 30 kGy, but remained stable when irradiation dosage increased to 60 kGy	Ocloo et al. (2014)
Potato[B]	~30%	Starch powder 3, 5, 10, 20, 35, 50 kGy at 0.83 kGy/h	~18%	Viscosity, rheological and textural parameters decreased, water solubility increased, and swelling power showed a maximum at 20 kGy	Atrous et al. (2017)
Potato[B]	~30%	Starch powder 5, 10, 20, 30kGy at 3.6 kGy/h	~12%	Water absorption, least gelling concentration, solubility, paste clarity, and susceptibility to a-amylolysis	Sujka et al. (2015)

(continued)

Table 7.4 (continued)

Starch samples		Irradiation treatment conditions		Results	
Starch/type	Amylose content	Material status Dosage/dose rate	Water content	Physical and digestion properties	Literatures
				increased, whereas fat absorption and viscosity decreased	
Potato[B]	ND	Starch powder 5, 10, 20, 30 kGy at 3.6 kGy/h	~14%	The gel structure became smooth rather than honey-comb like, swelling power decreased	Cieśla et al. (2015)
Potato[B]	~32%	Starch powder 5, 10, 20 kGy at 2 kGy/h	~10%	Pasting properties, swelling power, and syneresis decreased, whereas water absorption capacity and solubility increased	Gani et al. (2014)
Potato[B]	~31%	Tuber 0.1 kGy at 32.4 kGy/h	~75%	Rheological properties, pasting viscosities decreased, resistant starch content decreased	Lu et al. (2012a, b)
Potato[B]	~15%	Starch powder 0.01, 0.05, 0.1, 0.5 kGy	ND	Peak, trough, breakdown, final viscosities, and gel hardness decreased, whereas gel cohesiveness increased	Singh et al. (2011)
Potato[B]	~32%	Starch powder 10, 50 kGy at 2 kGy/h	~10%	Pasting viscosity decreased, resistant starch contents (determined with uncooked starch) increased	Chung and Liu (2010)
Elephant foot yam[B]	~28%	Starch powder 5, 10, 15, 20, 25 kGy at 2 kGy/h	~11%	Pasting parameters, swelling power, and syneresis decreased; solubility, light transmittance, water absorption capacity increased	Reddy et al. (2015a)
Banana[B]	~25%	5, 10, 15, 20, 25 kGy at 2 kGy/h	ND	Peak, hold, final and setback viscosities, swelling power, syneresis decreased, whereas in vitro digestibility, solubility, and water	Reddy et al. (2015b)

(continued)

Table 7.4 (continued)

Starch samples		Irradiation treatment conditions		Results	
Starch/type	Amylose content	Material status Dosage/dose rate	Water content	Physical and digestion properties	Literatures
				absorption capacity increased	
Cassava[B]	~22%	Starch powder 1.0, 2.5, 5.0, 7.5, 10.0 kGy	ND	Water solubility index increased with the irradiation dose, and swelling power increased to the highest at 5 kGy then decreased. Apparent viscosity was reduced significantly by gamma irradiation	Tran et al. (2022)
Ariá[C]	ND	Starch powder 1 kGy at 226.43 Gy/h, 5 kGy at 1131.49 Gy/h, 20 kGy at 4527.10 Gy/h, 50 kGy at 11,614.44 Gy/h	~8%	Pasting viscosities decreased, pasting temperature remained unchanged with irradiation dosage up to 20 kGy and reduced when irradiation dosage increased to 50 kGy	Pinto et al. (2021)
Broad bean[C]	~52%	Starch powder 5, 10, 15 kGy at 5 kGy/h	~8%	Water absorption capacity, freeze-thaw stability, and solubility increased, whereas pasting properties, syneresis decreased	Sofi et al. (2013)
Kidney beans[C]	36%-41%	Starch powder 5, 10, 20 kGy at 2 kGy/h	~13%	Solubility, water absorption capacity, and transmittance increased, whereas swelling power, syneresis, and pasting properties decreased	Gani et al. (2012)
Mung bean[C]	ND	Starch powder 0.5, 1, 3, 5 kGy	~10%	Irradiation process resulted in lower water retention ability, lower apparent viscosity, higher paste clarity, and, in general, harder and less elastic gels of starches	Castanha et al. (2019)
Bean[C]	~36%	Starch powder 10, 50 kGy at 2 kGy/h	~10%	Pasting viscosity decreased, resistant starch contents	Chung and Liu (2010)

(continued)

Table 7.4 (continued)

Starch samples		Irradiation treatment conditions		Results	
Starch/type	Amylose content	Material status Dosage/dose rate	Water content	Physical and digestion properties	Literatures
				(determined with uncooked starch) increased	
Bean[C]	ND	Starch powder 2.5, 5, 10, 20 kGy	~12%	Swelling power decreased, solubility increased	Duarte and Rupnow (1994)
Chickpea[C]	~30%	Flour powder 0.5, 1, 2.5, 5, 10 kGy at 0.5 kGy/h	ND	Pasting properties showed a significant decrease in peak viscosity, final viscosity, setback viscosity, trough viscosity, and pasting temperature in a dose- dependent manner. Swelling, solubility index, oil absorption capacity, and water absorption capacity increased significantly with dose, while syneresis decreased with dose	Bashir and Aggarwal (2017)
Chickpea[C]	~33%	Starch powder 4, 8, 12 kGy at 2 kGy/h	~12%	Swelling power, turbidity, syneresis, pasting parameters reduced, whereas solubility increased	Bashir and Haripriya (2016)
Chickpea[C]	ND	Grain 2, 5, 10, 20, 30, 50 kGy at 3.6 kGy/h	ND	Peak viscosity decreased significantly	Graham et al. (2002)
Cowpea[C]	ND	Flour powder and paste 2, 10, 50 kGy	ND	Pasting and swelling properties decreased significantly	Abu et al. (2006)

et al. 2012; Ashwar et al. 2014). However, some researcher reported that the swelling power increased at lower doses (Table 7.4). It was well known that the presence of amylose and lipids inhibited starch swelling, and gamma irradiation could fragment both amylose and lipid, which would promote starch swelling. Water solubility index of irradiated starches increased with irradiation dosage elevated in all works (Table 7.4); the increase in the solubility index is attributed to the radiation-induced depolymerization of the starch molecules resulting from breakage of glycosidic bonds and production of lower molecular weight fractions that can be dissolved by neighboring water molecules more easily, contributing to increases in both solubility index and light transmittance of irradiated starches. Syneresis refers to expulsion of liquid from the starch gel during storage; it is an index of gel stability. The syneresis of gels formed with irradiated treated starches decreased in all previous reports (Table 7.4); the decrease in the syneresis values was attributed to the production of lower molecular weight fractions that have higher tendency for water and also resulting from reduction in the magnitude of hydrogen bonding forces, interaction between amylose-amylose and amylose-amylopectin chains, in starch molecules upon irradiation (Bashir and Aggarwal 2017).

Pasting and rheological properties of starches could be characterized by Rapid Visco Analyzer (RVA) or rheometer; most pasting and rheological parameters of irradiated starches decreased continuously with irradiation dosage increasing (Table 7.4). The reduction in the pasting properties can be attributed to starch degradation resulting from irradiation induced changes. Pasting parameters can be easily correlated with the texture profile and quality of the product (Bashir and Aggarwal 2017). It was supposed that the degradation of amylopectin and amylose in starch induced by gamma irradiation treatment might cause a considerable decrease in water binding capacity and then reduced the swelling degree of starch granules, which resulted in a significant decrease in peak viscosity. The setback and cool paste viscosities are mainly attributed to reordering or polymerization of leached amylose and long linear amylopectin; accordingly, degradation or shortening of amylose and longer amylopectin branch chains after irradiation contributed to the decrease in setback and cool paste viscosities (Chung and Liu 2009).

Gamma irradiation can be employed to elevate the resistant starch fraction, which is useful for diabetic patients; however, some researchers observed that irradiation treatment could decrease resistant starch content (Table 7.4). The cleavage of anhydrous glycosidic linkages increases the accessibility of the digestive enzymes and thus decreases resistant starch content. The increase of resistant starch thus indicates that there were structural changes besides the chain cleavage; some authors attributed this to the increase in the proportion of β-bonded starch after irradiation as a result of transglucosidation (Rombo et al. 2004) or the increase in carboxyl groups which would result in inhibition of enzyme attack (Chung and Liu 2010). More studies on starch modified by gamma irradiation can provide knowledge on controlling the sugar produced in the starch for health purposes (Sunder et al. 2022).

7.3 Grafting and Cross-Linking Synthesis of Starch with Chemicals

Radiation-induced graft polymerization is a powerful method for preparation of polymeric materials with new properties (Suwanmala et al. 2012), such as superabsorbent polymer, biomaterials, and drug delivery systems. Radiation processing has many advantages over other conventional methods. In radiation processing, no catalysts or additives are needed to initiate the reaction, so that this process does not cause any further contamination associated with chemical initiators.

They are relatively simple and of high efficiency, and the degree of cross-linking and grafting can be controlled easily by varying the absorbed dose (Eid 2008). Zhang and Xu (2017) compared potassium persulfate and irradiation as initiator to graft acrylic acid onto starch and concluded that irradiation was more time-saving, energy-saving, and efficient. This kind of work has been done by many researchers (Ali and Abdel Ghaffar 2017; Abdel Ghaffar et al. 2016; Mahmoud et al. 2014; Sheikh et al. 2013; Lv et al. 2013; Abd El-Rehim et al. 2013; Suwanmala et al. 2012).

7.4 Conclusions

From literatures we listed in the tables, we could find that the research about starch irradiation boomed in the last decade, partly because this method is safe, inexpensive, and time-saving and has an enormous potential in various applications. Several review articles about the subject of current chapter are available for the readers (Sunder et al. 2022; Zhu 2016; Bhat and Karim 2009; Sokhey and Hanna 1993). Combination irradiation treatment with other modifications was performed to widen the applications of modified starch products and will attract more and more attention in future. Standardization of irradiation dosage under controlled conditions to achieve specific starch quality will have also potential interests in future.

References

Abd El-Rehim HA, Hegazy ESA, Diaa DA (2013) Radiation synthesis of eco-friendly water reducing sulfonated starch/acrylic acid hydrogel designed for cement industry. Radiat Phys Chem 85:139–146

Abdel Ghaffar AM, Radwan RR, Ali HE (2016) Radiation synthesis of poly(starch/ acrylic acid) pH sensitive hydrogel for rutin controlled release. Int J Biol Macromol 92:957–964

Abu JO, Duodu KG, Minnaar A (2006) Effect of γ-irradiation on some physicochemical and thermal properties of cowpea (*Vigna unguiculata* L. Walp) starch. Food Chem 95(3):386–393

Ali HE, Abdel Ghaffar AM (2017) Preparation and effect of gamma radiation on the properties and biodegradability of poly(styrene/starch) blends. Radiat Phys Chem 130:411–420

Ashwar BA, Shah A, Gani A, Rather SA, Wani SM, Wani IA, Masoodi FA, Gani A (2014) Effect of gamma irradiation on the physicochemical properties of alkali-extracted rice starch. Radiat Phys Chem 99:37–44

Atrous H, Benbettaieb N, Hosni F, Danthine S, Blecker C, Attia H, Ghorbel D (2015) Effect of γ-radiation on free radicals formation, structural changes and functional properties of wheat starch. Int J Biol Macromol 80:64–76

Atrous H, Benbettaieb N, Chouaibi M, Attia H, Ghorbel D (2017) Changes in wheat and potato starches induced by gamma irradiation: a comparative macro and microscopic study. Int J Food Prop 20(7):1532–1546

Baik BR, Yu JY, Yoon HS, Lee JW, Byun MW, Baik BK, Lim ST (2010) Physicochemical properties of waxy and normal maize starches irradiated at various pH and salt concentrations. Starch 62(1):41–48

Bao J, Ao Z, Jane JL (2005) Characterization of physical properties of flour and starch obtained from gamma-irradiated white rice. Starch 57(10):480–487

Barroso AG, del Mastro NL (2019) Physicochemical characterization of irradiated arrowroot starch. Radiat Phys Chem 158:194–198

Bashir K, Aggarwal M (2017) Physicochemical, thermal and functional properties of gamma irradiated chickpea starch. Int J Biol Macromol 97:426–433

Bashir M, Haripriya S (2016) Physicochemical and structural evaluation of alkali extracted chickpea starch as affected by γ-irradiation. Int J Biol Macromol 89:279–286

Ben Bettaïeb N, Jerbi MT, Ghorbel D (2014) Gamma radiation influences pasting, thermal and structural properties of corn starch. Radiat Phys Chem 103:1–8

Bertolini AC, Mestres C, Colonna P, Raffi J (2001) Free radical formation in UV- and gamma-irradiated cassava starch. Carbohydr Polym 44:269–271

Bhat R, Karim AA (2009) Impact of radiation processing on starch. Compr Rev Food Sci Food Saf 8:44–58

Castanha N, Miano AC, Sabadoti VD, Augusto PED (2019) Irradiation of mung beans (*Vigna radiata*): a prospective study correlating the properties of starch and grains. Int J Biol Macromol 129:460–470

Chaudhry MA, Glew G (1973) The effect of ionizing radiations on some physical and chemical properties of Pakistani rice: II. The effect on starch and starch fractions. Int J Food Sci Technol 8(3):295–303

Chung HJ, Liu Q (2009) Effect of gamma irradiation on molecular structure and physicochemical properties of corn starch. J Food Sci 74(5):353–361

Chung HJ, Liu Q (2010) Molecular structure and physicochemical properties of potato and bean starches as affected by gamma-irradiation. Int J Biol Macromol 47:214–222

Chung KH, Othman Z, Lee JS (2015) Gamma irradiation of corn starches with different amylose-to-amylopectin ratio. J Food Sci Technol 52:6218–6229

Cieśla K, Eliasson AC (2002) Influence of gamma radiation on potato starch gelatinization studied by differential scanning calorimetry. Radiat Phys Chem 64(2):137–148

Cieśla K, Sartowska B, Królak E (2015) SEM studies of the structure of the gels prepared from untreated and radiation modified potato starch. Radiat Phys Chem 106:289–302

Duarte PR, Rupnow JH (1993) Gamma-irradiation affects some physical properties of dry bean (*Phaseolus vulgaris*) starch. J Food Sci 58(2):389–394

Duarte PR, Rupnow JH (1994) Gamma-irradiated dry bean (*Phaseolus vulgaris*) starch: physicochemical properties. J Food Sci 59(4):839–843

Eid M (2008) In vitro release studies of vitamin B12 from poly N-vinyl pyrrolidone/starch hydrogels grafted with acrylic acid synthesized by gamma radiation. Nucl Inst Methods Phys Res B 266(23):5020–5026

El Saadany RMA, El Saadany FM, Foda YH (1974) Modification of rice starch by gamma irradiation to produce soluble starch of low viscosity for industrial purposes. Starch 26(12): 422–425

El Saadany RMA, El Saadany FM, Foda YH (1976) Transformation of tapioca starch by gamma irradiation. Starch 28(5):169–172

Ezekiel R, Rana G, Singh N, Singh S (2007) Physicochemical, thermal and pasting properties of starch separated from γ-irradiated and stored potatoes. Food Chem 105:1420–1429

Gani A, Bashir M, Wani SM, Masoodi FA (2012) Modification of bean starch by γ-irradiation: Effect on functional and morphological properties. LWT Food Sci Technol 49:162–169

Gani A, Nazia S, Rather SA, Wani SM, Shah A, Bashir M, Masoodi FA, Gani A (2014) Effect of γ-irradiation on granule structure and physicochemical properties of starch extracted from two types of potatoes grown in Jammu & Kashmir, India. LWT Food Sci Technol 58:239–246

Ghali Y, Ibrahim N, Gabr S, Aziz H (1979) Modification of corn starch and fine flour by acid and gamma irradiation. Part 1. Chemical investigation of the modified products. Starch 31(10): 325–328

Graham JA, Panozzo J, Lim PC, Brouwer JB (2002) Effects of gamma irradiation on physical and chemical properties of chickpeas (*Cicer arietinum*). J Sci Food Agric 82(14):1599–1605
Greenwood CT, Mackenzie S (1963) The irradiation of starch. Part I. The properties of potato starch and its components after irradiation with high-energy electrons. Starch 15(12):444–448
Gul K, Singh AK, Sonkawade RG (2016) Physicochemical, thermal and pasting characteristics of gamma irradiated rice starches. Int J Biol Macromol 85:460–466
Kanatt SR (2020) Irradiation as a tool for modifying tapioca starch and development of an active food packaging film with irradiated starch. Radiat Phys Chem 173:108873
Kertesz ZI, Schulz ER, Fox G, Gibson M (1959) Effects of ionizing radiations on plant tissues. IV. Some effects of gamma radiation on starch and starch fractions. J Food Sci 24(6):609–617
Komolprasert V, Morehouse KM (2004) Overview of irradiation of food and packaging. ACS Symposium Series. ACS, Washington
Kong X, Kasapis S, Bao J, Corke H (2009) Effect of gamma irradiation on the thermal and rheological properties of grain amaranth starch. Radiat Phys Chem 78(11):954–960
Kong X, Zhou X, Sui Z, Bao J (2016) Effects of gamma irradiation on physicochemical properties of native and acetylated wheat starches. Int J Biol Macromol 91:1141–1150
Lam ND, Quynh TM, Diep TB, Binh PT, Lam TD (2021) Effect of gamma irradiation and pyrolysis on indigestible fraction, physicochemical properties, and molecular structure of rice starch. J Food Process Preserv 45(10):e15880
Lee YJ, Kim SY, Lim ST, Han SM, Kim HM, Kang IJ (2006) Physicochemical properties of gamma-irradiated corn starch. J Food Sci Nutr 11:146–154
Lee JS, Ee ML, Chung KH, Othman Z (2013) Formation of resistant corn starches induced by gamma-irradiation. Carbohydr Polym 97(2):614–617
Liu T, Ma Y, Xue S, Shi J (2012) Modifications of structure and physicochemical properties of maize starch by γ-irradiation treatments. LWT Food Sci Technol 46:156–163
Lu ZH, Donner E, Yada RY, Liu Q (2012a) Rheological and structural properties of starches from γ-irradiated and stored potatoes. Carbohydr Polym 87(1):69–75
Lu ZH, Donner E, Yada RY, Liu Q (2012b) Impact of γ-irradiation, CIPC treatment, and storage conditions on physicochemical and nutritional properties of potato starches. Food Chem 133(4): 1188–1195
Lv X, Song W, Ti Y, Qu L, Zhao Z, Zheng H (2013) Gamma radiation-induced grafting of acrylamide and dimethyl diallyl ammonium chloride onto starch. Carbohydr Polym 92(1): 388–393
Mahapatra AK, Muthukumarappan K, Julson JL (2005) Applications of ozone, bacteriocins and irradiation in food processing: a review. Crit Rev Food Sci Nutr 45(6):447–461
Mahmoud GA, Abdel-Aal SE, Badway NA, Abo Farha SA, Alshafei EA (2014) Radiation synthesis and characterization of starch-based hydrogels for removal of acid dye. Starch 66(3–4):400–408
Mishina A, Nikuni Z (1959) Physical and chromatographical observations of γ-irradiated potato starch granules. Nature 184(4702):1867
Mukhtar R, Shah A, Noor N, Gani A, Wani IA, Ashwar BA, Masoodi FA (2017) γ-Irradiation of oat grain - effect on physico-chemical, structural, thermal, and antioxidant properties of extracted starch. Int J Biol Macromol 104:1313–1120
Mukisa IM, Muyanja CMBK, Byaruhanga YB, Schüller RB, Langsrud T, Narvhus JA (2012) Gamma irradiation of sorghum flour: Effects on microbial inactivation, amylase activity, fermentability, viscosity and starch granule structure. Radiat Phys Chem 81:345–351
Nene SP, Vakil UK, Sreenivasan A (1975) Effect of gamma radiation on physico-chemical characteristics of red gram (*Cajanus cajan*) starch. J Food Sci 40(5):943–947
Ocloo FCK, Minnaar A, Emmambux NM (2014) Effects of gamma irradiation and stearic acid, alone and in combination, on functional, structural, and molecular characteristics of high amylose maize starch. Starch 66:624–635

Ocloo FCK, Minnaar A, Emmambux NM (2016) Effects of stearic acid and gamma irradiation, alone and in combination, on pasting properties of high amylose maize starch. Food Chem 190: 12–19

Othman Z, Hassan O, Hashim K (2015) Physicochemical and thermal properties of gamma irradiated sago (*Metroxylon sagu*) starch. Radiat Phys Chem 109:48–53

Pinto CD, Sanches EA, Clerici MTPS, Pereira MT, Campelo PH, de Souza SM (2021) X-ray diffraction and Rietveld characterization of radiation-induced physicochemical changes in Ariá (*Goeppertia allouia*) C-type starch. Food Hydrocoll 117:106682

Polesi LF, Sarmento SBS, De Moraes J, Franco CML, Canniatti-Brazaca SG (2016) Physicochemical and structural characteristics of rice starch modified by irradiation. Food Chem 191:59–66

Polesi LF, Junior MDDM, Sarmento SBS, Canniatti-Brazaca SG (2017) Starch digestibility and physicochemical and cooking properties of irradiated rice grains. Rice Sci 24:48–55

Punia S, Dhull SB, Kunner P, Rohilla S (2020) Effect of γ-radiation on physico-chemical, morphological and thermal characteristics of lotus seed (*Nelumbo nucifera*) starch. Int J Biol Macromol 157:584–590

Raffi J, Agnel J (1983) Influence of the physical structure of irradiated starches on their electron spin resonance spectra kinetics. J Phys Chem 87:2369–2373

Rayas-Duarte P, Rupnow JH (1993) Gamma-irradiation affects some physical properties of dry bean (*Phaseolus vulgaris*) starch. J Food Sci 58(2):389–394

Reddy CK, Suriya M, Vidya PV, Vijina K, Haripriya S (2015a) Effect of γ-irradiation on structure and physico-chemical properties of Amorphophallus paeoniifolius starch. Int J Biol Macromol 79:309–315

Reddy KC, Vidya PV, Vijina K, Haripriya S (2015b) Modification of poovan banana (Musa AAB) starch by γ-irradiation: effect on in vitro digestibility, molecular structure and physico-chemical properties. Int J Food Sci Technol 50(8):1778–1784

Rombo GO, Taylor JRN, Minnaar A (2004) Irradiation of maize and bean flours: effects on starch physicochemical properties. J Sci Food Agric 84(4):350–356

Samec M (1960) Some properties of gamma-irradiated starches and their electrodialytic separation. J Appl Polym Sci 3(8):224–226

Sheikh N, Akhavan A, Ataeivarjovi E (2013) Radiation grafting of styrene on starch with high efficiency. Radiat Phys Chem 85:189–192

Shu X, Xu J, Wang Y, Rasmussen SK, Wu D (2013) Effects of gamma irradiation on starch digestibility of rice with different resistant starch content. Int J Food Sci Technol 48(1):35–43

Singh S, Singh N, Ezekiel R, Kaur A (2011) Effects of gamma-irradiation on the morphological, structural, thermal and rheological properties of potato starches. Carbohydr Polym 83(4): 1521–1528

Sofi BA, Wani IA, Masoodi FA, Saba I, Muzaffar S (2013) Effect of gamma irradiation on physicochemical properties of broad bean (*Vicia faba* L.) starch. LWT Food Sci Technol 54: 63–72

Sokhey AS, Hanna MA (1993) Properties of irradiated starches. Food. Structure 12:397–410

Stewart EM (2001) Food irradiation chemistry. In: Molins R (ed) Food irradiation: principles and applications. Wiley, New York, pp 37–76

Sudheesh C, Sunooj KV, George J, Kumar S, Vikas, & Sajeevkumar, V. A. (2019) Impact of γ– irradiation on the physico-chemical, rheological properties and in vitro digestibility of kithul (*Caryota urens*) starch; a new source of nonconventional stem starch. Radiat Phys Chem 162: 54–65

Sujka M, Ciesla K, Jamroz J (2015) Structure and selected functional properties of gamma-irradiated potato starch. Starch 67:1002–1010

Sunder M, Mumbrekar KD, Mazumder N (2022) Gamma radiation as a modifier of starch – physicochemical perspective. Curr Res Food Sci 5:141–149

Suwanmala P, Hemvichian K, Hoshina H, Srinuttrakul W, Seko N (2012) Preparation of metal adsorbent from poly(methyl acrylate)-grafted-cassava starch via gamma irradiation. Radiat Phys Chem 81(8):982–985

Tomasik P, Zaranyika MF (1995) Nonconventional methods of modification of starch. Adv Carbohydr Chem Biochem 51:243–320

Tran MQ, Nguyen VB, Tran XA (2022) Gamma radiation modification of cassava starch and its characterization. Polym Eng Sci 62(4):1197–1204

Wani IA, Jabeen M, Geelani H, Masoodi FA, Saba I, Muzaffar S (2014) Effect of gamma irradiation on physicochemical properties of Indian Horse Chestnut (*Aesculus indica* Colebr.) starch. Food Hydrocoll 35:253–263

Wani AA, Wani IA, Hussain PR, Gani A, Wani TA, Masoodi FA (2015) Physicochemical properties of native and γ-irradiated wild arrowhead (*Sagittaria sagittifolia* L.) tuber starch. Int J Biol Macromol 77:360–368

Wu D, Shu Q, Wang Z, Xia Y (2002) Effect of gamma irradiation on starch viscosity and physicochemical properties of different rice. Radiat Phys Chem 65(1):79–86

Yoon HS, Yoo JY, Kim JH, Lee JW, Byun MW, Baik BK, Lim ST (2010) In vitro digestibility of gamma-irradiated corn starches. Carbohydr Polym 81(4):961–963

Yu Y, Wang J (2007) Effect of γ-ray irradiation on starch granule structure and physicochemical properties of rice. Food Res Int 40:297–303

Zhang YN, Xu SA (2017) Effects of amylose/amylopectin starch on starch-based superabsorbent polymers prepared by γ-radiation. Starch 69(1–2):1500294

Zhu F (2016) Impact of γ-irradiation on structure, physicochemical properties, and applications of starch. Food Hydrocoll 52:201–212

Chapter 8
Microwave Treatment

Kao Wu and Zekun Xu

Abstract Microwave technology has been widely used in food processing and consumption industries for various purposes and can also be used for physical modification of starch. Unlike conventional heating methods, microwave heating has characteristics such as low cost, short start-up time, high heating rate, and efficiency. The modification effects of microwave treatment (MWT) on starch are influenced by many factors, like starch properties (starch type, moisture content, density, dielectric properties, temperature, etc.) and microwave processing conditions (frequency, power, radiation time, oven type, and geometry). This chapter gives detailed information about starch physical properties changes after MWT such as composition, pasting properties, gelation, micromorphology, swelling and gelatinization, and in vitro digestibility.

Keywords Microwave treatment · Starch composition · Starch pasting properties · Gelation

8.1 Introduction

Microwaves are defined as the electromagnetic waves with frequencies ranging from 300 MHz to 300 GHz, corresponding to wavelengths from 1 m to 1 mm (Chandrasekaran et al. 2013). Under microwave radiation, polar molecules (e.g., water molecules) absorb microwave energy, orient themselves rapidly with respect to the electric field, and quickly generate bulk heat by molecular friction (Román et al. 2015), quite different from conventional heating methods, which include three forms: conduction, convection, and radiation. Microwave technology has been widely used in food processing and consumption industries for various purposes

K. Wu (✉)
School of Food and Biological Engineering, Hubei University of Technology, Wuhan, China

Z. Xu
Department of Food Science and Technology, School of Agriculture and Biology, Shanghai Jiao Tong University, Shanghai, China

Z. Sui, X. Kong (eds.), *Physical Modifications of Starch*,
https://doi.org/10.1007/978-981-99-5390-5_8

(e.g., baking, dehydration, tempering, thawing, cooking, pasteurization, drying, and expansion) since the last century, and a frequency of 2450 MHz is most commonly used in food industry (Ma et al. 2015). The dielectric property of food is largely determined by the chemical composition, moisture content, and temperature (Jiang et al. 2018) and affects microwave heating through influencing the microwave response characteristics and penetration depth (Wu et al. 2022). Compared with conventional heating, microwave heating offers many advantages such as lower cost, less start-up time, high heating rate and efficiency, and convenient process control (Feng et al. 2012; Salazar-González et al. 2012). Though certain drawbacks in microwave heating such as the nonuniform heat distribution and high equipment costs exist, they do not hinder the high potentials of microwave heat treatment for the processing of agricultural products, as equipment costs can change with time and developing technology (Soni et al. 2020).

As dry starch (native and gelatinized) are thought to be electrically inert, starch–microwave radiation interaction has to be investigated at certain moisture content (Fan et al. 2015). Microwave irradiation has been reported to assist preparations of extruded starch pellets (Lee et al. 2000), cross-linked starch microsphere (Lin et al. 2013), and starch-based sustained release materials for drug delivery (Sriamornsak et al. 2010) and can benefit glucose extraction from starch-based wastes (Villière et al. 2015). It can also be used to reduce the cooking time by producing partial gelatinized noodles (Xue et al. 2008). On the other hand, microwave heating can be an effective drying method and will cause some morphological and physicochemical property changes (Chen et al. 2016), as a result of starch modification. At present, the microwave modification effects on starch are reported to be influenced by starch properties (starch type, moisture content, density, dielectric properties, temperature, etc.) and microwave processing conditions (frequency, power, radiation time, oven type, and geometry) (Braşoveanu and Nemţanu 2014).

8.2 Temperature-Time Curve During Microwave Treatment

Moisture and heating temperature are the key conditions for starch modifications based on hydrothermal methods and are mostly restricted to proper combinations for traditional methods. For examples, heat-moisture treatment (HMT) usually restricts the moisture content in a range of 10–30% with relatively high temperature (90–120 °C) to avoid starch gelatinization, and annealing treatment (ANN) adopts high water content (50–60%) and relatively low temperatures (<gelatinization point) (Maache-Rezzoug et al. 2008). However, studies about microwave treatment (MWT) on starches are not restricted by these rules (Table 8.1). For example, microwave irradiation with both high initial moisture content (IMC) (>90%) and high heating temperature (80 °C) were performed on rice starch, and a following freeze-drying method was adopted to obtain the irradiated samples (Fan et al.

Table 8.1 Starch–microwave related studies

Starch type	Initial moisture content	Microwave power and irradiation time	Heat attributes	Major findings	References
Bambara groundnut starch	30%	700 W, 0–60 s		MWT resulted in reduction in peak viscosity, hot paste viscosity, cold paste viscosity, pasting temperature, and set back ratio.	Oyeyinka et al. (2019)
				The native starch granules were smooth but had pinholes after MWT	
Barley	42.1–45.10%	330–713 W, 150–180 min	90 °C	Increased starch digestibility, lower melting enthalpy, and higher gelatinization temperature	Emami et al. (2012)
				Changes on pasting attributes varied with amylose content	
Canna edulis Ker	20%	1000 W, 300 W		Double higher resistant starch, increased molecular order and crystallinity during enzyme digestion	Zhang et al. (2010)
	20–45%	400–1000 W, 0–30 min	60 °C for35–45% moisture	Moisture content, irradiation power, and heating time had different impacts on heating rate, amylose leaching, enzymatic digestibility, crystallinity, and thermal stability	Zhang et al. (2009)
			110 °C for 20–30% moisture		
Cassava	20%	650 W, 3–15 min		Significant changes on pasting properties	Abraham (1993)
				Lower swelling power, water-binding capacity, paste translucency, enzyme susceptibility	
	30%	600–700 W, 0–60 s		MWT altered starch color, increased total color difference	Oyeyinka et al. (2021)
Lentil	25%	Half power of 650 W	85 °C, 6 min	Lower amylograph viscosities and limited retrogradation	Gonzalez and Perez (2002)
				Boat-shaped granule surface appeared roughened	

(continued)

Table 8.1 (continued)

Starch type	Initial moisture content	Microwave power and irradiation time	Heat attributes	Major findings	References
Indian horse chestnut	22%	15–45 s		Improved physicochemical, functional, and antioxidant properties	Shah et al. (2016)
Maize/corn		605 ± 17.5 W	30–80 °C, 4.52–5.29 °C/min	No differences in swelling behavior of common and modified maize starches between micro-wave and conventional heating	Casasnovas and Anantheswaran (2016)
	6.80%	450 W, 15 min		Deformed particles, decreased crystallinity	Szepes et al. (2005)
	10–35%	650 W,		Partial disruption of crystalline structure	Lee et al. (2007)
		3 min		Significant changes on pasting properties	
	15–40%	0.17 and 0.50 W/g, 1 h	48–65 °C for 0.17 W/g, 91–115 °C for 0.50 W/g	Irradiated power and IMC had significant impacts on gelatinization and pasting properties	Stevenson et al. (2005)
	30%	1 W/g, 20 min		Granule surface became porous with cavities in the center	Luo et al. (2006)
				Increased gelatinization temperature and Tc–to and decreased melting enthalpy	
				Modification changes varied with amylose content	
	30%	0.5 W/g	60 min	Reduced crystallinity, solubility and swelling characteristics, melting enthalpy and increased gelatinization temperature	Lewandowicz et al. (2000b, b)
				Modification changes depended on amylose content	
High amylose maize	30%	400 W/g,0–4 min		Viscosity reduced and resistant starch (RS) increased in 1 min treatment	Zhong et al. (2019)
				Viscosity increased and RS reduced in 2–4 min treatment	

Lotus seed native	70%	2.4–8.0 W/g	Until fully gelatinized	Increased starch–water interaction and crystalline region; decreased swelling power, amylose leaching, and molecular properties; and decreased digestibility	Zeng et al. (2016)
Potato	2–35%	10% output power of 800 W, 0–160 min	20–170 °C	Higher gelatinization temperature and less solubility	Lewandowicz et al. (1997)
				Pasting properties changing from tuber starch to typical cereal starch	
	9.60%	450 W, 15 min		Increased crystallinity, crystal structure changed from B to A	Szepes et al. (2005)
	10–90%	1.3 kw, 0–150 s		Proportion of damaged granules depended on moisture content rather than heating methods	Collison and Chilton (1974)
	97%	Modified power to simulate RHT	20 to over 80 °C, 27.2 °C/min	The microwave-specific effects occurred mainly at initially irradiation stage, and thermal effects began to dominate during and after gelatinization	Ma et al. (2015)
Rice	20%	270–1350 W,≈2–17 min	Melting temperature	Waxy and non-waxy had different pasting property changes after MWT	Anderson and Guraya (2006), Anderson et al. (2002)
				Slight digestibility changes	
	33.30%	800 W, 125 s	Completely gelatinized	Less damage to starch granules by MWT for non-waxy starch gels, compared with CHT	Jiang et al. (2011)
				Lower retrogradation rate and gel firmness and hydrolysis rate in non-waxy gel, compared with CHT	
	94%	Modified power to simulate RHT	20 to over 80 °C, 1.12–1.5 °C/s	Rapid heating accounted for the major property changes	Fan et al. (2012a, b, 2013a, b, c, 2014)
				Microwave effects occurred mainly in the early stage of MWT	

(continued)

Table 8.1 (continued)

Starch type	Initial moisture content	Microwave power and irradiation time	Heat attributes	Major findings	References
	15–35%	8 W/g, 3 min		MWT reduced starch paste viscosity	Guo et al. (2019)
				MWT strengthened starch paste stabilities during heating and cooling and increased the digestion rate and slowly digestible fraction of starch	
				MWT moderately disrupt glucan chain assembly in starch supramolecular structure	
	70–90%	8 W/g, 1–3 min		MWT destructed crystalline lamellae, changed the crystalline type from A to B + V, and decreased crystallinity and double helix content	Li et al. (2021)
				Water content had a stronger effect due to combined effects of water and heat for starch gelatinization	
				A highly porous material can be obtained simply upon MWT of indica rice starch for 3 min at a water content of 90 wt %	
Tapioca	1–35%	10% output power of 800 W, 0–160 min	20–170 °C	Higher gelatinization temperature and less solubility	Lewandowicz et al. (1997)
				Crystal structure changed	
Wheat	30%	0.5 W/g	60 min	Reduced crystallinity, solubility and swelling characteristics	Lewandowicz et al. (2000b, b)
				Increased gelatinization temperature, a drop in solubility and crystallinity for all starches, and the extent and type of these changes depended on variety of starches	

	50–67%	4300 W (absorbed: 1140 W), 10–30 s	94–98 °C	Lack of granular swelling and softer gel properties	Palav and Seetharaman (2007)
				Mechanism of MWT may be different with CHT	
	91%	2000 W,≈5–80 s	40–100 °C	No differences in gelatinization mechanism between MWT and CHT	Bilbao-Sáinz et al. (2007)
				Crystallinity disappeared at higher rate after MWT	
	92–99%	1.3 kw	55–95 °C, 90–350 °C/min	Granule swelling and leaching of polymers occurred only after complete loss of granule birefringence presenting evidence of an asynchronous process of gelatinization	Palav and Seetharaman (2006)
Sago	Starch: Water = 1: 5	20% output power of 900 W, 15 min		Apparent amylose content increase	Zailani et al. (2022)
				Higher RS content for their cooked sample after MWT	
				Higher oils binding capacities of starches after MWT	
Waxy maize	30%	160 W/g, 10–20 min		Lower population of A chain, higher proportion of B1, B2 and B3 chains were observed in MWT starch	Yang et al. (2017)
				α-(1,6) glycosidic linkages were destroyed easily than α-(1,4) glycosidic linkages in MWT	
Millet	30–50%	700 W,60 s		High moisture content enhances microwave effect on millet starch	Li et al. (2019a)
				High moisture promotes the appearance change of starch granule under microwave.	
				Microwave decreases the viscosity, swelling power and crystallinity, increases the T_O, ΔT and decreases the ΔH of millet starch	

MWT microwave treatment, *RHT* rapid conventional heat treatment, *CHT* conventional heat treatment

2012b). Another particular characteristic of MWT is the extremely high heating rate. For example, an oil bath with quite high temperature (200 °C) was reported to be able to reach the same heating speed with microwave irradiation power (1.2 kW) only before 65 °C, but afterward, the heating rate of oil bath began to fall behind significantly (Fan et al. 2013a). Unlike hours or days for most HMT and ANN, the treatment time of MWT has to be limited in a range (often in minutes), in order to prevent the irradiated starch to become a hard gel due to excess water loss (Cao et al. 2022). It is also reported that a long irradiation time at low power may not be suggested due to occurrence of starch clumps and peeled skins as for cornstarch (Stevenson et al. 2005).

The temperature-time curve of microwave-irradiated starch depends on the moisture content, starch type, microwave irradiation power, microwave frequency, and temperature measuring methods (Zhu and Guo 2017; Braşoveanu and Nemţanu 2014). Generally higher microwave power leads to higher heating temperatures. However, the temperature-time curve in MWT is often nonlinear though under constant power, and affected by moisture content, temperature, and degree of starch gelatinization (Fan et al. 2017a, b; Zhu and Guo 2017). Higher moisture content often contributes to higher heating rate (Zhang et al. 2009; Stevenson et al. 2005, as water friction determines the elevation of the temperature, and samples with abundant amounts of polarized water have a rapid increase in heating rates (Lee et al. 2007). Microwave heating of a material depends on its dielectric constant which determines how a material couples with microwaves and on the dielectric loss factor which expresses the ability of the material to absorb microwave energy and transform it into heat (Mudgett 1986). Zhu and Guo (2017) reported both dielectric constant and dielectric loss factor were affected by the moisture content microwave frequency and temperature, while dielectric constant and dielectric loss factor can also be influenced by the degree of starch gelatinization (Fan et al. 2013b, 2017a, b).

However, the contrary results were also reported in Lewandowicz et al. (1997) who found tuber starch samples with lower moisture content (1–5%) had much faster temperature rise than samples with high moisture content (7–15%), at microwave power 800 W with up to 200 min. They also reported a plateau in the curve (around 80 °C judged visually on the curve) with moisture content 17–35%, and the plateau length tended to grow with increased moisture content. For starch in sealed beakers, the plateau was observed to be higher. The authors considered some isothermal transformation occurred in the samples. Similarly, a temperature plateau emerged at 70 °C was observed for rice starch during MWT, accrediting to swelling and hydration in a gelatinization process slowing down the heating rate (Fan et al. 2012a, 2013a). In addition, plateau phenomena were also observed in *Canna edulis* Ker starch (microwave power: 400–1000 W, 10 min, maximum temperature 60 °C, MC = 40%) by Zhang et al. (2009), but the plateaus were not obvious with MC 20–30%. However, Stevenson et al. (2005) did not find any temperature plateaus on microwaved cornstarches and explained this phenomenon with four reasons. First, the plateaus may be caused by imprecisely control of microwave conditions due to domestic microwave oven they used, unlike sophisticated microwave oven in their study. Second, a constant wattage without any oscillation was adopted in their study and may be more stable than oscillating wattage in domestic microwave ovens. Third, they put temperature probe inside reaction vessels, while other researchers took the samples outside for temperature measurement and may not be accurate. Finally, the starch type may also have an impact on the temperature curve. For examples, Lewandowicz et al. (2000a) reported cereal starch was less affected by microwave irradiation compared with tuber starches. Anderson and Guraya (2006) found the heat transfer in waxy rice starch was higher than in non-waxy rice starch (microwave power: 270–1350 W, MC = 20%), and this was explained by the more bulky structure of amylopectin molecules in waxy starch facilitated rapid heat penetration. With moisture content 91%, Bilbao-Sáinz et al. (2007) found the temperature on irradiated wheat starch was gradually increased with time and no plateaus were found. These differences on time-temperature curve need further investigations.

8.3 Changes on Starch Composition and Functional Properties Under Microwave Irradiation

8.3.1 Composition

After MWT, only the minor impurities in the starches have some changes. It has been found that the chemical bonds, chemical groups, skeleton, and the way that they connect with each other was not destroyed, indicating no occurrence of chemical reaction (Fan et al. 2012a). A decrease of crude protein and crude fiber and an increase of ash, reducing sugars, and absolute density were found by Gonzalez and Perez (2002) for microwaved lentil starch (IMC = 25%, half power of 650 W, 6 min at 85 °C). An increase of amylose content was found in microwave-modified sago starch (Zailani et al. 2022) and groundnut starch (Oyeyinka et al. 2019). While Shah et al. (2016) found the amylose content showed no significant changes before and after MWT, but the phenolic content increased with antioxidant activity improvement. It is also reported some radicals are generated in microwaved starch (Przetaczek-Rożnowska et al. 2019) and can be influenced by the microwave power level, water content, and metal ions in the system (Fan et al. 2016, 2017a, b). However, the quite small concentration of these radicals and their rapid annihilation in water make them inoffensive (Braşoveanu and Nemţanu 2014).

8.3.2 Pasting Properties

Starch pasting properties are one of most important properties not only for direct industrial application but also for structure analysis. Starch type, amylose content (waxy or non-waxy), IMC, and irradiation dose can all have an impact on viscosity properties of microwave-modified starches. For examples, under irradiation condition of 10% output power of 800 W at IMC 20%, Lewandowicz et al. (1997) found the viscosity curves during heating in microwave-irradiated potato and tapioca starches were almost the same with their untreated starches, and the greater the IMC deviated from 20%, the lower the viscosity curves. After MWT (IMC = 20%, heated to melting temperature), the peak viscosity (PV), hot paste viscosity (HPV), cold paste viscosity (CPV), breakdown viscosity (BD), and setback viscosity (SB) had significant increase in non-waxy rice starches, whereas for waxy rice starches under the same treatment condition, all of these attributes decreased except SB increased (Anderson and Guraya 2006; Anderson et al. 2002), which were explained as reaggregation of amylopectin branches in waxy rice starches were much higher than non-waxy rice starches after MWT. The pasting properties of waxy barley starch were more likely to be influenced by microwave irradiation than high amylose and normal barley starches (Emami et al. 2012). The impact of MWT on Brabender viscosity curves appeared to be less in waxy cornstarch than in normal cornstarches (Lewandowicz et al. 2000a). The viscosity patterns of amylomaize V starch were changed after MWT, while waxy and normal maize starch did not change (Luo et al. 2006). Luo et al. (2006) found under greater irradiation doses generally led to greater changes on pasting properties (PV, HPV, BD, CPV, SB), but for peak temperature (PT), it was decreased after irritated at 0.17 W/g for 1 h with IMC 20–40% whereas increased after irradiation at 0.50 W/g for 1 h with IMC 25–40%. With increased treatment time, PV, HPV, FV, and SB were all significantly decreased in Indian horse chestnut starch (Shah et al. 2016).

However, though the impact of MWT on Brabender viscograph can be influenced by these factors, an increase in PT and/or a decrease in PV is mostly reported, e.g., wheat and corn starches (Lewandowicz et al. 2000a; Stevenson et al. 2005), lentil starch (Gonzalez and Perez 2002), cassava starch (Abraham 1993; Oyeyinka et al. 2021), maize starches (Luo et al. 2006; Yang et al. 2017)), and non-waxy rice starches (Anderson and Guraya 2006; Anderson et al. 2002), groundnut starch (Zailani et al. 2022), millet starch (Li et al. 2019b). But there are also some exceptions, e.g., Palav and Seetharaman (2007) reported a lower PT in microwave-modified wheat starch.

Though heating methods were reported to have no influence on starch viscosity values (at 60 °C, 200 s − 1) measured immediately after MWHT and conventional heat treatment (CHT) (Bilbao-Sáinz et al. 2007), the pasting properties of starches modified by the two methods (MWHT and CHT) can still have significant differences in most recent studies. For example, a significant lower PT and FV, but higher PV was found in microwave-heated wheat starch sample than conduction-heated sample, and this difference was considered to be incomplete disruption of granular integrity and less amylose leaching into the extragranular matrix, resulting in weaker amylose network formation (Palav and Seetharaman 2007). Therefore some internal structure changes may occur after MWT and can be supported by light microphotographs of wheat starches that showed that MWT can postpone the temperature point for symptoms of solubilization from 75 to 95 °C (Lewandowicz et al. 2000a). Moreover, an increase on X-ray diffraction (XRD) intensity is found in many microwave-irradiated starch, indicating more ordered crystalline array and increased interand intramolecular hydrogen bonding (Luo et al. 2006), which may also contribute to pasting property changes.

8.3.3 Gelation

With heating temperature above gelatinization temperature and adequate high moisture content, the starch formed a paste during MWT, and this paste turns to a gel once cooled. However, most studies show that this gel is much weaker and has significantly lower firmness than that formed after CHT, though they have lower moisture content (Palav and Seetharaman 2007; Jiang et al. 2011)). Even the paste did not form a gel at 8% starch concentration upon cooling, while other starches formed a gel at 6% concentration, which was explained by that amylose did not contribute to the continuous network needed for gel formation following microwave heating (Palav and Seetharaman 2006). Compared with CHT, starch samples after MWT had lower amount of reducing sugars produced by the action of a-amylase, suggesting greater granular disruption during conduction heating (Palav and Seetharaman 2007). This is in agreement with Luo et al. (2006) who suggested less separated amylose chain in irradiated starch than normal starch. Therefore this could contribute to the absence of a continuous network formation through the leaching of amylose after MWT, resulting in weaker gel formation (Palav and Seetharaman 2007; Jiang et al. 2011)). The tendency of retrogradation in rice starch was reduced after MWT (Jiang et al. 2011). Increasing microwave irradiation time and starch concentration led to increased gel firmness, and this may be explained as the rapid heating rate led to lack of granule swelling and less amount of leached amylose into extragranular matrix and a weaker amylose network formation, resulting in softer gel texture (Palav and Seetharaman 2007). It is also reported that compared with CHT, microwave irradiation could change the water distribution and result in a certain level of free water shifting to immobile water in starch gel, indicating improved starch–water interaction (Zeng et al. 2016). By investigating the freeze-thaw stabilities of three maize starches after MWT, Luo et al. (2006) found the decrease of syneresis was significant in normal and amylomaize V starch, while it was only marginal in waxy starches, suggesting that interaction between amylopectin chains did not occur to a significant extent during microwave irradiation. Therefore, the rapid heating rate during microwave heating likely results in lesser amount of leached amylose into the extragranular matrix and also a weaker amylose network formation (Palav and Seetharaman 2007), and its impact on waxy starches will not be significant.

8.3.4 Micromorphology

The effect of microwave treatment on starch granular morphology generally varied with microwave power, time, starch type, and moisture content. With appropriate irradiation dose and moisture to avoid gelatinization (low moisture content or heating temperature), the morphology of starch granule will have minor changes after MWT (Shah et al. 2016; Zeng et al. 2016; Zhong et al. 2019). For example, both granule morphology of waxy and non-waxy rice (IMC = 20%, heated to T_m) did not appear to be influenced by MWT except only if little more aggregation was found (Anderson and Guraya 2006). Potato starch had no significant micromorphology changes, while maize starch particles had some deformations (Szepes et al. 2005), indicating the granular morphology change may vary for different starch types. Many researches reported the presence of pores of starch surface after microwave was observed in cassava starch (Oyeyinka et al. 2021), millet starch (Li et al. 2019b), Bambara groundnut starch (Oyeyinka et al. 2019), indicating rearrangement of molecular structure (Luo et al. 2006), and the pores became bigger with the time extension and starch moisture content (Li et al. 2019b; Oyeyinka et al. 2021). After MWT (IMC = 25%, half power of 650 W, 85 °C), the boated-shape surface of lentil starch appeared to be roughened (Gonzalez and Perez 2002). Lewandowicz et al. (1997) found the microimages of conventional heated potato starch (IMC = 35%) heated at 68 °C were similar to microwaved heated at 90 °C, indicating starch–starch bonds were stronger after MWT. The microwave-induced micromorphology changes during gelatinization are included in the following swelling and gelatinization section.

8.3.5 Swelling and Gelatinization

Swelling and gelatinization are important properties of starches, and the detailed process may help explaining above modification changes. After microwave irradiation, most starches exhibit decreased starch swelling properties and solubility and increased gelatinization temperature. However, as MWT has been conducted with broad moisture range, these changes may have some differences. It is reported that gelatinization cannot occur with less than IMC 30%, but can partially and fully occur with IMC 30–55% and over 55%, respectively (Collison and Chilton 1974). Therefore the impact of microwave irradiation will be demonstrated in two sections as follows.

8.3.5.1 Crystalline and Thermal Property Changes at Low Moisture Content (<50%)

With relatively low moisture content, the microwave impact on starch are found to be similar to HMT, e.g., reduced swelling, increased gelatinization temperature, and crystalline rearrangement existence. During MWT, imperfect or weak crystallites may melt while the strong crystalline be reinforced, and this impact was more significant in type B starch than type A starch (Luo et al. 2006). The XRD pattern of microwaved type B starch shift to type A has been generally observed, e.g., potato starches (Lewandowicz et al. 1997; Szepes et al. 2005; Xu et al. 2019; Zhao et al. 2018), amylomaize V starch (Luo et al. 2006), canna edulis Ker starch (Zhang et al. 2009), and breadfruit starch (H. Marta et al. 2019), similar to the effect of HMT on type B starches like sweet potato starch (Genkina et al. 2004) and tuber and root starches (Gunaratne and Hoover 2002), and is explained similarly based on crystalline structure. As double helices of type A and B starches are packed in a pseudohexagonal array, type B starch has a large void in which 36 water molecules can be accommodated, while type A starch contains a helix in the center rather than a column of water. Thus the changes in X-ray pattern can be explained as dehydration (vaporization of the 36 water molecules in the central channel of the B-unit cell) and movement of a pair of double helices into central channel, which was originally occupied by vaporized water molecules. The pattern changes may also bring a higher increase on T_c–T_o (T_c, conclusion temperature; T_o, onset temperature) after MWT (Luo et al. 2006). Type A starch is found to have no changes on XRD pattern after MWT, e.g., wheat and cornstarch (Lewandowicz et al. 2000a), but crystalline rearrangements also occur and can be reflected by some changes on XRD peak intensities and thermal properties (Luo et al. 2006; Zhang et al. 2009). Appearance of V-type crystallinity pattern (peak at 20°2θ) of corn starch was observed after MWT indicated the increase of amylose-lipid complex (Herlina Marta et al. 2022), and Villanueva et al. (Villanueva et al. 2018) also reported the enhancement of V-type crystallinity pattern in rice flour after MWT.

An increase on gelatinization temperature and decrease on melting enthalpy have been mostly observed in microwave-irradiated starch, such as potato and tapioca starch (Lewandowicz et al. 1997), wheat starch (Lewandowicz et al. 2000a), corn starch (Lewandowicz et al. 2000a; Luo et al. 2006; Stevenson et al. 2005; Herlina Marta et al. 2022)), canna edulis Ker starch (Zhang et al. 2009), millet starch (Li et al. 2019b), rice starch (Villanueva et al. 2018), and waxy and normal barley starch (Emami et al. 2012). But lower values of T_o, T_p (peak temperature), and melting enthalpy after MWT have also been reported, e.g., Indian horse chestnut starch (Shah et al. 2016). A dropped solubility has been observed due to limited amylose leakage (Gonzalez and Perez 2002; Lewandowicz et al. 1997; Luo et al. 2006), and thus the increase on gelatinization temperature may be caused by the reduced destabilization of the amorphous areas induced by reduced granular swelling, and the reduction on gelatinization melting enthalpy may be explained as some of the double helices present in crystalline and in noncrystalline regions of the granule have been disrupted under the conditions prevailing during MWT (Luo et al. 2006). However, the extent of changes may be influenced by starch type and amylose content. For example, Lewandowicz et al. (2000a) found the decreasing extent of melting enthalpy followed the order wheat (significantly decreased) > corn (significantly decreased) > waxy corn (slightly decrease), and these results were consistent with the changes on crystallinities. Luo et al. (2006) reported that amylomaize V (type B starch) had the greatest reduction extent on swelling power, solubility, and melting enthalpy and greatest extent of increase on gelatinization temperature, compared with other waxy and normal starches. On the other hand, an increase on crystallinity has been observed on wheat (Lewandowicz et al. 2000a), normal cornstarch (Lewandowicz et al. 2000a; Luo et al. 2006), amylomaize V starches (Luo et al. 2006), potato starch (Szepes et al. 2005), and canna edulis Ker starch (Zhang et al. 2009). It had been postulated that MWT produced double helix chains with the starch crystallites shift and led to a crystalline array that was more randomly distributed and ordered than that in native starch, and the rearrangement of molecules in sections of starch that are part of the crystalline region and that irradiation had a more pronounced effect on amylose than on amylopectin (Luo et al. 2006). The crystallinity was increased in potato starch, while decreased in maize starch after MWT, indicating crystalline changes may also be affected by starch origin (Szepes et al. 2005).

The irradiation power and IMC are found to have significant impacts on changes on crystallinity and thermal properties. Stevenson et al. (2005) reported higher irradiation power and higher IMC led to higher gelatinization temperature and lower melting enthalpy on cornstarch, but this impact was relatively minimal with less than 30% IMC. Zhang et al. (2009) investigated the impact of microwave irradiation on *Canna edulis* Ker starch (IMC = 20–45%, 400–1000 W, 0–30 min), and the maximum temperature was set as 60 °C for IMC 35–45%, while it was 110 °C for IMC 20–30%. Interestingly, at IMC 40%, the increased irradiation time led to the XRD pattern remained unchanged as half-A/half-B type. The authors found increased IMC (20–40%) led to increased crystallinity but started to decrease at IMC 45%, and this explained this as surplus water molecules preventing perfection of the small-crystalline region and formation of additional crystallites. Increased IMC and decreased maximum temperature could lead to decreased amylose leaching. The impact of treatment time was rather complicated, leading to a first decrease but later increase on crystallinity and a first increase followed by decrease on amylose leaching. The author explained this as the structure of treated starch became strengthened with prolonged MWT due to leached amylose recombining with starch molecules. Besides, a two-phase transition containing one exothermic peak and one endothermic peak was observed in most of their samples.

8.3.5.2 Gelatinization at High Moisture Content (>50%)

Recent studies tend to support that some crystal disruption and formation occurs during MWT before or in the early gelatinization process. A proposed gelatinization process for MWT has been established by Palav and Seetharaman (2006), which is different compared to conventional heating. By using polarized microscopy, the authors found with increasing temperatures (55–95 °C, heating rate 90–350 °C/min) at IMC 92–99%, half of wheat starch granules after MWT exhibited birefringence though granule size appeared unchanged at 55 °C and melting enthalpies decreased by tenfold compared with raw starch and then at 60 °C only small B-type starch granule had birefringence with little swelling, in agreement with higher T_o. At 65 °C, all granules had size increases and no birefringence was observed. From 65 to 95 °C, increasing amount of leachate was observed, and finally all granules were completely ruptured. Thus it was considered that the polar molecule vibration during MWT impacted the amylopectin crystalline lamella and crystalline arrangement was destroyed before glass transition of the amorphous region of the granule, unlike CHT, in which the loss of birefringence and granule swelling typically occur semicooperatively. In addition, Palav and Seetharaman (2006) also pointed that the rupture of granules should be mainly attributed to the vibration motion. Moreover, the initial starch concentration (1–8%) was also found to greatly affect polymer leaching during MWT. The same authors (Palav and Seetharaman 2007) further investigated and characterized the changes at macroscopic level with higher starch concentrations (33–50%) after MWT and considered that the MWT-induced water molecule vibration in the crystalline regions destroyed the lamellar arrangement of the amylopectin crystals, even before the system reached gelatinization temperature. The rapid temperature rising and water molecular vibration caused that granule hydration fails to keep pace with granule expansion, and resultant stress generated caused granule collapse and rupture, forming film polymers coating the granule surface observed by SEM. These have been supported in a study on wheat starches (IMC = 91%) (Bilbao-Sáinz et al. 2007). The authors found intragranular population was significantly increased before gelatinization peak temperature, which came from mobilized starch protons, indicating partial gelatinization occurred at short heating time during MWT. Besides, an increase in melting enthalpy before swelling and starch granule hydration occurred during MWT, and the author confirmed that no annealing effects occurred by conducting experiments under similar condition using CHT. Therefore this was explained by that higher extent of hydration led to smaller exothermic portion of the gelatinization process, allowing more of the endothermic events to be recorded. Accordingly MWT may lead to more completed gelatinization, and this was supported by the higher rate of loss of crystallinity after MWT. However, these impacts seem to be rather little when heated to high temperature, as the viscosity of wheat starch samples after MWT and CHT at 85 °C, 90 °C, 95 °C was the same (at shear rate 200 s^{-1} at 60 °C), indicating the gelatinization process is not generally affected by the heating methods and also no unique structure differences during gelatinization were found between MWT and CHT. Both MWT and CHT exhibited similar changes to the double helix, V-type single helix, and amorphous structures with rising temperature (Fan et al. 2013a), and at controlled low and the same heating rate (≈5 °C/min), the swelling process was found to be the same for the three kinds of cornstarch suspensions after MWT and CHT (Casasnovas and Anantheswaran 2016).

A more specific and extensive gelatinization process under MWT at high IMC (>50%) has been established based on a series of studies on rice starches (IMC = 94%) by comparing MWT and rapid conventional heat treatment (RHT) at similar high heating rate, together with CHT (Fan et al. 2012b, c, 2013a, b, c, 2014), and a freeze-drying method was applied immediately to the microwaved starches. By changing the microwave power at appropriate time, the authors successfully got MWT to have almost the same heating curve with RHT (heating rate ≈60–90 °C/min). At the very beginning of MWT (e.g., <50 °C), the water molecules in the suspension absorb the microwave electromagnetic energy and became to vibrate fiercely with increased temperature. An initial increase on melting enthalpy at around 50 °C has been on observed on rice (Fan et al. 2012b, 2013c) and wheat starch (Bilbao-Sáinz et al. 2007), indicating structure changes. MWT and RHT had similar thermal results, and the activation energy and melting enthalpy were both higher than CHT; therefore, rapid heating had an enhanced effect on the stability of starch molecules in this initial heating stage (Fan et al. 2013c). In addition, Fan et al. (2013a) found the heating rate determined the differences in the proportions of amorphous starch, double helices, and V-type single helices, while the electromagnetic effects of microwave heating did not have a significant impact on the ordered structures in starch granules. Therefore this impact on initial treatment stage has been generally thought to be caused by the rapid heating. Bilbao-Sáinz et al. (2007) suggested the rapid heating rate led to a more complete swelling on wheat starch at this stage, resulting in smaller exothermic portion of the gelatinization process and more of the endothermic events to be recorded with higher melting enthalpy. By investigating rice starch with MWT, RHT, and CHT, Fan et al. (2014) considered the loose double helices through the fast heating might not be realistic though the crystalline lamellae exhibited increased thickness and the rapid heating seems to be the major factor accounting for the double helices of amylopectin densely packed, causing crystalline lamellae to become slightly ordered with strengthened H-bond in the initial heating of microwave (Fan et al. 2013c). This agrees with NMR and FTIR study in Fan et al. (2012a, 2013a), who suggested that microwave heating did not significantly change the ordered and disordered structures compared with RHT. Therefore these changes in the amorphous and double and single helix structures in starch granules are primarily related to the rapid increase in heat rather than electromagnetic effects of microwave heating (Fan et al. 2013a). However, starch with different types may have different structure changes. For example, the gelatinization enthalpy decreased rapidly for potato starch when heated to only 45 °C in all MWT, RHT, and CHT (Ma et al. 2015) and was explained as the starch polymorphs had become relatively loose by this point and the energy required for further destruction of the crystalline regions was diminished. This was also explained as type B starch (potato) has lowered density and stability at original status. Besides, RHT led to the largest melting enthalpy, indicating that fast heating led to insufficient expansion of the crystalline regions and therefore total heat required for destruction of this part of the granules was relatively large, while microwave could have some vibration impact, leading to a loose crystalline regions to some extent (Ma et al. 2015).

With further heating below onset gelatinization temperature (e.g., 50–60 °C), crystalline starts to be melted, and a progressive shift of the endotherm to higher temperatures and decrease on melting enthalpies was observed once the onset of the endothermic transition was reached in both microwaved and conduction-heated sample, but the former had higher gelatinization peak and melting enthalpy like around 50 °C, e.g., potato starch (Ma et al. 2015), rice starch (Fan et al. 2013c), and wheat starch (Bilbao-Sáinz et al. 2007). The decrease on gelatinization melting enthalpy may be explained as the rapid heating can destroy the ordered lamellar arrangements within starch semicrystalline region before T_o and molecular vibration of the microwaves could accelerate this impact (Fan et al. 2014). The higher gelatinization enthalpy in microwaved sample than conventional heated sample could be due to the more perfect structure of the remaining crystals (Bilbao-Sáinz et al. 2007). On the other hand, the starch–water interactions at this stage are quite different between RHT and MWT. Based on starch–water interaction analysis by using H-1 NMR, Fan et al. (2013b) reported that water absorption was the major process that occurred before 60 °C for samples under RHT and CHT, and before 55 ° C for samples under MWT. Thus, the water-binding capability was lower when reaching the gelatinization temperature in microwaved sample. Heating rate directly affects the strength of the hydrogen bonds in amylopectin with respect to their molecular and submicroscopic lamellar structure (Fan et al. 2013c). The starch slurry started to from hydrocolloid gels at 55 °C for microwaved sample and at 60 °C for convention heating and rapid conventional heating, according to Fan et al. (2013b). The authors also found the glass transition temperature followed a descending order RHT, CHT, and MWT. They suggested the rapid heating inhibited the disruption of hydrogen bonds and thus possessed the stronger hydrogen bonds and lowest water mobility, while the microwave effect of vibration contributed to the hydrogen bonds disruption, but its impact was significantly stronger than former, leading to the weakest hydrogen bonds and highest water mobility. Zeng et al. (2016) found higher irradiation power led to increased extent on reducing swelling power of lotus starch. Besides, the authors also found unlike CHT, the granule surface of microwave treated lotus seed starch granules was collapsed and wrinkled before the beginning of gelatinization, suggesting the mobility of the starch protons may occur at the beginning of gelatinization. Similarly Bilbao-Sáinz et al. (2007) found the mobilization of the wheat starch protons was slightly greater for microwaves heated samples before any swelling and hydration of the starch granules occurred through NMR.

With further heating approaching or above gelatinization temperature, the ordered structure of starch granules disrupts and gelatinizes. It has been confirmed the vibrational motion of polar molecules did not impact the gelatinization of rice starch (Fan et al. 2012b). The high thermal impact becomes predominant, and all treatments gradually reached the same gelatinization status. However, a lower amorphous content in microwaved starch was observed, compared with traditional and rapid heated starches at the range of 65–70 °C (Fan et al. 2013a). Zeng et al. (2016) reported the amylose leaching was significantly lower in microwaved lotus seed starches than conventional heating from 50 to 90 °C, and the higher the irradiation power, the more extent of the restricting impact. The authors explained this as the increased amylose–amylose interaction, and reduced branching degree in amylopectin as a result of partially degraded dextrins, as the molar weight was significantly decreased. In addition, the authors also found the proportion of crystalline region in microwaved starch was higher than that of conventional heating, indicating that the existence of more efficient packing of double helices within the crystalline lamella and the intensive interaction between amylose–amylose interaction occur in microwave heating.

8.3.6 In Vitro Digestibility

Starch is the major carbohydrate source as daily energy source, and thus its digestibility is important. Usually native starch belongs to the high glycemic index (GI) food and has low resistant starch content. The starch digestibility is mostly reduced after MWT. For example, the resistant starch content of *Canna edulis* Ker starch increased from 27% to 55% after MWT (IMC, 20%, irradiated at 1000 W for 30 min and freeze-dried) (Zhang et al. 2010). After MWT, the enzyme susceptibility of cassava starch lowered from 30.00% to 27.50% (Abraham 1993). The extent of reduction on digestibility is found to vary with different irradiation doses, amylose content, and moisture. For example, Zeng et al. (2016) found the digestibility of lotus seed starch had higher resistant starch and slowly digestible starch with greater microwave power. Zhang et al. (2009) found IMC and maximum heating temperature, irradiation time, and power all could influence the enzymatic hydrolysis on canna edulis Ker starch, and an appropriate combination led to a lowest digestibility. The enzyme susceptibility of wheat starch after MWT decreased with increased irradiation time and decreased solids concentration (Palav and Seetharaman 2007). Zhong et al. (2019) found maize starch with 1 min MWT showed highest RS, compared to native starch and starches treated with 2–4 min MWT. There are also a few exceptions, e.g., both Anderson et al. (2002) and Anderson and Guraya (2006) found the digestibility of waxy and non-waxy rice starch had slight increases after MWT. The digestibility of barley starch was increased after MWT, and the higher the irradiation power, the greater the digestion rate (Emami et al. 2012). The digestibility of microwave cooked rice had initially higher digestibility than conductive cooked rice, while after certain storage time, it became lower (Li et al. 2014). It was attributed to that microwave cooking-storage induced more amorphous-crystalline structures, which was believed to be related to the high-frequency movement of polar groups by microwave (Li et al. 2019b; Xiong et al. 2022).

As molecular arrangement, degree of crystallinity, and retrogradation of starch influence the enzymatic availability and rate of digestion (Lopez-Rubio et al. 2008), the explanations for the digestibility changes relate to previous described internal structure changes caused by MWT. Zeng et al. (2016) suggested decreased branching degree after MWT contributed to amylose–amylose interaction and the retrogradation of lotus seed starch, promoting the formation of resistant starch during cooling. A predicted two-stage in vitro digestion mechanism for microwaved starch has been drawn by Zhang et al. (2010). The authors considered the enzymes initially eroded selected zones of granule surface, rapidly forming pits which became canals as observed by SEM. Then the enzyme would hydrolyze the whole granule through the canals. The author found the crystalline structure of treated starch was not destroyed in the process of digestion. Besides, the analysis on Fourier transform infrared spectroscopy (FTIR) and XRD results suggested that during digestion, the hydrogen bonds were strengthened and ordered content and crystallinity were increased within the starch granule, indicating the enzymatic resistance of microwaved *Canna edulis* starches may be mainly caused by the specific structures formed during digestion.

8.4 Conclusion and Application Perspectives

The main advantages of microwave processing are the high heating speed, short time cost, and simple operation steps. Generally microwave irradiation causes internal molecular rearrangements inside starch granules, resulting in property changes such as pasting, swelling, and gelatinization properties, and the rapid heating speed may be the most important factor accounting for the changes rather than polar molecular vibration. The modification effects depend on the starch type (amylose content), initial moisture content, heating temperature, heating rate, and irradiation time. Its effects may also be influenced by the sample storing conditions before and/ or after MWT (Emami et al. 2012). The common uneven heating caused by microwave heating should also be noticed (Fan et al. 2012c). Therefore, the comparison between different literatures is difficult to some extent due to different starches used and treatment conditions adopted, and systematic research on microwave modification are in still need to reach full understandings on the modification mechanisms. MWT can improve molecule manipulation and modification (Zhongdong et al. 2005), and recently it is reported to assist other treatment/modification methods with synergistic effects to further change starch properties, e.g., vacuum treatment (Mollekopf et al. 2011), ultrasonic treatment (Jiang et al. 2011), and conventional heat-moisture treatment (Deka and Sit 2016). It can help some chemical modifications (Rivero et al. 2009) and significantly reduce the reaction time for starch esterification (Lewandowicz et al. 2000b; Lukasiewicz and Kowalski 2012). Microwave treatment does have high potential for the processing of agricultural products (Vadivambal and Jayas 2007), but some other potential application areas of microwave modification products has been reported, e.g., microwave-modified rice starch can be used to control drug release in the simulated gastric fluid (Luu et al. 2015). However, generally it has not yet been implemented for industrially purpose (Braşoveanu and Nemţanu 2014).

References

Abraham TE (1993) Stabilization of paste viscosity of cassava starch by heat moisture treatment. Starch 45:131–135

Anderson AK, Guraya HS (2006) Effects of microwave heat-moisture treatment on properties of waxy and non-waxy rice starches. Food Chem 97:318–323

Anderson AK, Guraya HS, James C, Salvaggio L (2002) Digestibility and pasting properties of rice starch heat-moisture treated at the melting temperature (tm). Starch 54:401–409

Bilbao-Sáinz C, Butler M, Weaver T, Bent J (2007) Wheat starch gelatinization under microwave irradiation and conduction heating. Carbohydr Polym 69:224–232

Braşoveanu M, Nemţanu MR (2014) Behaviour of starch exposed to microwave radiation treatment. Starch 66:3–14

Cao H, Sun R, Liu Y, Wang X, Guan X, Huang K, Zhang Y (2022) Appropriate microwave improved the texture properties of quinoa due to starch gelatinization from the destructed cyptomere structure. Food Chemistry X 14:100347

Casasnovas J, Anantheswaran RC (2016) Dynamic measurement of starch granule swelling during microwave heating. Carbohydr Polym 151:1052–1057

Chandrasekaran S, Ramanathan S, Basak T (2013) Microwave food processing—a review. Food Res Int 52:243–261

Chen X, Li X, Mao X, Huang H, Miao J, Gao W (2016) Study on the effects of different drying methods on physicochemical properties, structure, and in vitro digestibility of Fritillaria thunbergii Miq. (Zhebeimu) flours. Food Bioprod Process 98:266–274

Collison R, Chilton W (1974) Starch gelation as a function of water content. Int J Food Sci Technol 9:309–315

Deka D, Sit N (2016) Dual modification of taro starch by microwave and other heat moisture treatments. Int J Biol Macromol 92:416–422

Emami S, Perera A, Meda V, Tyler RT (2012) Effect of microwave treatment on starch digestibility and physico-chemical properties of three barley types. Food Bioprocess Technol 5:2266–2274

Fan D, Li C, Ma W, Zhao J, Zhang H, Chen W (2012c) A study of the power absorption and temperature distribution during microwave reheating of instant rice. Int J Food Sci Technol 47: 640–647

Fan D, Lin L, Wang L, Huang L, Hu B, Gu X, Zhao J, Zhang H (2017b) The influence of metal ions on the dielectric enhancement and radical generation of rice starch during microwave processing. Int J Biol Macromol 94:266–270

Fan D, Liu Y, Hu B, Lin L, Huang L, Wang L, Zhao J, Zhang H, Chen W (2016) Influence of microwave parameters and water activity on radical generation in rice starch. Food Chem 196: 34–41

Fan D, Ma W, Wang L, Huang J, Zhang F, Zhao J, Zhang H, Chen W (2013a) Determining the effects of microwave heating on the ordered structures of rice starch by NMR. Carbohydr Polym 92:1395–1401

Fan D, Ma W, Wang L, Huang J, Zhao J, Zhang H, Chen W (2012a) Determination of structural changes in microwaved rice starch using Fourier transform infrared and Raman spectroscopy. Starch 64:598–606

Fan D, Ma S, Wang L, Zhao J, Zhang H, Chen W (2012b) Effect of microwave heating on optical and thermal properties of rice starch. Starch 64:740–744

Fan D, Ma S, Wang L, Zhao H, Zhao J, Zhang H, Chen W (2013b) 1 H NMR studies of starch–water interactions during microwave heating. Carbohydr Polym 97:406–412

Fan D, Shen H, Huang L, Gao Y, Lian H, Zhao J, Chen W (2015) Microwave-absorbing properties of rice starch. Polymers 7(9):1895–1904

Fan D, Wang L, Chen W, Ma S, Ma W, Liu X, Zhao J, Zhang H (2014) Effect of microwave on lamellar parameters of rice starch through small-angle X-ray scattering. Food Hydrocoll 35: 620–626

Fan D, Wang L, Ma S, Ma W, Liu X, Huang J, Zhao J, Zhang H, Chen W (2013c) Structural variation of rice starch in response to temperature during microwave heating before gelatinisation. Carbohydr Polym 92:1249–1255

Fan D, Wang L, Zhang N, Xiong L, Huang L, Zhao J, Wang M, Zhang H (2017a) Full-time response of starch subjected to microwave heating. Sci Rep 7:3967

Feng H, Yin Y, Tang J (2012) Microwave drying of food and agricultural materials: basics and heat and mass transfer modeling. Food Eng Rev 4:89–106

Genkina NK, Wasserman LA, Noda T, Tester RF, Yuryev VP (2004) Effects of annealing on the polymorphic structure of starches from sweet potatoes (Ayamurasaki and Sunnyred cultivars) grown at various soil temperatures. Carbohydr Res 339:1093–1098

Gonzalez Z, Perez E (2002) Evaluation of lentil starches modified by microwave irradiation and extrusion cooking. Food Res Int 35:415–420

Gunaratne A, Hoover R (2002) Effect of heat–moisture treatment on the structure and physicochemical properties of tuber and root starches. Carbohydr Polym 49:425–437

Guo Y, Xu T, Li N, Cheng Q, Qiao D, Zhang B, Zhao S, Huang Q, Lin Q (2019) Supramolecular structure and pasting/digestion behaviors of rice starches following concurrent microwave and heat moisture treatment. Int J Biol Macromol 135:437–444

Jiang H, Liu Z, Wang S (2018) Microwave processing: effects and impacts on food components. Crit Rev Food Sci Nutr 58:2476–2489

Jiang Q, Xu X, Jin Z, Tian Y, Hu X, Bai Y (2011) Physico-chemical properties of rice starch gels: effect of different heat treatments. J Food Eng 107:353–357

Lee EY, Lim KI, Lim JK, Lim S-T (2000) Effects of gelatinization and moisture content of extruded starch pellets on morphology and physical properties of microwave-expanded products. Cereal Chem 77:769–773

Lee S, Sandhu KS, Lim S (2007) Effect of microwave irradiation on crystallinity and pasting viscosity of corn starches different in amylose content. Food Sci Biotechnol 16:832–835

Lewandowicz G, Fornal J, Walkowski A (1997) Effect of microwave radiation on physicochemical properties and structure of potato and tapioca starches. Carbohydr Polym 34:213–220

Lewandowicz G, Fornal J, Walkowski A, Mączyński M, Urbaniak G, Szymańska G (2000b) Starch esters obtained by microwave radiation—structure and functionality. Ind Crop Prod 11:249–257

Lewandowicz G, Jankowski T, Fornal J (2000a) Effect of microwave radiation on physicochemical properties and structure of cereal starches. Carbohydr Polym 42:193–199

Li N, Cai Z, Guo Y, Xu T, Qiao D, Zhang B, Zhao S, Huang Q, Niu M, Jia C, Lin L, Lin Q (2019b) Hierarchical structure and slowly digestible features of rice starch following microwave cooking with storage. Food Chem 295:475–483

Li J, Han W, Xu J, Xiong S, Zhao S (2014) Comparison of morphological changes and in vitro starch digestibility of rice cooked by microwave and conductive heating. Starch 66:549–557

Li Y, Hu A, Wang X, Zheng J (2019a) Physicochemical and in vitro digestion of millet starch: effect of moisture content in microwave. Int J Biol Macromol 134:308–315

Li N, Zhao S, Qiao D, Lin Q, Zhang B, Xie F (2021) Multiscale structural disorganization of *indica* rice starch under microwave treatment with high water contents. ACS Food Sci Technol 1:45–53

Lin Q, Pan J, Lin Q, Liu Q (2013) Microwave synthesis and adsorption performance of a novel crosslinked starch microsphere. J Hazard Mater 263(Part 2):517–524

Lopez-Rubio A, Flanagan BM, Shrestha AK, Gidley MJ, Gilbert EP (2008) Molecular rearrangement of starch during in vitro digestion: toward a better understanding of enzyme resistant starch formation in processed starches. Biomacromolecules 9:1951–1958

Lukasiewicz M, Kowalski S (2012) Low power microwave-assisted enzymatic esterification of starch. Starch 64:188–197

Luo Z, He X, Fu X, Luo F, Gao Q (2006) Effect of microwave radiation on the physicochemical properties of normal maize, waxy maize and amylomaize V starches. Starch 58:468–474

Luu TD, Phan NH, Tran TT-D, Van Vo T, Tran PH-L (2015) Use of microwave method for controlling drug release of modified sprouted rice starch. In: 5th international conference on biomedical engineering in Vietnam. Coference. Springer, Cham

Ma S, Fan D, Wang L, Lian H, Zhao J, Zhang H, Chen W (2015) The impact of microwave heating on the granule state and thermal properties of potato starch. Starch 67:391–398

Maache-Rezzoug Z, Zarguili I, Loisel C, Queveau D, Buléon A (2008) Structural modifications and thermal transitions of standard maize starch after DIC hydrothermal treatment. Carbohydr Polym 74:802–812

Marta H, Cahyana Y, Arifin HR, Khairani L (2019) Comparing the effect of four different thermal modifications on physicochemical and pasting properties of breadfruit (Artocarpus altilis) starch. Int Food Res J 26:269–276

Marta H, Cahyana Y, Bintang S, Soeherman GP, Djali M (2022) Physicochemical and pasting properties of corn starch as affected by hydrothermal modification by various methods. Int J Food Prop 25:792–812

Mollekopf N, Treppe K, Fiala P, Dixit O (2011) Vacuum microwave treatment of potato starch and the resultant modification of properties. Chem Ing Tech 83:262–272

Mudgett RE (1986) Microwave properties and heating characteristics of foods. Food Technol 40: 99–105

Oyeyinka SA, Akinware RO, Bankole AT, Njobeh PB, Kayitesi E (2021) Influence of microwave heating and time on functional, pasting and thermal properties of cassava starch. Int J Food Sci Technol 56:215–223

Oyeyinka SA, Umaru E, Olatunde SJ, Joseph JK (2019) Effect of short microwave heating time on physicochemical and functional properties of Bambara groundnut starch. Food Biosci 28:36–41

Palav T, Seetharaman K (2006) Mechanism of starch gelatinization and polymer leaching during microwave heating. Carbohydr Polym 65:364–370

Palav T, Seetharaman K (2007) Impact of microwave heating on the physico-chemical properties of a starch–water model system. Carbohydr Polym 67:596–604

Przetaczek-Rożnowska I, Fortuna T, Wodniak M, Łabanowska M, Pająk P, Królikowska K (2019) Properties of potato starch treated with microwave radiation and enriched with mineral additives. Int J Biol Macromol 124:229–234

Rivero IE, Balsamo V, Müller AJ (2009) Microwave-assisted modification of starch for compatibilizing LLDPE/starch blends. Carbohydr Polym 75:343–350

Román L, Martínez MM, Rosell CM, Gómez M (2015) Effect of microwave treatment on physicochemical properties of maize flour. Food Bioprocess Technol 8:1330–1335

Salazar-González C, San Martín-González MF, López-Malo A, Sosa-Morales ME (2012) Recent studies related to microwave processing of fluid foods. Food Bioprocess Technol 5:31–46

Shah U, Gani A, Ashwar BA, Shah A, Wani IA, Masoodi FA (2016) Effect of infrared and microwave radiations on properties of Indian horse chestnut starch. Int J Biol Macromol 84: 166–173

Soni A, Smith J, Thompson A, Brightwell G (2020) Microwave-induced thermal sterilization- a review on history, technical progress, advantages and challenges as compared to the conventional methods. Trends Food Sci Technol 97:433–442

Sriamornsak P, Juttulapa M, Piriyaprasarth S (2010) Microwave-assisted modification of arrowroot starch for pharmaceutical matrix tablets. Adv Mater Res 93-94:358–361

Stevenson DG, Biswas A, Inglett GE (2005) Thermal and pasting properties of microwaved corn starch. Starch 57:347–353

Szepes A, Hasznos-Nezdei M, Kovács J, Funke Z, Ulrich J, Szabó-Révész P (2005) Microwave processing of natural biopolymers—studies on the properties of different starches. Int J Parm 302:166–171

Vadivambal R, Jayas DS (2007) Changes in quality of microwave-treated agricultural products – a review. Biosyst Eng 98:1–16

Villanueva M, Harasym J, Muñoz JM, Ronda F (2018) Microwave absorption capacity of rice flour. Impact of the radiation on rice flour microstructure, thermal and viscometric properties. J Food Eng 224:156–164

Villière A, Cravotto G, Vibert R, Perrier A, Lassi U, Lévêque J-M (2015) Production of glucose from starch-based waste employing ultrasound and/or microwave irradiation. In: Production of biofuels and chemicals with ultrasound. Springer, Dordrecht, pp 289–315

Wu Y, Mu R, Li G, Li M, Lv W (2022) Research progress in fluid and semifluid microwave heating technology in food processing. Compr Rev Food Sci Food Saf 21:3436–3454

Xiong Q, Qiao D, Niu M, Xu Y, Jia C, Zhao S, Li S, Zhang B (2022) Microwave cooking enriches the nanoscale and short/long-range orders of the resulting *indica* rice starch undergoing storage. Foods 11:501

Xu X, Chen Y, Luo Z, Lu X (2019) Different variations in structures of A- and B-type starches subjected to microwave treatment and their relationships with digestibility. LWT 99:179–187

Xue C, Sakai N, Fukuoka M (2008) Use of microwave heating to control the degree of starch gelatinization in noodles. J Food Eng 87:357–362

Yang Q, Qi L, Luo Z, Kong X, Xiao Z, Wang P, Peng X (2017) Effect of microwave irradiation on internal molecular structure and physical properties of waxy maize starch. Food Hydrocoll 69: 473–482

Zailani MA, Kamilah H, Husaini A, Awang Seruji AZR, Sarbini SR (2022) Functional and digestibility properties of sago (*Metroxylon sagu*) starch modified by microwave heat treatment. Food Hydrocoll 122:107042

Zeng S, Chen B, Zeng H, Guo Z, Lu X, Zhang Y, Zheng B (2016) Effect of microwave irradiation on the physicochemical and digestive properties of lotus seed starch. J Agric Food Chem 64: 2442–2449

Zhang J, Chen F, Liu F, Wang Z-W (2010) Study on structural changes of microwave heat-moisture treated resistant *Canna edulis* Ker starch during digestion in vitro. Food Hydrocoll 24:27–34

Zhang J, Wang ZW, Shi XM (2009) Effect of microwave heat/moisture treatment on physicochemical properties of *Canna edulis* Ker starch. J Sci Food Agric 89:653–664

Zhao K, Li B, Xu M, Jing L, Gou M, Yu Z, Zheng J, Li W (2018) Microwave pretreated esterification improved the substitution degree, structural and physicochemical properties of potato starch esters. LWT 90:116–123

Zhong Y, Liang W, Pu H, Blennow A, Liu X, Guo D (2019) Short-time microwave treatment affects the multi-scale structure and digestive properties of high-amylose maize starch. Int J Biol Macromol 137:870–877

Zhongdong L, Peng L, Kennedy JF (2005) The technology of molecular manipulation and modification assisted by microwaves as applied to starch granules. Carbohydr Polym 61:374–378

Zhu Z, Guo W (2017) Frequency, moisture content, and temperature dependent dielectric properties of potato starch related to drying with radio-frequency/microwave energy. Sci Rep 7:9311

Chapter 9
Ultrahigh Pressure Treatment

Huayin Pu, Junrong Huang, and Simin Guo

Abstract Ultrahigh pressure (UHP), also called high hydrostatic pressure (HHP), is often defined as pressure exceeding 100 MPa. The implementation of UHP technology depends on the UHP equipment. A typical laboratory-scale UHP equipment includes a pressure vessel, closures for sealing the vessel, pumps to intensify the high pressure, and a controlling system, in general with cavity volume of 10 mL to 10 L. A commercial UHP equipment has a product handling system to transfer the product without stirring apparatus and difficult to realize the continuous monitoring of structure and properties changes (Bolumar et al., Electron beam pasteurization & complementary food processing technologies. Elsevier, London, pp. 127–155, 2015) (Fig. 9.1). Depending on the equipment, the pressure settings vary from 100 to 1000 MPa but frequently in the range of 400–600 MPa in an industrial environment (Bolumar et al., Electron beam pasteurization & complementary food processing technologies. Elsevier, London, pp. 127–155, 2015). Pressure transmits through a specific medium, which in most cases is water but is replaced by oil at a higher pressure (usually exceeds 600 MPa).

Keywords Ultrahigh pressure · Gelatinization · Crystalline structure

9.1 Introduction

Ultrahigh pressure (UHP), also called high hydrostatic pressure (HHP), is often defined as pressure exceeding 100 MPa. The implementation of UHP technology depends on the UHP equipment. A typical laboratory-scale UHP equipment includes a pressure vessel, closures for sealing the vessel, pumps to intensify the high pressure and a controlling system, in general with cavity volume of 10 mL–10 L. A commercial UHP equipment has a product handling system to transfer the product without stirring apparatus and difficult to realize the continuous monitoring of

H. Pu · J. Huang (✉) · S. Guo
School of Food Science and Engineering, Shaanxi University of Science and Technology, Xian, China
e-mail: huangjunrong@sust.edu.cn

Z. Sui, X. Kong (eds.), *Physical Modifications of Starch*,
https://doi.org/10.1007/978-981-99-5390-5_9

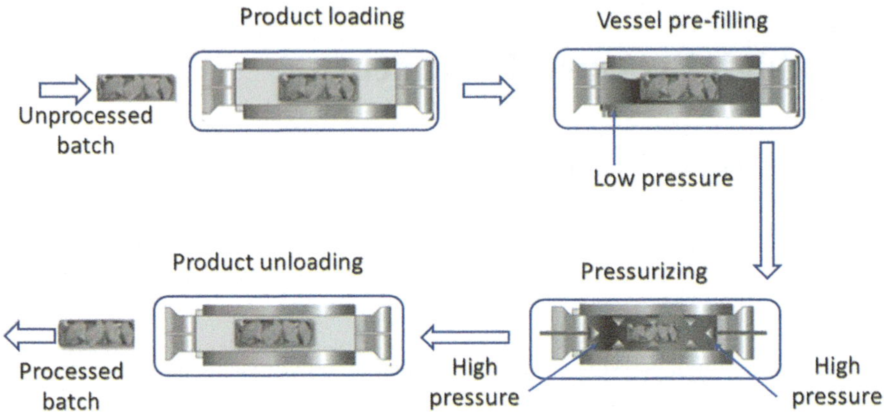

Fig. 9.1 Scheme of high pressure processing

structure and properties changes (Bolumar et al. 2015) (Fig. 9.1). Depending on the equipment, the pressure settings vary from 100 to 1000 MPa but frequently in the range of 400–600 MPa in an industrial environment (Bolumar et al. 2015). Normally, pressure is transmitted through water, but oil is used instead at higher pressures (usually above 600 MPa).

UHP is traditionally applied in ceramics, synthetic materials, steel, and super alloy production (Mota et al. 2013). The original application of UHP in food industry was reported in increasing shelf life of milk in 1899. Currently, UHP is mainly used for food nonthermal sterilization, inactivation of enzymes, macromolecular modification, shucking of aquatic products, and quality improvement for end-use products such as meats and wines. Since 1981, as the gelatinization of starch at room temperature is reported by Thevelein and his co-workers, the UHP treatment of starch attracts more and more attention (Thevelein et al. 1981).

Starches are typically treated with UHP as water-based suspensions. The information on dry (lower moisture content) starch is less comprehensive. Starch granules exhibit significant pressure resistance when there is not enough moisture present. Only extreme UHP (>600 MPa) treatment can alter the shape and surface appearance of dry starch granules, as well as destroy the crystalline structure (Liu et al. 2008; Słomińska et al. 2015; Kudla and Tomasik 1992a, b).

Earlier reports were about pressure-induced gelatinization process and comparison with heat-induced gelatinization. The relationship between the structure and properties of starch and the potential use of UHP have been the subject of considerable research in recent years. Heat-induced gelatinization can be represented directly by using hot-stage polarized light microscope (HS-PLM), rapid visco analyzer (RVA), and differential scanning calorimeter (DSC). However, for pressure-induced gelatinization, it is difficult to carry out the pressure treatment and properties determination simultaneously due to instrumental limitation. The diamond anvil cell (DAC) combined with synchrotron radiation technology realizes the online analysis for lamellar structure change during pressure-induced gelatinization process, but fails to obtain macroscopic properties because of the smaller sample capacity for this device. Almost all studies on pressure-induced gelatinization must prepare UHP-treated samples first. The gelatinization process is commonly evaluated by the degree of gelatinization, calculated by gelatinization enthalpy in DSC results. Besides, Bauer and Knorr characterized the gelatinization according to the electrical conductivity of UHP-treated starch and found the electrical conductivity correlated well with the degree of gelatinization (Bauer and Knorr 2004). This is applicable for the quick and simple determination of pressure-induced starch gelatinization.

The two most crucial elements in pressure-induced gelatinization are botanical source and pressure. In addition, starch gelatinization is also influenced by concentration, holding duration, temperature, and pH. Starches with different botanical sources vary in gelatinization pressure, usually evaluated by the gelatinization pressure, which is defined to be the initial pressure (or a pressure range) to induce starch gelatinization and generally determined by using polarizing microscope (PLM) or DSC. In general, without heating or addition of other components, a complete gelatinization usually needs 600 MPa (Liu et al. 2010). Potato starch (with B-type crystal) exceeds 600 MPa and is considered to have a higher pressure resistance (Kawai et al. 2007). Additionally, although botanical source (especially crystal type) can affect the gelatinization pressure of starch, pretreatment may change the pressure gelatinization characteristics of starch. For example, the suitable annealing treatment can increase the pressure resistance of corn starch, or delay the UHP gelatinization process (Pu et al. 2021). At present, it is still controversial whether or not UHP treatment at the pressure lower than gelatinization pressure can affect starch structure. Therefore, it would be worthwhile to research how UHP affects starch structure at lower pressures (lower than gelatinization pressures), which would facilitate the comprehensive understanding of starch performance during UHP treatment processing.

9.2 The Effects of UHP on Starch Physicochemical Properties

9.2.1 Gelatinization Properties of Starch

The gelatinization properties are mainly reflected in the change of gelatinization enthalpy/temperature and viscosity. In general, UHP treatment lowers the gelatinization enthalpy in a certain pressure range, while gelatinization temperature decreases with increasing pressure (Table 9.1), indicating UHP treatment can weaken the ordered structure of starch, which leads to the reduction of thermal stability of starch. Another explanation on the decrease in onset temperature is the formation of new starch crystal with lower gelatinization temperature due to the retrogradation of starch molecules. However, Thevelein and Muhr reported that a

Table 9.1 The effects of ultrahigh pressure (UHP) treatment on gelatinization enthalpy (△H) and onset gelatinization temperature (To) of various starches

Starch	Pressure (gelatinization pressure)/MPa	Starch content/ %	Time/ min	△H	To	References
Potato	51–253	0.4	4	Decline	Rise	Thevelein et al. (1981)
Wheat, potato, smooth pea	58–401	0.4	–	Decline	Rise and decline	Muhr and Blanshard (1982)
Barley	450–600	10, 25	15, 30	Decline	–	Stolt et al. (2000)
Maize, waxy maize, tapioca, rice, potato, high amylose maize	690	1:1, 1:2 (w/V)	5, 60	Decline	Decline (rise for waxy maize, rice and high amylose maize)	Katopo et al. (2002)
Potato	600	10	2, 3	Decline	Decline	Błaszczak et al. (2005b)
Potato	600–1000	10–70	60–3960	Decline	–	Kawai et al. (2007)
Sorghum	200–600 (600)	25	10	Decline	Rise	Vallons and Arendt (2009b)
Buckwheat	200–600 (600)	25	10	Decline	Rise	Vallons and Arendt (2009a)
Mung bean	120–600 (600)	20	30	Decline	Decline	Li et al. (2011b)
Rice	120–600 (600)	20	30	Decline (120–480 MPa)	Decline	Li et al. (2011a)
Lotus seed	100–600 (600)	15	30	Decline (100–500 MPa)	Decline	Guo et al. (2015b)
Sorghum	120–600	20(w/v)	20	Decline	Decline	Liu et al. (2016a)
Quinoa, maize	100–600 (600)	10 (w/v)	5	Rise and decline	Decline	Li and Zhu (2018)
Waxy rice	100–600 (600)	10	20	Decline	Rise and decline	Zeng et al. (2018)
Pea	150–600	15	25	Decline	Decline	Liu et al. (2018)

(continued)

Table 9.1 (continued)

Starch	Pressure (gelatinization pressure)/MPa	Starch content/ %	Time/ min	ΔH	To	References
Proso millet grains	150–600(600)	30(w/v)	15	Decline	Rise	Li et al. (2018))
Mango	300–650	–	10	Decline	Rise	Kaur et al. (2019)
Lily	100–600 (600)	15	30	–	Decline and rise	Zhang et al. (2021)
Litchi kernel	300–600	20%	10	Decline	Decline	Sandhu et al. (2021)

treatment of potato starch in dilute (0.4%) suspension with pressures up to a relatively low pressure (<150 MPa) increased its gelatinization temperature (Thevelein et al. 1981; Muhr and Blanshard 1982). Under UHP treatment, the gelatinization temperature (onset temperature) of rice, waxy maize, high amylose maize, sorghum, buckwheat lily, mango, and proso millet grains starches also increase (Table 9.1). The increase in onset temperature may be related to the preferential gelatinization of starch granules with low pressure resistance. Kweon and his co-workers propose that the reason is annealing of amylopectin (Kweon et al. 2008b), while Li and his co-workers attribute this to the formation of amylose–lipid complexes (Li et al. 2018).

UHP, a type of nonthermal processing technique, is mainly applied at room temperature. However, in some case, it is interesting to discuss the gelatinization induced by pressure-heat combinations. For pressure-induced gelatinization at a constant holding time, either increasing temperature or pressure can promote starch gelatinization. In other words, the higher the temperature, the lower pressure to realize the complete gelatinization of starch (Bauer and Knorr 2005; Tan et al. 2009; Dominique et al. 2019; Okur et al. 2019). On the other hand, at a constant temperature and pressure, the degree of gelatinization increases with increasing holding time. However, if the temperature and/or pressure is unable to induce starch gelatinization, it is invalid to prolong holding time (Stolt et al. 2000; Bauer and Knorr 2005). In order to further understand the heat-pressure combination effect, the phase diagrams of various starches have been provided and used to estimate the degree of gelatinization after applying a certain pressure and temperature on a starch–water mixture with starch concentrations in the range of 5% and 60% w/w (Baks et al. 2008). Besides, NMR relaxometry is also used to analyze the gelatinization pressure (Okur et al. 2019).

Just as heat-induced gelatinization, pressure-induced gelatinization is also influenced by solutes in the starch suspensions. The addition of sugar (20%) reduced gelatinization pressure (wheat starch/350 MPa (tapioca starch/530 MPa, potato starch/700 MPa), 5% suspension, 15 min at 29 °C), whereas the degree of gelatinization is linearly correlated with the number of equatorial hydroxyl groups for different sugars (fructose, glucose, sucrose, trehalose) (Rumpold and Knorr 2005). Another results also suggested that the tea polyphenols could also accelerate HHP-induced gelatinization and changed the structural and physicochemical properties of rice starch under sufficiently high pressure (≥400 MPa) at room temperature (Du et al. 2019). Different hydrocolloids had different effects on UHP-induced gelatinization process of starch. The gelatinization degree of starch added with xanthan gum (400 MPa, 10% suspension, 35 min at 40 °C) was significantly lower than that of λ- carrageenan and guar gum, indicating that xanthan gum may play a role in stabilizing the starch granular structure in the pressure-induced gelatinization process (Schneider Teixeira et al. 2018). Similarly, during the UHP treatment, wheat starch's crystalline and glass transitions were significantly defended by salt. (Kweon et al. 2008a). Additionally, the influence of salts on the gelatinization pressure varies, and the extent of effect on the gelatinization pressure depends not only on the solute added but also on the source of starch (700 MPa, 15% suspension, 15 min at 29 °C) (Rumpold and Knorr 2005). At high chloride concentrations (>2 M), the impact of the salts on starch gelatinization augmentation followed the order $Na^+ < K^+ < Li^+ < Ca^{2+}$, which corresponds to the order of the lyotropic series. At concentrations above 1 M, the effect of potassium salts on starch gelatinization upon pressurization also followed the order of the Hofmeister series ($Cl^- < Br^- < I^- < SCN^-$). This conclusion is helpful in the practical application of UHP-treated starch (Rumpold and Knorr 2005).

After UHP treatment, the change of viscosity in gelatinization process is mainly measured by RVA or the rheometer. If the UHP-treated starch sample still appears, a drastic increase in viscosity during heating indicates the treatment pressure is lower than the gelatinization pressure of the starch. In contrast, the treatment pressure is higher than gelatinization pressure. Thus, the gelatinization pressure can be obtained by comparing the RVA curves of samples treated at different pressures. Oh and his co-workers determined the gelatinization pressures (10% suspension, 30 min at 20 °C) of normal rice, waxy rice, normal corn, waxy corn, tapioca and potato starches by using this method (Oh et al. 2008). More information can also be obtained by using RVA. The pasting properties of mung bean (20% suspension, 30 min at room temperature), rice (20% suspension, 30 min at room temperature), red adzuki bean (20% suspension, 15 min at 25 °C), lotus seed (15% suspension, 30 min at room temperature), lily (15% suspension, 30 min at room temperature), lentil (20% suspension, 10 min at room temperature), sorghum (20% suspension, 20 min at room temperature), pea (15% suspension, 15 min at 25 °C), mango (10 min at 26 °C), tartary buckwheat (20% suspension, 20 min at room temperature), and proso millet (30% suspension, 15 min at room temperature) starches were reported (Li et al. 2011a, b, 2018; Guo et al. 2015b; Zhang et al. 2021; Ahmed et al. 2016; Liu et al. 2016a, b; Leite et al. 2017; Kaur et al. 2019). Gelatinized starch (treated at 600 MPa) exhibits the lowest peak viscosity (PV), breakdown (BD) and setback (SB) (except for mung bean starch). PV is an indicator of early and rapid swelling of starch granules, and BD represents the stability and resistance of starch granules to shear stress, while SB represents the rapid retrogradation of leached amylose in starch. In comparison to native starch, UHP-gelatinized starch has a generally stronger starch aggregation, more stable hot paste, and reduced retrogradation tendency, although some starches exist a varying trend. This could be due to some structural factors such as granular swelling, amylose leaching, starch–water and amylose–lipid interaction.

9.2.2 Rheological Properties of Starch Paste and Gel

The form of gelatinized starch is mostly applied in the food industry. Therefore, the properties of starch paste and starch gel are widely studied. In steady-state rheological behavior, the consistency coefficient (K) value is an approximate measurement of the viscosity of the starch paste at rest. The elevating K value reflects the granular swelling and the increasing in degree of gelatinization. After UHP treatment for starch suspension, the K of starch paste increased with increasing pressure or holding time, however existing a limited K value during the increasing of holding time (Stolt et al. 1999, 2000; Guo et al. 2015a; Jiang et al. 2014). Furthermore, increasing temperature could further increase the K value in a constant pressure (Tan et al. 2009).

The dynamic rheological behavior and textural parameters reflect the properties of gels, which are related to the retrogradation of starch molecules. The dynamic rheological test is generally carried out according to strain sweep measurement. The storage modulus G' is a measure of the energy stored in the sample. For a gel, it reflects the cross-link density of the network (Stolt et al. 1999) . In most reports, G' value increases with increasing pressure or holding time; however, continuous increasing pressure or holding time could show an opposite effect for starch gels, indicating that excessive pressurization can produce weaker gel (Stolt et al. 1999, 2000; Tan et al. 2009; Guo et al. 2015a).

The gel prepared in a higher concentration (related to the source of starch) is usually measured by textural analyzer. Theoretically, starch gel formation mainly depends on the amount of swollen starch granules. A reduction in the levels of leached amylose for pressure-induced gels plays an important role in reducing gel hardness. Therefore, UHP treatment can result in the formation of relatively weaker gels.

According to the above discussion, the properties of pressure-induced gels will undoubtedly differ from that of the heat-induced gels. By comparing the whole process of pressure-induced and heat-induced gelatinization, it was found that the hardness, gumminess and chewiness of gels increased first and then decreased with increasing pressure/temperature. However, the degree of gelatinization corresponding to the maximal value of the above-mentioned three texture parameters for pressure-induced gels was lower than that for heat-induced gels (Liu et al. 2020b).

Stute and his co-workers reported a rapid retrogradation peak in DSC curve for UHP treated (wheat, corn, pea, waxy rice) starches (450–500 MPa, 25%, 15 min at 20 °C), suggesting that amylose significantly retrogrades and mainly occurs within the granules, which is associated with the limited swelling of starch granules and a limited releasing of amylose during UHP gelatinization (Stute et al. 1996). Therefore, UHP (600 MPa, 25% suspension, 10 min at 30–80 °C/20, 30 min at 30 °C) resulted in the formation of harder gels than thermal processing (25% suspension, 20 min at 90 °C) (Vittadini et al. 2007). Meanwhile, longer UHP treatments (600 MPa, 25% suspension, 10–30 min at 30 °C) caused only a slight decrease in hardness, and was significant only at longer processing times (30 min). These can be explained by the different water-starch and/or starch–starch molecular interactions due to partial preservation of the granular structure in gel after UHP-induced gelatinization (Vittadini et al. 2007).

However, the contradictory results were reported for wheat starch. Pressure-induced wheat starch gels were softer but dense compared with heat-induced gel, indicating that wheat starch gels obtained under pressure would be less sensitive to retrogradation (Douzals et al. 1998). This difference may also be related to the difference on storage time of the starch gel. After 28 days of storage, the pressure-induced gel showed a slower increase in hardness compared with heat-induced gel (Vittadini et al. 2007).

9.2.3 Starch Digestibility

The amylase digestibility of pressure-treated starch was first discussed in 1989 by Hayashi and Hayashida (Hayashi and Hayashida 1989). The author indicates that the amylase digestibility increases with increasing pressure because of the starch gelatinization. However, the formation of new structure during long-period pressure treatment decreases the digestibility. The object in this report is starch paste obtained by UHP treatment, while the research in recent years is more focused on the dried samples after UHP treatment.

General, more slowly digestible starch (SDS) and resistant starch (RS) were observed in UHP-gelatinized starches than in native (Liu et al. 2016c,c) and heat-gelatinized starches (Bauer et al. 2005; Tian et al. 2014). Linsberger-Martin and his co-workers reported that increasing pressure, holding time or temperature led to increases in RS for amaranth, quinoa and wheat starches (Linsberger-Martin et al. 2012). Bajaj and his co-workers proposed that UHP (20% suspension, 300–600 MPa, 30 min at room temperature) could induce the increase in both SDS and RS for various starches including wheat, corn, waxy corn starch, potato, sweet potato and kidney bean starches (Bajaj et al. 2021). However, Deng and his co-workers (20% suspension, 30 min at 25 °C) proposed that excessive pressurization (600 MPa) and cycle UHP treatment (15 + 15 min) decreased the RS, in spite of increased the SDS content of rice starch (Deng et al. 2014). The similar results have been reported by Zeng and his co-workers (waxy rice starch, 10% suspension, 400 MPa, 20 min at 25 °C) (Zeng et al. 2018) and Shen and his co-workers (high amylose maize starch, 30%–70% suspension, ≥600 MPa, 15 min at 30 °C) (Shen et al. 2018). The above-mentioned differences may be related to starch varieties and processing conditions (especially for drying conditions).

Additionally, when a variety of starch processing technologies are combined, the effect of UHP treatment on starch digestion characteristics is also significant. The addition of UHP treatment to annealing (Bauer and Knorr 2005) or retrogradation (Colussi et al. 2018) treatment can further induce increasing RS in starch paste. Besides, the re-cooking of the UHP-treated samples led to lower levels of glucose production compared to the thermally-treated samples (Papathanasiou et al. 2015).

The gels prepared by UHP treatment have the potential application in drug release. The drug release rate depends on the starch source. Gel-forming polymer containing potato starch exhibits faster drug dissolution, while the pressurization of maize starch results in a gel exhibiting sustained drug release (Szepes et al. 2008). Overall amylose content, pressure, and starch source are important factors in affecting the digestibility of starch treated with UHP (Dupuis et al. 2014).

9.3 The Effects of UHP on Starch Structure

9.3.1 Granular Micrograph Structure

Granular micrograph variation can intuitively reflect the effects of UHP on various starches. The primary techniques for observing starch granules are the light microscope (LM) and scanning electron microscopy (SEM). To investigate the surface and internal structure in depth, confocal laser scanning microscopy (CLSM) and atomic force microscopy (AFM) are also used.

Starch granules in pressure-induced gelatinization are inclined to show restricted swelling and lower release of amylose in UHP treatment, differed from heat-induced gelatinization (Stute et al. 1996). This is possibly caused by the absence of shear forces and no hot paste is formed in UHP treatment (BeMiller and Huber 2015). Another explanation could be that amylose somehow stabilizes the starch granule structure under pressure (Stolt et al. 2000; Douzals et al. 1998).

Corn (300–600 MPa, 30% suspension, 15 min at room temperature) and red adzuki bean (150–600 MPa, 20% suspension, 15 min at 25 °C) starch show a granular maintaining with a rough surface followed by complete disintegrating (Li et al. 2015, 2016; Stute et al. 1996). The surface of chestnut starch granules show a minor crack (400–600 MPa, 20% suspension, 10 min at 26–38 °C) (Ahmed and Al-Attar 2017). Mung bean (120–600 MPa, 20% suspension, 30 min at room temperature), lotus seed (100–600 MPa, 15% suspension, 30 min at room temperature), buckwheat (200–600 MPa, 25% suspension, 10 min at 20 °C), sorghum (200–600 MPa, 25% suspension, 10 min at 20 °C), lentil (400–600 MPa, 20% suspension, 10 min at 26–38 °C), mango kernel (300–600 MPa, 10 min at 26–38 ° C) and Litchi (Litchi chinensis) kernel (300–600 MPa, 20% suspension, 10 min at 26–38 °C) starches are inclined to form a doughnut-shaped structure (Fig. 9.2a–c)

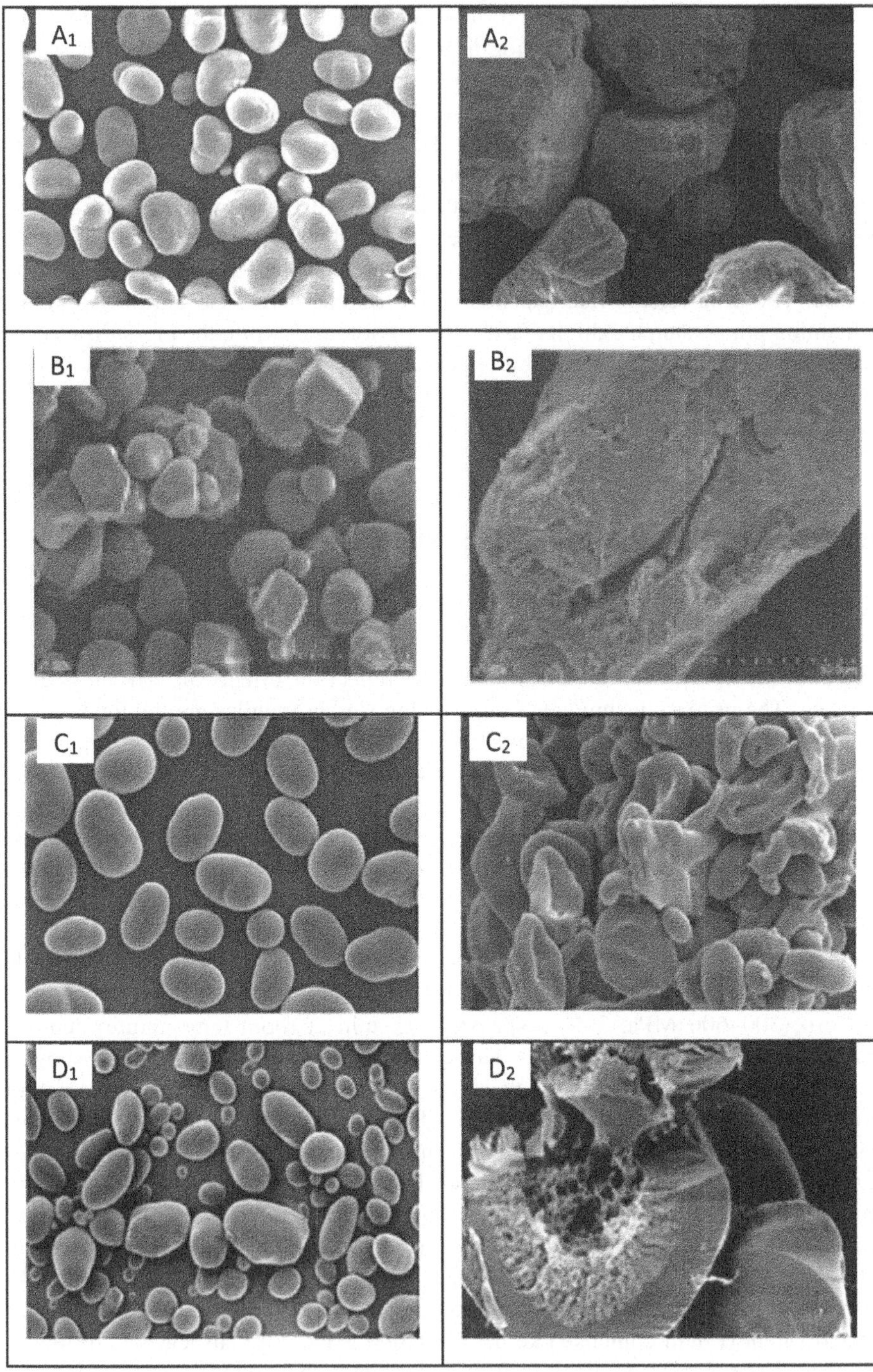

Fig. 9.2 Scanning electron micrographs of typical native ($\mathbf{a}_1$-red adzuki bean starch, $\mathbf{b}_1$-corn starch, $\mathbf{c}_1$-mung bean starch, $\mathbf{d}_1$-potato starch) and UHP treated starch granules ($\mathbf{a}_2$–600 MPa, 30% suspension, 15 min at room temperature; $\mathbf{b}_2$–480 MPa, 20% suspension, 30 min at room

(Vallons and Arendt 2009a, b; Li et al. 2011b; Guo et al. 2015b; Ahmed et al. 2016; Kaur et al. 2019; Sandhu et al. 2021).

The inner structure destruction (gel-like network formation) along with maintaining of external structure for potato starch granules during UHP treatment (600 MPa, 10% suspension, 3 min at 20 ± 2 °C) was observed by Blaszczak and his co-workers, indicating the special granular structure of potato starch, which may be related to the characteristic surface organization of potato starch (Fig. 9.2d) (Błaszczak et al. 2005b). The results of Gebhardt and his co-workers obtained by micro small-angle and wide-angle X-ray scattering (SAXS/WAXS) also support this conclusion (Gebhardt et al. 2007). On the basis of AFM results, the number of surface blocklets of rice starch granules increases and the diameter decreases after treatment at a relatively low pressure (200 MPa, 20% suspension, 30 min at 25 °C). When the pressure reaches 600 MPa, starch granules show totally different structure with more fine and flat surfaces (Deng et al. 2014).

9.3.2 *Order and Disorder Structure*

As a kind of semi-crystalline polymer, the spherocrystal structure of starch induces the polarization cross under polarizing microscope. During common heat gelatinization, the crystalline structure is disrupted, the polarization cross vanishes, the gelatinization enthalpy drops, and these signs point to an order-disorder transformation. Polarizing microscope (PLM), differential scanning calorimetry (DSC) and X-ray diffraction (XRD) are frequently applied to study on order and disorder structure change of starch after UHP treatment.

Starch crystal exists in four different forms and shows different XRD patterns. In general, cereal starches generate A-type XRD patterns, B-type patterns exist in tuber and high amylose starches, and legume, root, some fruit and stem starches show C-type patterns. V-type patterns are also present in high amylose starch or retrograded starch, and are in form of the amylose single helices co-crystallized with compounds such as iodine, dimethyl sulfoxide (DMSO), alcohols, or fatty acids (Buleon et al. 1998). Most A-type and C-type starches can be completely gelatinized at 600 MPa except some rare starches such as taro, wrinkled pea and babassu starches (5% suspension, 15 min at 20 °C). This pressure cannot induce the gelatinization of starch with B-type pattern (potato starch, 5% suspension, 15 min at 20 ° C) (Oh et al. 2008; Stute et al. 1996; Yang et al. 2016a; Rubens et al. 1999), suggesting a better pressure resistance of B-type starches. Potato starch can just be gelatinized completely when the pressure reaches 800 MPa at low starch concentration (10–20%, 1 h at 40 °C) and the required pressure to realize complete

Fig. 9.2 (continued) temperature; **c$_2$**–600 MPa, 10% suspension, 30 min at room temperature; **d$_2$**–600 MPa, 10% suspension, 3 min at 20 °C) (Błaszczak et al. 2005b; Li et al. 2011b, 2015, 2016). (Reused with permission from Elsevier Publications)

gelatinization increases with increasing starch concentration (Kawai et al. 2007). A- and C-type patterns could be converted into B-type patterns during UHP treatment, which is attributed to water being introduced into the crystalline packing unit under UHP and/or the rearrangement of double helices (Liu et al. 2010, 2016b). Nevertheless, there are still some exceptions (unchanged), as shown in Table 9.2.

Studies on the lamellar and double helix structure for starch granules after UHP treatment provide further information on the order and disorder structures. The influence of UHP treatment (waxy corn and Hylon VII starches, 650 MPa, 30% suspension, 9 min at 20 °C; rice starch, 20% at 25 °C, 200 MPa, 30 min/200 MPa, 15 + 15 min/600 MPa, 30 min/600 MPa, 15 + 15 min) on C1 and C4 peaks is obvious as determined by using solid-state ^{13}C CP/MAS NMR, indicating the decrease of double helix and the increase of single helix (Deng et al. 2014; Błaszczak et al. 2005a). This is attributed to the unwinding of double helix during UHP treatment.

For lamellar structure, the thickness changes of lamellar thickness (≈9 nm) might depend on starch source and/or treatment condition. During UHP treatment, increasing pressure induced the increase in lamellar thickness for quinoa, waxy corn, normal corn and G80 starches (Yang et al. 2016a; Li et al. 2019), but almost no change for G50 and maize starches (Yang et al. 2016a; Li et al. 2019). The increasing in lamellar thickness could be related to the the entry of water being forced into the starch lamellae during UHP. Therefore, this may be demonstrated by observing whether the granule swells (Yang et al. 2016a).

However, The synchrotron SAXS results proved the average thickness of long period and amorphous layers decreased with increasing pressure for waxy corn and potato starches, indicating the pressure-induced compression, the compression effect is identified as the "shock absorbers" of amorphous layer and acting as a protection for crystalline layer during compressive forces (Yang et al. 2016b). Interestingly, the thickness of the crystalline layer first increased and then decreased. The initial increase of the thickness of the crystalline layer is likely to indicate the annealing of starch (Yang et al. 2016b). Additionally, compared with waxy corn starch, the decrease of long period with pressure (up to 750 MPa) is less for waxy potato starch, indicating the B-type starch (waxy potato starch) is much less compressible compared to A-type starch (waxy corn starch), which is identified to be caused by the different amylopectin structure. These conclusions are benefit to the understanding of the high pressure resistance of B-type crystal starches (Yang et al. 2016b).

Furthermore, studies on the order and disorder structure variation are helpful to further understand the difference between heat-induced and pressure-induced gelatinization. Liu and his co-workers found that the long-range and short-range structure of UHP-treated corn starch were more orderly than those of heat-treated corn starch under the same gelatinization degree. The amorphous lamellar thickness increased with increasing the gelatinization degree, but heat-induced gelatinization showed a faster reduction in amorphous lamellar thickness compared with pressure-induced gelatinization. The mechanism behind these differences could suggest that compared with UHP-treated starch, the amorphous lamellae of heat-treated corn starch were easier to absorb water, or more amorphous material was transformed to

Table 9.2 The effects of ultrahigh pressure (UHP) treatment on polarization cross and crystal form of starches

Starch	Pressure (gelatinization pressure)/MPa	Starch content/ %	Time/ min	PLM	XRD	References
				Polarization cross	Crystal form	
A-type						
Maize, waxy maize, tapioca, rice	690	1:1,1:2	5,60	–	A → B	(Katopo et al. 2002)
Waxy corn	650	30	3–9	Disappeared (9 min)	Unchanged	(Błaszczak et al. 2005a)
Sorghum	200–600 (600)	25	10	Disappeared (500 MPa)	–	(Vallons and Arendt 2009b)
Buckwheat	200–600 (600)	25	10	Disappeared (400 MPa)	–	(Vallons and Arendt 2009a)
Rice	120–600 (600)	20	30	–	A→B	(Li et al. 2011a)
Rice	200, 600	20	30	–	A → B	(Deng et al. 2014)
Waxy corn, corn	100–600	40	30	Partially disappeared (600 MPa)	A → B	(Yang et al. 2016a)
Sorghum	120–600 (600)	20	20	–	A → (B)	(Liu et al. 2016a)
Tartary buckwheat	120–600	20	20	–	A → B	(Liu et al. 2016c)
Common buckwheat	120–600	20	20	–	A → B	(Liu et al. 2016b)
Waxy rice	100–600 (600)	10	20	Disappeared (600 MPa)	Unchanged	(Zeng et al. 2018)
Proso millet	150–600	30	15	Disappeared (600 MPa)	A → (B)	(Li et al. 2018)
Maize	100–500	20	15–30	–	A → (B)	(Rahman et al. 2020)
Corn, waxy corn, kidney bean, wheat, rice, sweet	300–600	20	30	–	Unchanged	(Bajaj et al. 2021)
B-type						
Potato, High amylose maize	690	1:1,1:2	5		Unchanged	(Katopo et al. 2002)
Potato	100–500	20	15–30	–	B → B + V	(Rahman et al. 2020)
Potato	300, 600	20	30	–	Unchanged	(Bajaj et al. 2021)

(continued)

Table 9.2 (continued)

Starch	Pressure (gelatinization pressure)/MPa	Starch content/ %	Time/ min	PLM	XRD	References
				Polarization cross	Crystal form	
B + V type						
[a]Hylon VII	650	30	3–9	Unchanged (9 min)	Unchanged	(Błaszczak et al. 2005a)
[a]G50, G80	100–600	40	30	Unchanged (600 MPa)	Unchanged	(Yang et al. 2016a)
[a]Hi-maize™ 260	200–1000	30–50	15	–	Unchanged	(Shen et al. 2018)
C-type						
Mung bean	120–600 (600)	20	30	–	C → B	(Li et al. 2011b)
Lotus seed	100–600 (600)	15	30	–	C → B	(Guo et al. 2015b)
Red adzuki bean	150–600 (600)	20	15	Disappeared (600 MPa)	Unchanged	(Li et al. 2015)
Lentil	400–600(600)	20	10	–	C → B	(Ahmed et al. 2016)
Pea	150–600(600)	15	25	–	C → B	(Liu et al. 2018)

[a]Hylon VII is high amylose (70%) corn starch
G50 and G80 are high amylose corn starches, containing 50% and 80% amylose, respectively
Hi-maize™ 260 is high amylose (73%) maize starch

ordered structure in heat-treated one (Liu et al. 2020a). Wang and his co-workers proposed a similar conclusion in the study of rice starch. UHP-treated rice starch had greater crystallinity, larger R(1047/1022), and more double helix and V-type single helix structures as compared to heat-treated rice starch at a similar DG. The authors believe that UHP simultaneously induced annealing and pressure-induced gelatinization until achieving a certain degree of gelatinization (Wang et al. 2020).

9.3.3 Molecular Structure

Blaszzak and his co-workers found that corn amylopectin formed a polydispersed product after UHP treatment (690 MPa, 30% suspension, 3 min at 20 °C), measured by using HPLC, indicating molecular degradation (Błaszczak et al. 2005a). However, it is not valid for high amylose corn starch (Hylon VII). Guo and his co-workers utilized high-performance size-exclusion chromatography and multiangle laser-light scattering and refractive index detectors (HPSEC–MALLS-RI) to study the influence of UHP (15% suspension, 30 min at room temperature) on molecular weight distribution of lotus seed starch and suggested that the *Mw* and *Mn* values decreased with increase of pressure, indicating that lotus seed starch was slightly degraded during UHP treatment and formed molecular chains with a low degree of polymerization (Guo et al. 2015b).

Besides, Szwengiel and his co-workers believes that the structural changes found for pressurized starches were more strongly determined by the starch origin than by the processing applied. The combination of SEC and FTIR data showed that α-1,6-glycosidic bonds are more frequently split in pressurized amaranth, Hylon VII, and waxy maize starch, while in sorghum and maize starches, the α-1,4 bonds are most commonly split (Szwengiel et al. 2018). More studies in molecular structure of pressure-treated starch need to be carried out.

9.4 The Potential Applications of UHP

The potential application of UHP is necessarily related to the superiority of UHP treatment and modified starch. Generally, UHP treatment can induce starch gelatinization at room temperature, providing a method for the preparation of pregelatinized starch and cold water-swelling (CWS) starch. However, considering the requirement of special equipment and the discontinuous product preparation process, UHP treatment shows no advantages as compared to the thermal process. Nevertheless, application of UHP treatment is feasible in starch-containing high value-added or heat-sensitive system. On the other hand, compared with heat-gelatinized starch, UHP-gelatinized starch shows a diverse interior structure organization and has better granular maintaining. This is very important because many potential applications are based on this characteristic, including preparation of resistant starch (*RS*). Additionally, some reports indicate that UHP-modified starch can be applied in the binding of aroma compounds and textural improvement (Błaszczak et al. 2007; Zhang et al. 2014).

Recently, the utilization of UHP in preparation of chemical modified starch was introduced. Combining chemical modification with UHP technology has been applied in acid hydrolysis (Lee et al. 2006; Choi et al. 2009b), hydroxyl propylation (Chotipratoom et al. 2015; Chun et al. 2016), acetylation (Choi et al. 2009a; Colussi et al. 2014; Kim et al. 2010), cationization (Chang et al. 2014) and cross-linking (Hwang et al. 2009; Kim et al. 2012) of starches. In comparison, the maintenance of granular structures along with the improvement of reaction efficiency by UHP-assisted modification facilitates the industry application, although the structure-properties relationship needs to be further studied and more chemical modified starches will be prepared.

In general, starch properties can generally be influenced by a wide range of variables, including the starch source, pressure, temperature, concentration, and holding time; therefore, the application of UHP depends on the reasonable selection of these parameters and further illustrating the structure-properties relationship. Investigations on the structural recombination of starch treated at lower pressure (lower than gelatinization pressure) and the performance of dry starch in UHP treatment will be helpful to broaden the application of UHP-treated starch.

References

Ahmed J, Al-Attar H (2017) Structural properties of high-pressure-treated chestnut flour dispersions. Int J Food Prop 20(sup1):S766–S778

Ahmed J, Thomas L, Taher A, Joseph A (2016) Impact of high pressure treatment on functional, rheological, pasting, and structural properties of lentil starch dispersions. Carbohydr Polym 152: 639–647

Bajaj R, Singh N, Ghumman A, Kaur A, Mishra HN (2021) Effect of high pressure treatment on structural, functional, and in-vitro digestibility of starches from tubers, cereals, and beans. Starch 74(1–2):2100096

Baks T, Bruins ME, Janssen AEM, Boom RM (2008) Effect of pressure and temperature on the gelatinization of starch at various starch concentrations. Biomacromolecules 9(1):296–304

Bauer BA, Knorr D (2004) Electrical conductivity: a new tool for the determination of high hydrostatic pressure-induced starch gelatinisation. Innovative Food Sci Emerg Technol 5(4): 437–442

Bauer BA, Knorr D (2005) The impact of pressure, temperature and treatment time on starches: pressure-induced starch gelatinisation as pressure time temperature indicator for high hydrostatic pressure processing. J Food Eng 68(3):329–334

Bauer BA, Wiehle T, Knorr D (2005) Impact of high hydrostatic pressure treatment on the resistant starch content of wheat starch. Starch 57(3–4):124–133

BeMiller JN, Huber KC (2015) Physical modification of food starch functionalities. Annu Rev Food Sci Technol 6:19–69

Błaszczak W, Fornal J, Valverde S, Garrido L (2005a) Pressure-induced changes in the structure of corn starches with different amylose content. Carbohydr Polym 61(2):132–140

Błaszczak W, Misharina TA, Yuryev VP, Fornal J (2007) Effect of high pressure on binding aroma compounds by maize starches with different amylose content. LWT Food Sci Technol 40(10): 1841–1848

Błaszczak W, Valverde S, Fornal J (2005b) Effect of high pressure on the structure of potato starch. Carbohydr Polym 59(3):377–383

Bolumar T, Georget E, Mathys A (2015) High pressure processing (HPP) of foods and its combination with electron beam processing. Woodhead Publishing, Sawston, pp 127–155

Buleon A, Colonna P, Planchot V, Ball S (1998) Starch granules: structure and biosynthesis. Int J Biol Macromol 23(2):85–112

Chang YJ, Choi HW, Kim HS, Lee H, Kim W, Kim DO et al (2014) Physicochemical properties of granular and non-granular cationic starches prepared under ultra high pressure. Carbohydr Polym 99:385–393

Choi H-S, Kim H-S, Park C-S, Kim B-Y, Baik M-Y (2009a) Ultra high pressure (UHP)-assisted acetylation of corn starch. Carbohydr Polym 78(4):862–868

Choi H-W, Lee J-H, Ahn S-C, Kim B-Y, Baik M-Y (2009b) Effects of ultra high pressure, pressing time and HCl concentration on non-thermal starch hydrolysis using ultra high pressure. Starch 61(6):334–343

Chotipratoom S, Choi J-H, Bae J-E, Kim BY, Baik M (2015) Freeze-thaw stability, glass transition, and retrogradation of high hydrostatic pressure-assisted hydroxypropylated corn starch. Food Sci Biotechnol 24(4):1327–1333

Chun EH, Oh SM, Kim HY, Kim BY, Baik MY (2016) Effect of high hydrostatic pressure treatment on conventional hydroxypropylation of maize starch. Carbohydr Polym 146:328–336

Colussi R, Kaur L, Zavareze EDR, Dias ARG, Stewart RB, Singh J (2018) High pressure processing and retrogradation of potato starch: influence on functional properties and gastro-small intestinal digestion in vitro. Food Hydrocoll 75:131–137

Colussi R, Pinto VZ, El Halal SL, Vanier NL, Villanova FA, Marques ESR et al (2014) Structural, morphological, and physicochemical properties of acetylated high-, medium-, and low-amylose rice starches. Carbohydr Polym 103:405–413

Deng Y, Jin Y, Luo Y, Zhong Y, Yue J, Song X et al (2014) Impact of continuous or cycle high hydrostatic pressure on the ultrastructure and digestibility of rice starch granules. J Cereal Sci 60(2):302–310

Dominique LW, Gipsy TM, Giovanna F (2019) Potato starch hydrogels produced by high hydrostatic pressure (HHP): A first approach. Polymers (Basel) 11(10):1673

Douzals JP, Perrier Cornet JM, Gervais P, Coquille JC (1998) High-pressure gelatinization of wheat starch and properties of pressure-induced gels. J Agric Food Chem 46(12):4824–4829

Du J, Yang Z, Xu X, Wang X, Du X (2019) Effects of tea polyphenols on the structural and physicochemical properties of high-hydrostatic-pressure-gelatinized rice starch. Food Hydrocoll 91:256–262

Dupuis JH, Liu Q, Yada RY (2014) Methodologies for increasing the resistant starch content of food starches: A review. Compr Rev Food Sci Food Saf 13(6):1219–1234

Gebhardt R, Hanfland M, Mezouar M, Riekel C (2007) High-pressure potato starch granule gelatinization: synchrotron radiation micro-SAXS/WAXS using a diamond anvil cell. Biomacromolecules 8(7):2092–2097

Guo Z, Zeng S, Lu X, Zhou M, Zheng M, Zheng B (2015b) Structural and physicochemical properties of lotus seed starch treated with ultra-high pressure. Food Chem 186:223–230

Guo Z, Zeng S, Zhang Y, Lu X, Tian Y, Zheng B (2015a) The effects of ultra-high pressure on the structural, rheological and retrogradation properties of lotus seed starch. Food Hydrocoll 44: 285–291

Hayashi R, Hayashida A (1989) Increased amylase digestibility of pressure-treated starch. Agric Biol Chem 53(9):2543–2544

Hwang D-K, Kim B-Y, Baik M-Y (2009) Physicochemical properties of non-thermally cross-linked corn starch with phosphorus oxychloride using ultra high pressure (UHP). Starch 61(8): 438–447

Jiang B, Li W, Hu X, Wu J, Shen Q (2014) Rheology of mung bean starch treated by high hydrostatic pressure. Int J Food Prop 18(1):81–92

Katopo H, Song Y, Jane JL (2002) Effect and mechanism of ultrahigh hydrostatic pressure on the structure and properties of starches. Carbohydr Polym 47(3):233–244

Kaur M, Punia S, Sandhu KS, Ahmed J (2019) Impact of high pressure processing on the rheological, thermal and morphological characteristics of mango kernel starch. Int J Biol Macromol 140:149–155

Kawai K, Fukami K, Yamamoto K (2007) Effects of treatment pressure, holding time, and starch content on gelatinization and retrogradation properties of potato starch–water mixtures treated with high hydrostatic pressure. Carbohydr Polym 69(3):590–596

Kim HS, Choi HS, Kim BY, Baik MY (2010) Characterization of acetylated corn starch prepared under ultrahigh pressure (UHP). J Agric Food Chem 58(6):3573–3579

Kim H-S, Hwang D-K, Kim B-Y, Baik M-Y (2012) Cross-linking of corn starch with phosphorus oxychloride under ultra high pressure. Food Chem 130(4):977–980

Kudla E, Tomasik P (1992a) The modification of starch by high pressure. Part I: air-and oven-dried potato starch. Starch 44(5):167–173

Kudla E, Tomasik P (1992b) The modification of starch by high pressure. Part II: compression of starch with additives. Starch 44(7):253–259

Kweon M, Slade L, Levine H (2008a) Effect of sodium chloride on glassy and crystalline melting transitions of wheat starch treated with high hydrostatic pressure: prediction of solute-induced Barostability from nonmonotonic solute-induced thermostability. Starch 60(3–4):127–133

Kweon M, Slade L, Levine H (2008b) Role of glassy and crystalline transitions in the responses of corn starches to heat and high pressure treatments: prediction of solute-induced barostabilty from solute-induced thermostability. Carbohydr Polym 72(2):293–299
Lee J-H, Choi H-W, Kim B-Y, Chung M-S, Kim D-S, Choi SW et al (2006) Nonthermal starch hydrolysis using ultra high pressure: I. effects of acids and starch concentrations. LWT Food Sci Technol 39(10):1125–1132
Leite TS, de Jesus ALT, Schmiele M, Tribst AAL, Cristianini M (2017) High pressure processing (HPP) of pea starch: effect on the gelatinization properties. LWT Food Sci Technol 76:361–369
Li W, Bai Y, Mousaa SAS, Zhang Q, Shen Q (2011a) Effect of high hydrostatic pressure on physicochemical and structural properties of Rice starch. Food Bioprocess Technol 5(6): 2233–2241
Li W, Gao J, Saleh ASM, Tian X, Wang P, Jiang H et al (2018) The modifications in physicochemical and functional properties of Proso millet starch after ultra-high pressure (UHP) process. Starch 70(5–6):1
Li W, Tian X, Liu L, Wang P, Wu G, Zheng J et al (2015) High pressure induced gelatinization of red adzuki bean starch and its effects on starch physicochemical and structural properties. Food Hydrocoll 45:132–139
Li W, Tian X, Wang P, Saleh AS, Luo Q, Zheng J et al (2016) Recrystallization characteristics of high hydrostatic pressure gelatinized normal and waxy corn starch. Int J Biol Macromol 83:171–177
Li W, Zhang F, Liu P, Bai Y, Gao L, Shen Q (2011b) Effect of high hydrostatic pressure on physicochemical, thermal and morphological properties of mung bean (Vigna radiata L.) starch. J Food Eng 103(4):388–393
Li G, Zhu F (2018) Effect of high pressure on rheological and thermal properties of quinoa and maize starches. Food Chem 241:380–386
Li G, Zhu F, Mo G, Hemar Y (2019) Supramolecular structure of high hydrostatic pressure treated quinoa and maize starches. Food Hydrocoll 92:276–284
Linsberger-Martin G, Lukasch B, Berghofer E (2012) Effects of high hydrostatic pressure on the RS content of amaranth, quinoa and wheat starch. Starch 64(2):157–165
Liu Y, Chao C, Yu J, Wang S, Wang S, Copeland L (2020b) New insights into starch gelatinization by high pressure: comparison with heat-gelatinization. Food Chem 318:126493
Liu H, Fan H, Cao R, Blanchard C, Wang M (2016a) Physicochemical properties and in vitro digestibility of sorghum starch altered by high hydrostatic pressure. Int J Biol Macromol 92: 753–760
Liu H, Guo X, Li Y, Li H, Fan H, Wang M (2016c) In vitro digestibility and changes in physicochemical and textural properties of tartary buckwheat starch under high hydrostatic pressure. J Food Eng 189:64–71
Liu PL, Hu XS, Shen Q (2010) Effect of high hydrostatic pressure on starches: a review. Starch 62(12):615–628
Liu Y, Selomulyo VO, Zhou W (2008) Effect of high pressure on some physicochemical properties of several native starches. J Food Eng 88(1):126–136
Liu H, Wang L, Cao R, Fan H, Wang M (2016b) In vitro digestibility and changes in physicochemical and structural properties of common buckwheat starch affected by high hydrostatic pressure. Carbohydr Polym 144:1–8
Liu Z, Wang C, Liao X, Shen Q (2020a) Measurement and comparison of multi-scale structure in heat and pressure treated corn starch granule under the same degree of gelatinization. Food Hydrocoll 108:106081
Liu M, Wu N-N, Yu G-P, Zhai X-T, Chen X, Zhang M et al (2018) Physicochemical properties, structural properties, and in vitro digestibility of pea starch treated with high hydrostatic pressure. Starch 70(1–2):1700082
Mota MJ, Lopes RP, Delgadillo I, Saraiva JA (2013) Microorganisms under high pressure—adaptation, growth and biotechnological potential. Biotechnol Adv 31(8):1426–1434

Muhr A, Blanshard J (1982) Effect of hydrostatic pressure on starch gelatinisation. Carbohydr Polym 2(1):61–74
Oh HE, Pinder DN, Hemar Y, Anema SG, Wong M (2008) Effect of high-pressure treatment on various starch-in-water suspensions. Food Hydrocoll 22(1):150–155
Okur I, Ozel B, Oztop MH, Alpas H (2019) Effect of high hydrostatic pressure in physicochemical properties and in vitro digestibility of cornstarch by nuclear magnetic resonance relaxometry. J Food Process Eng 42(6):e13168
Papathanasiou MM, Reineke K, Gogou E, Taoukis PS, Knorr D (2015) Impact of high pressure treatment on the available glucose content of various starch types: A case study on wheat, tapioca, potato, corn, waxy corn and resistant starch (RS3). Innovative Food Sci Emerg Technol 30:24–30
Pu H, Liu G, Huang M, Zhang C, Niu W, Chen X et al (2021) Effects of annealing on ultra-high pressure induced gelatinization of corn starch. Innovative Food Sci Emerg Technol 74:102849
Rahman MH, Mu TH, Zhang M, Ma MM, Sun HN (2020) Comparative study of the effects of high hydrostatic pressure on physicochemical, thermal, and structural properties of maize, potato, and sweet potato starches. J Food Process Preserv 44(11):14852
Rubens P, Snauwaert J, Heremans K, Stute R (1999) In situ observation of pressure-induced gelation of starches studied with FTIR in the diamond anvil cell. Carbohydr Polym 39(3): 231–235
Rumpold BA, Knorr D (2005) Effect of salts and sugars on pressure-induced gelatinisation of wheat, tapioca, and potato starches. Starch 57(8):370–377
Sandhu KS, Kaur M, Punia S, Ahmed J (2021) Rheological, thermal, and structural properties of high-pressure treated litchi (Litchi chinensis) kernel starch. Int J Biol Macromol 175:229–234
Schneider Teixeira A, Deladino L, García MA, Zaritzky NE, Sanz PD, Molina-García AD (2018) Microstructure analysis of high pressure induced gelatinization of maize starch in the presence of hydrocolloids. Food Bioprod Process 112:119–130
Shen X, Shang W, Strappe P, Chen L, Li X, Zhou Z et al (2018) Manipulation of the internal structure of high amylose maize starch by high pressure treatment and its diverse influence on digestion. Food Hydrocoll 77:40–48
Słomińska L, Zielonka R, Jarosławski L, Krupska A, Szlaferek A, Kowalski W et al (2015) High pressure impact on changes in potato starch granules. Pol J Chem Technol 17(4):65–73
Stolt M, Oinonen S, Autio K (2000) Effect of high pressure on the physical properties of barley starch. Innovative Food Sci Emerg Technol 1(3):167–175
Stolt M, Stoforos NG, Taoukis PS, Autio K (1999) Evaluation and modelling of rheological properties of high pressure treated waxy maize starch dispersions. J Food Eng 40(4):293–298
Stute R, Klingler R, Boguslawski S, Eshtiaghi M, Knorr D (1996) Effects of high pressures treatment on starches. Starch 48(11–12):399–408
Szepes A, Makai Z, Blümer C, Mäder K, Kása P, Szabó-Révész P (2008) Characterization and drug delivery behaviour of starch-based hydrogels prepared via isostatic ultrahigh pressure. Carbohydr Polym 72(4):571–578
Szwengiel A, Lewandowicz G, Gorecki AR, Blaszczak W (2018) The effect of high hydrostatic pressure treatment on the molecular structure of starches with different amylose content. Food Chem 240:51–58
Tan F-J, Dai W-T, Hsu K-C (2009) Changes in gelatinization and rheological characteristics of japonica rice starch induced by pressure/heat combinations. J Cereal Sci 49(2):285–289
Thevelein JM, Van Assche JA, Heremans K, Gerlsma SY (1981) Gelatinisation temperature of starch, as influenced by high pressure. Carbohydr Res 93(2):304–307
Tian Y, Li D, Zhao J, Xu X, Jin Z (2014) Effect of high hydrostatic pressure (HHP) on slowly digestible properties of rice starches. Food Chem 152:225–229
Vallons KJR, Arendt EK (2009a) Effects of high pressure and temperature on buckwheat starch characteristics. Eur Food Res Technol 230(2):343–351
Vallons KJR, Arendt EK (2009b) Effects of high pressure and temperature on the structural and rheological properties of sorghum starch. Innovative Food Sci Emerg Technol 10(4):449–456

Vittadini E, Carini E, Chiavaro E, Rovere P, Barbanti D (2007) High pressure-induced tapioca starch gels: physico-chemical characterization and stability. Eur Food Res Technol 226(4): 889–896

Wang C, Xue Y, Yousaf L, Hu J, Shen Q (2020) Effects of high hydrostatic pressure on the ordered structure including double helices and V-type single helices of rice starch. Int J Biol Macromol 144:1034–1042

Yang Z, Gu Q, Lam E, Tian F, Chaieb S, Hemar Y (2016b) In situ study starch gelatinization under ultra-high hydrostatic pressure using synchrotron SAXS. Food Hydrocoll 56:58–61

Yang Z, Swedlund P, Hemar Y, Mo G, Wei Y, Li Z et al (2016a) Effect of high hydrostatic pressure on the supramolecular structure of corn starch with different amylose contents. Int J Biol Macromol 85:604–614

Zeng F, Li T, Gao QY, Liu B, Yu SJ (2018) Physicochemical properties and in vitro digestibility of high hydrostatic pressure treated waxy rice starch. [article]. Int J Biol Macromol 120:1030–1038

Zhang L, Ji H, Yang M, Ma H (2014) Effects of high hydrostatic pressure treated mung bean starch on characteristics of batters and crusts from deep-fried pork nuggets. Int J Food Eng 10(2):261

Zhang D, Xu H, Jiang B, Wang X, Yang L, Shan Y et al (2021) Effects of ultra-high pressure on the morphological and physicochemical properties of lily starch. Food Sci Nutr 9(2):952–962

Chapter 10
Ultrasonic Treatment

Zhaofeng Li

Abstract Starch is an indispensable nutritional resource for human beings. However, the applications of starch are limited because of its low solubility, high viscosity, and high retrogradation tendency in its native form. These limitations can be overcome by starch modification techniques, particularly physical methods. Ultrasonic treatment (UT) is a physical method applied in the modification of starch because it generates strong shear force, high temperature, and free radicals locally, which can change the structure and properties of starch. It has been widely studied using different methods to explain the influence of frequency and intensity of ultrasound, duration and settings of ultrasonication, temperature and moisture content of the system on the properties of different starches. This review discusses the characterisation of UT-modified starches by differential scanning calorimetry (DSC), scanning electron microscopy (SEM), X-ray diffraction (XRD), pasting properties (RVA), and rheometer. The potential applications of UT-modified starches are also reviewed.

Keywords Ultrasonic treatment · Starch system · Granular structure of starch · Physicochemical properties of starch

10.1 Overview of Ultrasonic Treatment in Starch System

Ultrasound defines the mechanical waves at a frequency above the upper value of normal human hearing range (>16 kHz). It can be divided into three frequency ranges including power ultrasound (16—100 kHz), high frequency ultrasound (100 kHz—1 MHz), and diagnostic ultrasound (1—10 MHz) (Patist and Bates 2008). Ultrasonic equipment is environmentally friendly, easy to operate and control, and also simple to realize automation (Bartsch and Schmidt-Naake 2006). What's more, ultrasonic treatment exhibited positive impacts on food processing

Z. Li (✉)
School of Food Science and Technology, Jiangnan University, Wuxi, China
e-mail: zfli@jiangnan.edu.cn

Z. Sui, X. Kong (eds.), *Physical Modifications of Starch*,
https://doi.org/10.1007/978-981-99-5390-5_10

and preservation, such as improving the product yields, shortening processing time and cost, improving t product grade and safety, reducing pathogens, and helping to create novel products with desired properties. As a promising new technology, it is supposed to be full of potentialities in future development of food processing (Awad et al. 2012; Jiang et al. 2020; Knorr et al. 2004; Patist and Bates 2008; Singla and Sit 2021).

Most of the ultrasonic application in starch is in starch–water system. When the sinusoidal ultrasound waves pass through the liquid, the acoustic energy of ultrasound cannot be absorbed by the molecules, and propagates via a series of compression and rarefaction waves induced in the medium molecules, and is then transformed to chemically usable forms via the cavitation phenomenon (Chan et al. 2010; Knorr et al. 2004). This phenomenon is caused by ultrasonic treatment when the amplitude is high enough, leading to the formation of bubbles in the system. The pressure from the wave motion makes the bubbles grow during the expansion cycle and shrink during the compression cycle. The bubbles keep growing over a lot number of cycles, and when they reach the point that the oscillation frequency of the ultrasound waves equals that of the bubble wall, the bubbles explode during the compression cycle (Chan et al. 2010; Kerboua and Hamdaoui 2018; Patist and Bates 2008). The explosion of bubbles produces extremely high shearing force and turbulence with large energy waves, resulting in drastically rise of local temperature, consequently changing the physical and chemical conditions of the system. Bubble collapsing causes a water jet shooting onto the granule surface. These "micro-jets", as well as the "shear forces" induced by cavitation, can availably damage the starch granules (Zuo et al. 2012). Besides, the cavitation dissociate the water molecules, leading to the release of free radicals such as hydroxide (OH) and hydrogen (H) radicals. These radicals may chemically modify the system in turn (Czechowska-Biskup et al. 2005; Riesz and Kondo 1992).

Early in 1933, Szent-Györgyi (Szent-Györgyi 1933) firstly reported the action of ultrasound on starch, which decreased the viscosity and cut off the starch chains. During the past decades, large number of efforts have been devoted to further investigate the impacts of ultrasonic treatment on different aspects of starches, which are expounded in this chapter. And many factors have been found to influence the effects of ultrasound on starch granules, such as the frequency and power of ultrasonic treatment, temperature and time of the treatment, and the properties of starch dispersion systems including concentration and botanical origin of starch. Although it is difficult to directly compare the experimental results from different research because their methods and materials varied greatly, this chapter describes the trends of the effect of ultrasonic treatment on starch.

10.2 Effect of Ultrasonic Treatment on the Starch Yield and Composition

10.2.1 Starch Yield

Cereals, tubers, and roots are major materials to produce starch (Kringel et al. 2020). Starch granules, especially starch granules present in the tubers and roots, are usually embedded in cellulosic fibres and combined together by pectin substrates. Hence, the raw materials need to be grated firstly, breaking the cells and releasing the starch (Daiuto et al. 2005; Kringel et al. 2020). The yield of starch is decreased, however, since it is impossible to grate the plant cell walls completely by mechanical disintegration. Enzymatic methods have been used to increase the recovery of starch from roots and tubers, but the methods are ineluctably time-consuming and costly, because long incubation periods and pure enzyme are required (Dzogbefia et al. 2008).

Ultrasounication can increase the yield of starch (Patist and Bates 2008; Zhu 2015). Compared to conventional method, ultrasonic treatment increased the recovery of corn and sorghum starch (Benmoussa and Hamaker 2011). As both cycle and amplitude were constant, the longer treatment time caused higher breakdown of plant cells, thus releasing more starch and increasing the starch yield. Ultrasound can be applied to facilitate starch–protein separation during ultrasound-assisted laboratory-scale corn wet-milling (Liu et al. 2021). Sit et al. (2014) carried out a study on the effect of sonication parameters on yield of taro (a tropical tuber crop) starch, and found that the starch yield significantly increased under the control of all ultrasonic treatment combinations. Compared with starch from sole wet-milling, ultrasonication increased the yield and purity of starch from yellow dent maize (Zhang et al. 2005). Wang and Wang (2004b) reported that a higher yield of rice starch was achieved at lower amplitude rather than at 100% amplitude, indicating that high amplitude with more energy might not only disintegrate the cellulosic materials but also break the starch molecules, resulting in the dissolution of partial starch in water and the difficult for recovering starch by centrifugation in the later process. Thus, the conditions of ultrasonic treatment should be carefully controlled in order to avoid the damage to granules.

Ultrasonic treatment has been combined with other additives such as sodium dodecyl sulphate (SDS) (Park et al. 2006; Wang and Wang 2004a) and protease (Cameron and Wang 2006; Jadwong and Therdthai 2018; Wang and Wang 2004b) for the improvement in the efficiency of starch extraction from various cereals. When coupled with SDS, both the starch recovery and purity from different cereals (such as rice, sorghum, and barley) were enhanced by ultrasonic treatment without the use of alkaline solution (Park et al. 2006; Wang and Wang 2004a). With the aid of protease, ultrasonic treatment reduced the steeping time during wet-milling process of degermed corn flour without the use of SO_2 (Cameron and Wang 2006), and evidently increased the starch recovery from rice flour while the morphology and molecular structure of the granules were not influenced (Wang and Wang 2004b).

Ultrasonic treatment has also been coupled with sucrose density separation for better extraction of starch from a small amount of flour. This strategy could be significantly useful in cases when plenty of raw materials are difficult to get, for example, genetic breeders and crop molecular biologists (Benmoussa and Hamaker 2011). A series of complex processes to apply the lab-scale results to the industrial scale are still needed.

10.2.2 Starch Composition

Amylose content is a decisive factor affecting the properties and applications of starch (Li 2022; Srichuwong and Jane 2007). Researchers used iodine complexation to determine the amylose content of native and sonicated corn starches. The result showed an increase in the amylose content after ultrasonic treatment, implying that partial amylose may depolymerize during the ultrasonication process (Li et al. 2016). Chan et al. (2021) reported that both high- and low-temperature ultrasonication did not increase the gelatinization degree of semigelatinized high amylose maize starch, but generated cracks and pores on the starch granule surface and enhanced the short-range ordered molecular structure and apparent amylose content. Ultrasound may scissor the starch chains and produced more linear molecules, as reported by Chan et al. (2010), focusing on the increase in amylose content of corn, potato, mung bean, and sago starches after ultrasonic treatment.

Minor components such as protein and lipid in starch granules can influence the functionality of starch. Besides, the residual protein are combined with the starch granules tightly and may inhibit the enzymolysis of starch (Dhital et al. 2019; Srichuwong and Jane 2007). Ultrasonic treatment can reduce the residue contents in starch during extraction. Chan et al. (2010) revealed that ultrasonic treatment in a bath improved the enzyme susceptibility of mung bean and corn starches because the protein content of starch was decreased by ultrasonic treatment. Ultrasonic treatment has also been employed in wet-milling of corn flour and hominy feed (Zhang et al. 2005. Compared to starch extracted by single wet-milling treatment, the starch recovered with the involvement of ultrasonic treatment had fewer residual impurities such as proteins and lipids (Zhang et al. 2005.

10.3 Effect of Ultrasonic Treatment on the Structure of Starch

10.3.1 Effect of Ultrasonic Treatment on the Granular Structure of Starch

Ultrasonic treatment often results in the morphological changes of starch granules by forming cracks and pores through damaging the granules. Various techniques have been used to observe the morphology of starch granules caused by ultrasonic treatment, such as light microscopy (LM), polarized light microscopy (PLM), scanning electron microscopy (SEM), transmission electron microscopy (TEM), and laser light scattering (LLS) technique (Sujka and Jamroz 2013; Yang et al. 2019; Zhu et al. 2012; Zuo et al. 2009, 2012). The differences in morphology of starch before and after ultrasonic treatment can be seen clearly (Fig. 10.1).

Plenty of factors can affect the effect of ultrasonic treatment on the morphology of starch granules (Zuo et al. 2009). Firstly, the type and structure of starch, as well as the concentration of starch slurry, are the main factors because they determine the susceptibility of starch to ultrasonication. For example, the ultrasonication treatment made the surface of normal and waxy corn starches become porous, while made the amylomaize V starch become cracked (Luo et al. 2008). Treated with the same ultrasonic action, the surface fractures and depressions of potato and wheat starches were more obvious than those of corn and rice starches (Sujka and Jamroz 2013). Besides, the damage of granules decreased with the increase of starch concentration (Amini et al. 2015; Gallant et al. 1972). Secondly, the parameters of ultrasonication settings including power, frequency, and duration of treatment are also the main factors. The starch granules tended to agglomerate when increasing the power input (Jambrak et al. 2010). Increasing the duration time could cause more damage on starch granules (Huang et al. 2007). Dual-frequency ultrasound (25 and 80 kHz) caused more dents and pores on granules' surface than single frequency ultrasound

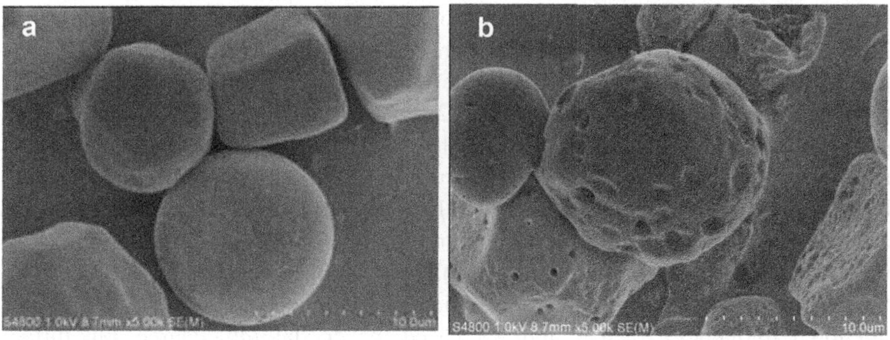

Fig. 10.1 SEM figures of native (**a**) and sonicated (**b**) corn starches (Li et al. 2016). (Reused with permission from *Wiley Publications*)

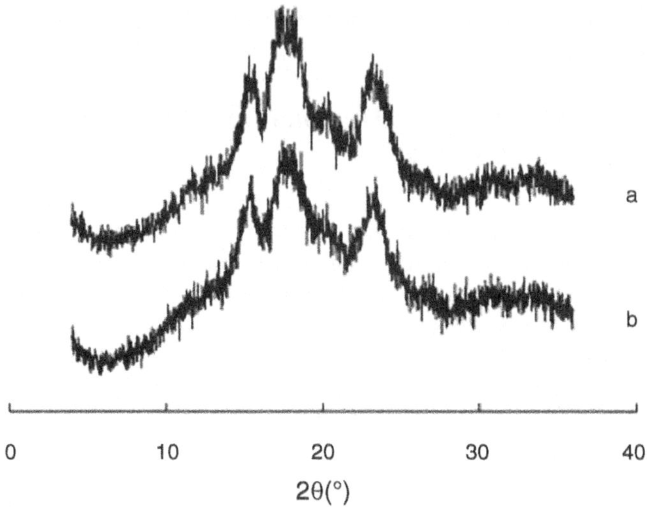

Fig. 10.2 X-ray diffraction patterns of native corn starch (**a**) and sonicated corn starch (**b**) (Li et al. 2016). (Reused with permission from *Wiley Publications*)

(25 kHz or 80 kHz) (Zheng et al. 2013). The treatment sequence also showed a significant effect on the physicochemical properties of normal maize and potato starches (Wang et al. 2022). Finally, other conditions such as temperature, the type of ultrasonication equipment (probe or bath), and the gas type of the atmosphere can affect the results as well. Heating can make starch granules more vulnerable to ultrasonic destruction (Zuo et al. 2009). Different ultrasonication baths and probes led to different outcomes for the same kind of starch (Jambrak et al. 2010). The gas type can affect the damage degree of starch granules. The hydrogen atmosphere was the most harmful, followed by air and oxygen, carbon dioxide, and vacuum (Gallant et al. 1972). Furthermore, the solubility of atmosphere gases had negative effect on the pit size (Degrois et al. 1974).

Ultrasonication treatment can change the morphology of starch granules, the crystallinity of starch is consequently changed (Bonto et al. 2021; Zhang et al. 2021). The conditions of ultrasonication are one of the decisive factors that changing the crystallinity degree of starch granules. Li et al. (2016) found that ultrasonic treatment (starch concentration of 40%, 20 kHz, 800 W, 4 h) did not change the crystalline structural type of the corn starch (Fig. 10.2), but caused the destruction of approximately 25.1% of the crystalline structure of the starch granules, which could be attributed to excessively small crystallites or a decreased degree of crystalline arrangement. Simliar results could be seen in another study where ultrasonic treatment was applied to isolate starches from brown rice grains (Park and Han 2016). Huang et al. (2007) reported that ultrasonication (starch concentration 30%, 500 W, unknown frequency) up to 3 min increased the crystallinity degree, but further treatment for 15 min decreased it. It could be inferred that ultrasonication disrupted the amorphous parts in the granules during the initial process and increased the crystallinity degree. Further treatment decreased the crystallinity degree, probably due to the damage of crystals in the granules by further ultrasonication (Huang et al. 2007).

Other deterction technologies also can help determine the effect of ultrasonication on starch granules in more details. Small-angle X-ray scattering (SAXS) analysis showed that the repeated lamellae thickness was unaffected, while the background material (electron density as ρ_u) was more affected than the amorphous material (electron density as ρ_a) (Zhu et al. 2012). It seems that the amorphous region of the granules is more susceptible to ultrasonication than the crystalline part. Thus, starches with a higher content of amylose (most amorphous in the granules) were easier to be influenced in structure and properties by ultrasonication (Luo et al. 2008; Zhu 2015; Zhu et al. 2012). Fourier transform infrared spectroscopy (FTIR) analysis revealed that ultrasonication affected the crystal structure of sweet potato starch, but did not change the functional chemical groups, and the effect of dual-frequency treatment was more obvious than that of single ones (Zheng et al. 2013). Thus, to get systematic and comprehensive data, it is advantageous to combine and use various methods together, such as FTIR, solid-state nuclear magnetic resonance spectroscopy (NMR), WAXS, as well as SAXS. For example, the aggregation structure and physicochemical characteristics of ultrasonicated sweet potato starches, sweet potato starch modified with different sonication time (15, 20, 25, and 30 min) were studied using combined analytical techniques including SEM, XRD, FTIR, Raman, and DSC. The results showed that crystalline structure, short-range ordered, and ordered molecular structure were partially destroyed after ultrasonication (Wang et al. 2020, b).

10.3.2 Effect of Ultrasonication on the Molecular Structure of Starch

The influence of ultrasonication on molecular structure of starch chains had been examined by different methods (Chan et al. 2021; Iida et al. 2008; Isono et al. 1994; Wang et al. 2021; Zheng et al. 2013). Starch chains of amylose/amylopectin can form inclusion complexes with iodine, and exhibit different colours, reflecting the length and branching pattern of starches. It has been reported that ultrasonication converted the colour of starch–iodine complexes from blue to red (Szent-Györgyi 1933). Ultrasonication could reduce the blue value, $_{max}$ value, and A_{680}/A_{545} ratio (A_{680} and A_{545} represent the absorbance of the starch solution at the wave lengths of 680 and 545 nm, respectively) of starches from different origins (e.g., potato, wheat, corn, and rice) (Sujka and Jamroz 2013). These phenomena indicated that the ultrasonic treatment induced the chain scission and weakened the ability of starch to form complexes with iodine.

The effect of ultrasonication on molecular structure of starch chains has been studied in more details by diverse methods. Chains of starches from various sources were disrupted by ultrasonication. The sonicated starch showed smaller molecular size, lower weight average molar mass, and narrower molecule weight with the use of chromatography (Iida et al. 2008; Isono et al. 1994; Li et al. 2016). The ^1H and ^{13}C NMR spectrogram of sonicated starch showed an increased proportion of highly mobile fraction (Iida et al. 2008). When the molecular weight of sonicated starch approached to a certain limit (e.g., the limit for rice starch was 1.1×10^4 in number-average molecular weight when using pullulan as standards), the effect of ultrasonication decreased significantly (Isono et al. 1994). As mentioned earier, ultrasonication decreased the polydispersity of polymers through the competing degradation processes and recombination processes (Lorimer and Mason 1987). During the degradation process, the high pressure induced by cavitation, the dramatically elevated temperature, and the shear force led to the scission of the molecular chains. It was reported that the stronger the ultrasound, the shorter the starch chains. The minimum length of starch chain was obtained when the ultrasound cannot degrade the chains anymore (Czechowska-Biskup et al. 2005; Jambrak et al. 2010). Besides, ultrasonication also can produce free radicals that may break the chains of starch (Kardos and Luche 2001).

Further study remains to be carried out for better understanding about the fine structure of ultrasound treated starch, such as the cluster and internal molecular structure of amylopectin. This is because amylopectin is the main component of the amorphous part of starch granules, and amylopectin is more vulnerable to ultrasonication than the crystalline region of starch granules. In some other research studies, ultrasonication hardly affected the molecular size of starch chains from various origins (Chung et al. 2002; Jackson et al. 1988; Zuo et al. 2009). It is thus clear that the degree of the starch chain degradation is dominated by experimental conditions and parameters of ultrasonication. Hence, by setting appropriate parameters, ultrasonication could be used to promote the dissolution of starch samples for HPSEC analysis without any molecular degradation and involvement of alkaline or DMSO which may either result in starch degradation or environmental pollution (Jackson et al. 1988).

10.4 Effect of Ultrasonication on the Physicochemical Properties of Starch

10.4.1 Solubility and Swelling

Ultrasonication has been reported to improve the swelling of potato, wheat, rice, and corn starches from diverse sources with different amylose contents (Jambrak et al. 2010; Luo et al. 2008; Sujka and Jamroz 2013). Ultrasonication can also enhance the solubility of starches in water because the physical and chemical damage of starch granules by ultrasonic wave promoted the water penetration and starch hydration (Chan et al. 2010; Jambrak et al. 2010; Kaur and Gill 2019; Luo et al. 2008; Sujka and Jamroz 2013; Zheng et al. 2013). Recently, a kind of granular cold water swelling starch was prepared from corn starch by using ultrasound-assisted alcoholic-alkaline treatment. The maximum cold water solubility reached 93.33% after optimization (Zhu et al. 2016).

The type and composition of starch can influence the change in swellability and solubility of granules after ultrasonication treatment (Wang et al. 2020, b). For example, the swelling power and solubility of corn starches with different amylose contents (waxy, normal, and high amylose genotypes) changed in different degrees after ultrasonication (Luo et al. 2008). The results showed that the most significant increase in swelling and solubility appeared in high amylose genotypes, followed by normal and waxy corn starches, confirming that amylose in starch granules was more susceptible to ultrasonication (Luo et al. 2008). Apart from the internal factors of granules, varying ultrasonication parameters can also lead to different results. For example, an increase in ultrasound power and intensity led to the increase in the swelling power of corn starch (Jambrak et al. 2010). The ultrasonication with dual-frequency (25 kHz and 80 kHz) as more efficient than that of single frequency (25 kHz or 80 kHz) in improving the solubility of sweet potato starch (Zheng et al. 2013).

10.4.2 Gelatinization

Differential scanning calorimetry (DSC) is the most typical technique to measure the gelatinization properties of starch treated with ultrasonication. The enthalpy change (ΔH) and the gelatinization temperatures including onset (T_o), peak (T_p), and conclusion (T_c) temperatures, are common parameters to refelect the gelatinization (Rahaman et al. 2021; Zhu 2015). The ΔH indicates the loss of molecular order, and the T_p represents the crystallinity related to double helix length (Sang et al. 2008). However, various studies on the effect of ultrasound on starch gelatinization seem to show contradictory results. It has been reported that ultrasonication decreased the ΔH of rice starch (Yu et al. 2013) and increased/decreased that of corn starches with different amylose contents (Huang et al. 2007; Jambrak et al. 2010; Luo et al. 2008). And it was also observed that the gelatinization temperaturesof rice starch was decreased (Yu et al. 2013), while that of corn starch was increased or hardly increased (Huang et al. 2007; Jambrak et al. 2010), or the gelatinization temperature range of corn starches containing different amount of amylose was reduced (Luo et al. 2008). These discrepancies from various reports may be caused by the different types and composition of starches and the discrepant operational conditions of ultrasonication. The initial erosion of the amorphous part in the granules caused the increase in the crystallinity of starch, resulting in the raise of gelatinization temperatures and ΔH of corn starch (Huang et al. 2007). The futher ultrasonication treatment gradually damaged the crystallinity and relatively the ΔH and gelatinization temperatures of starch were reduced (Jambrak et al. 2010; Luo et al. 2008; Yu et al. 2013). In some research studies, the ultrasonication conditions were so intensive that the whole granules were damaged quickly and consequently the initial increases in ΔH and gelatinization temperatures can be ignored. Moreover, some studies have found that the temperature of starch slurry increased during the ultrasonication process without thermostat, leading to partial gelatinization of starch granules and the decrease in ΔH and gelatinization temperatures (Yu et al. 2013).

10.4.3 Properties of the Paste

10.4.3.1 Pasting

The effect of ultrasonication on pasting properties of starch can be commonly determined by the rapid visco-analyser (RVA) (Chan et al. 2010; Zuo et al. 2009), the Brabender viscograph (Herceg et al. 2010; Huang et al. 2007; Isono et al. 1994; Li et al. 2016; Luo et al. 2008; Zheng et al. 2013), and other viscosity analysers (Azhar and Hamdy 1979; Chung et al. 2002; Sujka and Jamroz 2013). Figure 10.3 showed the amylograms of native and sonicated corn starches mesured by Brabender viscograph. During the pasting process, the starch slurry was stirred under fixed shearing force and the granules were heated and cooled with a set procedure. The recorded viscosity cahnges revealed that ultrasonication caused a decrease in viscosity of the starch–water system during pasting. Similar outcomes has been reported by different studies using starches from diverse origins (Herceg et al.

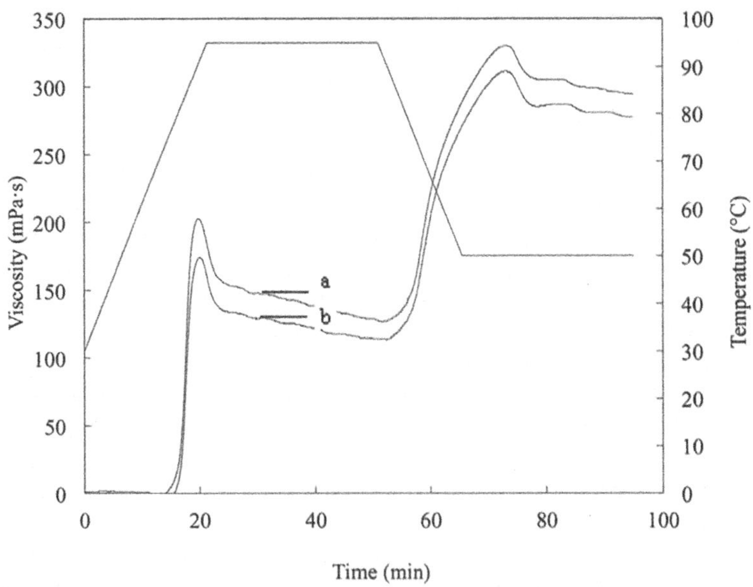

Fig. 10.3 Brabender viscosity curves of native (**a**) and sonicated (**b**) corn starch at 6% (w/w) (Li et al. 2016). (Reused with permission from *Wiley Publications*)

2010; Isono et al. 1994; Li et al. 2016; Luo et al. 2008; Zheng et al. 2013; Zuo et al. 2009). The treatment of ultrasonication damaged the starch granules, resulting in worse swellability, and these disrupted granules were further permeated by water, which facilitated the hydration and caused lower pasting viscosity of starch. By controlling the ultrasonication conditions, a lower viscosity can be achieved through the physical disruption of the starch granules without scission of molecular chains (Zuo et al. 2009). However, contrary results were observed in another study where the pasting temperature, peak viscosity, hold viscosity, final viscosity, and setback viscosity of the taro starches extracted by ultrasound pretreatment were significantly higher than those of control (Sit et al. 2014). Chan et al. (2010) also reported that ultrasonication increased the peak viscosity of potato starch during pasting. As mentioned in the above section, these contradictories may be resulted from the different damage degree of the amorphous and crystalline regions of the starch granules under different ultrasonic conditions.

Obviously, the ultrasonication conditions influence the ultrasound effect on pasting properties of starch. The dual-frequency ultrasonication was more effective than a single frequency ultrasonication (Zheng et al. 2013). The degree of changes was enlarged with the increase of temperature and duration (Park and Han 2016; Zuo et al. 2009). Apart from the experimental conditions, the starch composition can also affect the degree of changes in pasting behaviours. In a study, the pasting properties of corn starches (native/sonicated) with different amylose contents were measured. Under the same ultrasonication conditions, the pasting viscosity of high amylose starch was much more influenced by ultrasound than that of the waxy starch, indicating that the amylose in starch granules was more sensitive (Luo et al. 2008). However, it is notable that when starches were imposed of high-power ultrasond and the starch pastes were depolymerised, the apparent viscosity was decreased by molecular scission occurred at the C-O-C bond of α-1,6 glycosidic linkage as the FTIR showed, and the extent of molecular scission was inversely correlated with amylose content (Kang et al. 2016). The better resistance to ultrasonication of high amylose starch pastes could be attributed to the aggregation of amylose. The effect of ultrasonication on other rheological properties including dynamic oscillation and small deformation rheology of starch are also investigated. Carmona-García et al. (2016) showed that ultrasound treatment of plantain starch resulted in a more pronounced decrease in the storage modulus (*G*') compared with its native counterpart. Dhull et al. (2021) also observed a decrease in both storage modulus and loss modulus in the dynamic shear test for the sample of black rice starch pastes after ultrasonic treatment when compared with that of the native starch.

10.4.3.2 Gel Transparency

When using the gelatinized starch paste for many industrial application, the transparency is one of the important qualities that should be considered. It has been reported that ultrasonication can help to improve the transparency of the starch paste/gel from various sources including potato, sweet potato, mung bean, and rice (Chung et al. 2002; Sujka and Jamroz 2013). The reason for the transparency enhancement may be that ultrasonication can increase the solubility of starch in water and facilitate the hydration of starch chains, leading to the improvement of light transmittance. However, in another study using taro starch, a decrease in transparency was found and this decrease trend was consistent with the increase in viscosity during pasting. It was speculated that under the certain experimental conditions, ultrasonication allowed more water to be absorbed by the starch granules and made the granules swell more, resulting in a more viscous starch paste and a decrease of the transmittance (Sit et al. 2014).

10.4.3.3 Flow Behaviour

The flow behaviour of starch gels treated with ultrasonication can be desicribed by power law equation. The consistency coefficient, which was dependent on the ultrasound power, was decreased by ultrasonication acting on corn starch, and the flow index was higher than 1 (Jambrak et al. 2010).

10.4.4 Retrogradation and Freeze-Thaw Stability

When the gelatinized starch paste is gradually cooled, the crystallinity, gel firmness and turbidity of starch gel begin to increase, and B-type polymorph appears in the system accompaning by other changes. These changes are the characteristic phenomenon and is defined as the retrogradation of starch (Hoover 1995). Short-term retrogradation is mainly attributed to the recombination of amylose, while the long-term one is much related to the recrystallization of amylopectin (Hoover 1995; Miles et al. 1985). The difference in retrogradation of the starch gels from sonicated or non-sonicated starches had been reported. Results showed that the sonicated starches (stored at 4 °C for 24 h after gelatinization) exhibited higher hardness, cohesiveness, and adhesiveness than that of the control ones (Herceg et al. 2010). Ultrasonication (before the gelatinization of starch) caused a increase in onset melting temperature and a decrease in ΔH of the retrograded rice starch, and the increase in power input had a greater impact on rice starch (Yu et al. 2013). As for the freeze-thaw stability, ultrasonication increased the gel syneresis of high amylose corn starch (amylose content of 50%) while reduced that of the waxy corn starch during freezee-thaw cycle (Luo et al. 2008). Therefore, it is difficult to directly compare the data from various studies because the effect of ultrasonication is influenced by the type and composition of starch. Better freeze-thaw stability was also obtained in another study where the taro starch was the ultrasonically extracted starches, making the starch more acceptable for application in frozen foods (Sit et al. 2014).

10.5 Effect of Ultrasonication on the Modification of Starch

As the concept of "green chemistry and technology" is emerging, non-polluting processes and products are becoming more and more popular in synthetic chemistry. Ultrasonication has been widely employed in order to optimize the use of materials (especially renewable biologic resources) and decrease the wastes and pollutions (Kardos and Luche 2001). As for starch modification, native starches are generally modified by physical, chemical, and enzymatical methods and novel starches with new properties are created for various applications. Ultrasonication is not only a physical treatment method, but also an auxiliary method to promote the reactivity and efficiency of other kinds of modifications, which will be reviewed later.

10.5.1 Physical Modification

Comparing with starches modified chemically and enzymatically, physically modified products are more popular because they are considered healthy and green. Ultrasonication has been employed to improve the melting process of high amylose corn starch (70% amylose content) using glycerol as the plasticizer. Ultrasound treatment caused a significant reduction in intrinsic viscosity for the sample previously processed with the highest glycerol content, probably because of its higher solubility in water (Lima and Andrade 2010).

The influence of ultrasonication on other kinds of physical modification of starch including extrusion and hydrothermal treatments are also reported in some studies. The combination of ultrasonic treatment, extrusion, and hydrothermal treatment were usually used to produce triple modified starch (TMS) (Dey and Sit 2017). An example of homogeneous triple physical modification was a combination of extrusion/ultrasonication/ANN (Ashogbon 2021).

10.5.2 Chemical Modification

By controlling experimental conditions, ultrasonication can shorten the modification time and improve the degree of substitution (e.g., acetylation, hydroxypropylation, octenyl-succinylation, and carboxymethylation) of starches from various origin, such as corn (Chen et al. 2014; Huang et al. 2007), potato (Čížová et al. 2008), cassava (Gao et al. 2011), and yam (Dioscorea zingiberensis) (Zhang et al. 2012). Here comes some examples, simultaneous ultrasonication enhanced the reaction efficiency and increased the contents of carbonyl and carboxyl groups of corn starch due to the expedited oxidation process (Chong et al. 2013). Ultrasonication was applied to the carboxymethylated potato starch, which can reduce the time of octenyl-succinylation from conventionally-spent 24 h to a few minutes, and also can make the modification become free of ptoluenesulphonic acid (a kind of potential pollutant to environment) (Čížová et al. 2008). During the carboxymethylation of corn starch, a degree of substitution up to 38.7% could be achieved when the ultrasonication was employed before alkalization process (Shi and Hu 2013). Obviously, it is of great significance to study the effect of ultrasonication on other chemical modification or their combinations.

Two factors may account for the effect of ultrasonication on these modifications. On the one hand, as mentioned in anterior sections, ultrasonication can make starch granules produce pores, cracks, and holes on their surface and improve the swelling power and water solubility of starch. These changes make the structure of granules more open for the chemical agents and catalysts to permeate into the granules, thus increasing the probabilities for chemical reactions (Zhang et al. 2012). On the other hand, the cavitation and sonolysis can produce free radicals (e.g., hydroxide radicals (·OH) and hydrogen (·H)) from water, and also facilitate the chemical modifications (Chong et al. 2013), although the detailed reaction mechanism needs futher exploration. Obviously, in order to obtain desired results, the optimization of ultrasonication conditions is necessary because the conditions of ultrasonic treatment can determine the degree of substitution enhancement (Shi and Hu 2013).

10.5.3 Enzymatic Modification

The enzymatic modification offers lower energy cost, better control of net products and higher yields (Lan et al. 2016). The enzyme reactions are also mild and very specific for starch substrates present in complex food system (Ashogbon 2021). Ultrasonication has been used to increase the efficiency of enzymatic modification of starch from diverse sources such as corn (Apar et al. 2006; Wu et al. 2011), mung bean (Hu et al. 2013), rice, and wheat (Apar et al. 2006). For example, ultrasonication facilitated the hydrolysis of mung bean starch by α-amylase, and ultrasonication with dual-frequency (25 kHz and 40 kHz) showed greater effect on starch than that of ultrasonication with single frequency (25 kHz or 40 kHz) (Hu et al. 2013). Using the glucoamylase as a catalyzer to prepare microporous corn starch, the production efficiency was greatly enhanced by ultrasonication (Wu et al. 2011).

The increased efficiency of enzymatic modification of starch could be attributed to the enhanced enzyme susceptibility of starch by ultrasonication. As mentioned before, ultrasonication can create pores and cracks on the starch granules, making it easier for enzymes to permeate into the granules and thus increasing the interfacial area of enzyme-substrate as well as the chance for modification. Wang et al. (2020, b) reported that sonicating the mixed enzymatic reaction system below 65 °C promoted starch hydrolysis significantly, inducing more than five-fold growth in the degree of starch hydrolysis as much as the ultrasound pretreatment caused. This was because ultrasound could promote the enzymatic hydrolysis of amylopectin, which was harder for glucoamylase to hydrolyze compared to amylose. The molecular mechanism of ultrasound which promoted enzymolysis includes the decrease in the relative crystallinity, viscosity, enthalpies, molecular weight and polymerization, and changes in the surface structures (Li et al. 2018). However, it is necessary to optimize the experimental conditions to avoid excessive disruption of granules and negative effects on the activity and stability of some enzymes (Özbek and Ülgen 2000), and to maximize the efficiency of ultrasonication on enzymatic modifications of starch at the meanwhile.

10.6 Effect of Ultrasonication on the Application of Starch

Ultrasonication has been widely employed in different industries and has proved successful in commercialized applications of a varity of food and other industries, such as defoaming and extrusion (Awad et al. 2012; Knorr et al. 2004; Patist and Bates 2008). Thus, the applications of ultrasonication in starch commercialization have great potential.

10.6.1 Food Application

Among diverse applications of ultrasonication in starch-related industries, the most reported one should be the improvement of starch extraction from various botanical sources, which has been mentioned in the starch yield section. Recently, ultrasonication was employed to assist the production of glutinous rice flour from broken rice by extracting amylose. By utilizing ultrasonication in alkaline solution, amylose could be extracted by washing at lower temperature within shorter time compared to the traditional aqueous leaching technique. The amylose content of the obtained starch decreased to less than 2% with a yield more than 80% (Setyawati et al. 2016).

Ultrasonication has been used in rough rice parboiling because of its facilitation in gelatinization of native starch (Wambura et al. 2008). With the help of ultrasond treatment, a better reservation of the micro-nutrients in rice could be achieved because of the lower temperature, which greatly improved the efficiency of one-step combined soaking and gelatinization for rice parboiling with decreased energy consumption and duration of the whole process (Wambura et al. 2008).

Some micro-nutrients are faced with poor water solubility that has great influence on their bioavailability. Nano-emulsion is a feasible technique to solve this problem (Abbas et al. 2014). During the preparation of food grade oil-water nano-emulsion of curcumin stabilized by OSA-modified starch (corn starch octenyl succinate), ultrasonication treatment was successfully applied to not only aid the formation of this nano-emulsion but also effectively reduce the energy consumption through the whole process (Abbas et al. 2014). The Nile Red fluorescent probe dyed oil droplets of PGU-stabilized emulsions were observed by confocal laser scanning microscopy

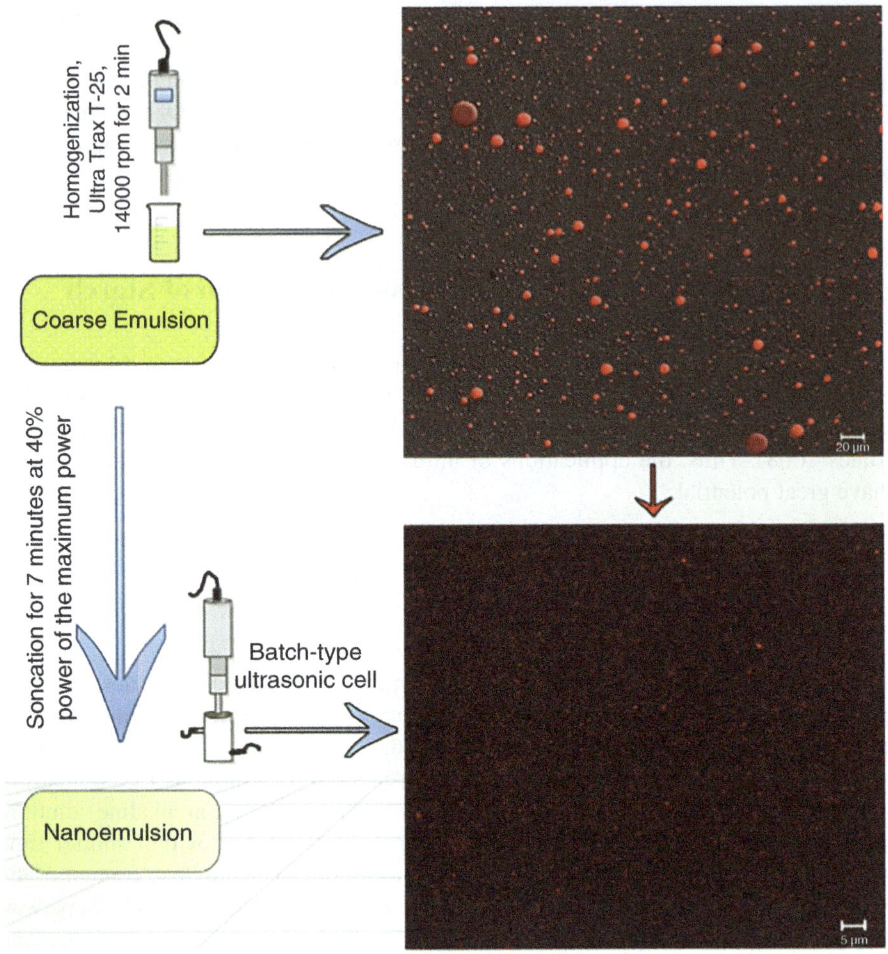

Fig. 10.4 Confocal laser scanning microscopy (CLSM) of conventional emulsions and ultrasound-assisted nano-emulsion (Abbas et al. 2014). (Reused with permission from *Elsevier Publications*)

(CLSM), as shown in Fig. 10.4. Therefore, it is promising to employ ultrasonication for nano-emulsification of other nutrients.

As substitutes for fossil oil based materials, biodegradable packaging biomaterials have attracted many researchers and entrepreneurs for their great potentials for sustainable development. Starch-based edible films are typic representatives of these developing environmental biomaterials (Cheng et al. 2010). When treated the gelatinized normal corn starch (amylose content of 28%) solution with ultrasound, the resulting films had better transparency, enhanced moisture resistance, and tougher structure (Cheng et al. 2010). However, plasticizers such as glycerol were needed to overcome the decrease in strain at break, which was caused by ultrasonication.

The preparation of V-type amylose inclusion complexes with small bioactive compounds is commonly used in the preparation of encapsulation and sustained releasing products (Tian et al. 2013). For example, cinnamaldehyde (an type of anti-antimicrobial agent) was compounded successfully using high amylose corn starch when treated with ultrasond, and the controlled release of cinnamaldehydein food was beneficial for food preservation (Tian et al. 2013). Ultrasonication is also expected to improve the encapsulation efficiency and loading of other small hydrophobic bioactive molecules for sustained releasing applications. For instance, Liu et al. (2017) prepared the starch nanoparticles self-assembled via a rapid ultrasonication method and achieved the encapsulation of peppermint oil. The encapsulation efficiency of peppermint oil-loaded starch nanoparticles reached 87.7%.

10.6.2 Industrial Applications

10.6.2.1 Preparation of Functional Nanoparticles

Nanoparticles with special properties have been applied in many aspects. The preparation of nanoparticles from biomacromolecules including cellulose and starch has been the focus of both manufacturers and scholars (Bel Haaj et al. 2013; Konwarh et al. 2011). With the aid of ultrasonication, nanoparticles of different sizes (around a few nm to hundreds nm) can be produced from diverse starches.

Nanoparticles made from pinhão (*A. angustifolia*) seeds had a size around 453 nm and showed great potentials to serve as the materials for coating and film formation (Gonçalves et al. 2014). Using waxy and normal corn starches (30–100 nm) as raw materials, starch nanoparticles with smaller size were produced without the involvement of chemicals. The efficiency and yield of starch nanoparticles were enhanced by the assistance of ultrasonication (Bel Haaj et al. 2013). The antimicrobial nanoparticles with size less than 5.4 nm were produced by ultrasonication using the mixture of starch and aqueous extract from leaves of Mesua ferrea Linn (Konwarh et al. 2011). The ultrasonication can be used to produce nanoparticles mainly because of its capacity of damaging the starch granules and scissoring the molecular chains in the amorphous region, which has been mentioned before.

Recently, retrograded starch (RS III) and resistant starch type IV nanoparticles (chemically modified starch, RS IV) with sizes around 600–700 nm were prepared from a novel and convenient synthesis route via employing ultrasonication combined with water in-oil miniemulsion cross-linking technique (Ding and Kan 2016; Ding et al. 2016). It appeared that ultrasonication increased the amylose content and decreased the RS content, but the resulting nanoscale RS III and RS IV showed high adsorption capacities and retained good antidigestibility. Thus, they would show great potentials in biomedical applications and in the development of new medical materials such as the drug-carrier materials. In another study, ultrasonic treatment was utilized to prepare amylose nanoparticles (ANPs) by nanoprecipitation. The

ANPs obtained with ultrasonic treatments were smaller and more uniform than those prepared without ultrasonic treatment, indicating that ultrasonication was a method of higher efficiency and lower cost (Chang et al. 2017).

10.6.2.2 Production of Starch Sugar and Bioethanol

It has been reported that ultrasonication can be employed in liquefaction and saccharification (Jin et al. 2020; Li et al. 2018). For example, ultrasonic pretreatment was used to improve the high-temperature liquefaction of corn starch at high concentrations. The results showed that compared with the control, the viscosity of sonicated starch paste was lower and the reducing sugar content was higher, indicating more efficient liquefaction of starch (Li et al. 2016). It could be mainly attributed to the rough and holey surface, looser granule structure, lower crystallinity, and higher amylose content caused by ultrasonication, leading to the enhancement of enzymatic reaction. The application of ultrasonication during enzymatic hydrolysis can also improve the release rate of glucose, but the experimental conditions need to be optimized in order to avoid the loss of enzyme activity (Nikolić et al. 2010).

Bioethanol has been considered as a promising alternatives of the major energy resources, especially the fossil fuels which are becoming in short supply and unfriendly to the environment. Among various substrates for bioethanol production, starch-based feedstocks are the most commom ones in industries because they are easy to obtain and inexpensive (Nikolić et al. 2010). Ultrasonication has been reported to expedite the saccharification and conversion to ethanol from diverse sources. Usually, the feedstocks containing starch were treated with ultrasound in advance and then they were liquefied and saccharified for fermentation-derived ethanol production (Montalbo-Lomboy et al. 2010; Nikolić et al. 2010; Nitayavardhana et al. 2010; Pejin et al. 2012). It was found that the pretreatment with optimized ultrasonication conditions can not only improve the conversion rate and ethanol yield, but also decrease the cost of the whole processes (Pejin et al. 2012).

10.6.2.3 Waste Recycling

The combination of ultrasonication and microwave was employed in a complex facility to recycle the starch-based industrial waste in sulphuric acid solution (Hernoux-Villière et al. 2013). During the composite treatment, microwave expedited the heat transfer while ultrasonication enforced the mass transfer. Thus, the granules in starch-based industrial waste was disrupted and converted into sugars with a conversion rate of 46% (Hernoux-Villière et al. 2013). More studies about the effect of ultrasonication on starch-based industrial waste are still needed to support the environmentally friendly development.

References

Abbas S, Bashari M, Akhtar W, Li WW, Zhang X (2014) Process optimization of ultrasound-assisted curcumin nanoemulsions stabilized by OSA-modified starch. Ultrason Sonochem 21: 1265–1274

Amini AM, Razavi SMA, Mortazavi SA (2015) Morphological, physicochemical, and viscoelastic properties of sonicated corn starch. Carbohydr Polym 122:282–292

Apar DK, Turhan M, Özbek B (2006) Enzymatic hydrolysis of starch by using a sonifier. Chem Eng Commun 193:1117–1126

Ashogbon AO (2021) The recent development in the syntheses, properties, and applications of triple modification of various starches. Starch 73(3–4):2000125

Awad TS, Moharram HA, Shaltout OE, Asker D, Youssef MM (2012) Applications of ultrasound in analysis, processing and quality control of food: a review. Food Res Int 48:410–427

Azhar A, Hamdy MK (1979) Sonication effect on potato starch and sweet potato powder. J Food Sci 44:801–804

Bartsch M, Schmidt-Naake G (2006) Application of nitroxide-terminated polymers prepared by sonochemical degradation in the synthesis of block copolymers. Macromol Chem Phys 207: 209–215

Bel Haaj S, Magnin A, Pétrier C, Boufi S (2013) Starch nanoparticles formation via high power ultrasonication. Carbohydr Polym 92:1625–1632

Benmoussa M, Hamaker BR (2011) Rapid small-scale starch isolation using a combination of ultrasonic sonication and sucrose density separation. Starch 63:333–339

Bonto AP, Tiozon RN Jr, Sreenivasulu N, Camacho DH (2021) Impact of ultrasonic treatment on rice starch and grain functional properties: a review. Ultrason Sonochem 71:105383

Cameron DK, Wang YJ (2006) Application of protease and high-intensity ultrasound in corn starch isolation from degermed corn flour. Cereal Chem J 83:505–509

Carmona-García R, Bello-Pérez LA, Aguirre-Cruz A, Aparicio-Saguilán A, Hernández-Torres J, Alvarez-Ramirez J (2016) Effect of ultrasonic treatment on the morphological, physicochemical, functional, and rheological properties of starches with different granule size. Starch 68(9–10):972–979

Chan HT, Bhat R, Karim AA (2010) Effects of sodium dodecyl sulphate and sonication treatment on physicochemical properties of starch. Food Chem 120:703–709

Chan CH, Wu RG, Shao YY (2021) The effects of ultrasonic treatment on physicochemical properties and in vitro digestibility of semigelatinized high amylose maize starch. Food Hydrocoll 119:106831

Chang Y, Yan X, Wang Q, Ren L, Tong J, Zhou J (2017) Influence of ultrasonic treatment on formation of amylose nanoparticles prepared by nanoprecipitation. Carbohydr Polym 157: 1413–1418

Chen HM, Huang Q, Fu X, Luo FX (2014) Ultrasonic effect on the octenyl succinate starch synthesis and substitution patterns in starch granules. Food Hydrocoll 35:636–643

Cheng W, Chen J, Liu D, Ye X, Ke F (2010) Impact of ultrasonic treatment on properties of starch film-forming dispersion and the resulting films. Carbohydr Polym 81:707–711

Chong WT, Uthumporn U, Karim AA, Cheng LH (2013) The influence of ultrasound on the degree of oxidation of hypochlorite-oxidized corn starch. LWT Food Sci Technol 50:439–443

Chung KM, Moon TW, Kim H, Chun JK (2002) Physicochemical properties of sonicated mung bean, potato, and rice starches. Cereal Chem 79:631–633

Čížová A, Sroková I, Sasinková V, Malovíková A, Ebringerová A (2008) Carboxymethyl starch octenylsuccinate: microwave- and ultrasound-assisted synthesis and properties. Starch 60:389–397

Czechowska-Biskup R, Rokita B, Lotfy S, Ulanski P, Rosiak JM (2005) Degradation of chitosan and starch by 360-kHz ultrasound. Carbohydr Polym 60:175–184

Daiuto É, Cereda M, Sarmento S, Vilpoux O (2005) Effects of extraction methods on yam (dioscorea alata) starch characteristics. Starch 57:153–160

Degrois M, Gallant D, Baldo P, Guilbot A (1974) The effects of ultrasound on starch grains. Ultrasonics 12:129–131

Dey A, Sit N (2017) Modification of foxtail millet starch by combining physical, chemical and enzymatic methods. Int J Biol Macromol 95:314–320

Dhital S, Brennan C, Gidley MJ (2019) Location and interactions of starches in planta: effects on food and nutritional functionality. Trends Food Sci Technol 93:158–166

Dhull SB, Punia S, Kumar M, Singh S, Singh P (2021) Effect of different modifications (physical and chemical) on morphological, pasting, and rheological properties of black rice (Oryza sativa L. Indica) starch: a comparative study. Starch 73(1–2):2000098

Ding Y, Kan J (2016) Characterization of nanoscale retrograded starch prepared by a sonochemical method. Starch 68:264–273

Ding Y, Zheng J, Xia X, Ren T, Kan J (2016) Preparation and characterization of resistant starch type IV nanoparticles through ultrasonication and miniemulsion cross-linking. Carbohydr Polym 141:151–159

Dzogbefia V, Ofosu G, Oldham J (2008) Physicochemical and pasting properties of cassava starch extracted with the aid of pectin enzymes produced from *Saccharomyces cerevisiae* ATCC52712. Sci Res Essays 3:406–409

Gallant D, Degrois M, Sterling C, Guilbot A (1972) Microscopic effects of ultrasound on the structure of potato starch preliminary study. Starch 24:116–123

Gao W, Lin X, Lin X, Ding J, Huang X, Wu H (2011) Preparation of nano-sized flake carboxymethyl cassava starch under ultrasonic irradiation. Carbohydr Polym 84:1413–1418

Gonçalves PM, Noreña CPZ, da Silveira NP, Brandelli A (2014) Characterization of starch nanoparticles obtained from *Araucaria angustifolia* seeds by acid hydrolysis and ultrasound. LWT Food Sci Technol 58:21–27

Herceg IL, Jambrak AR, Šubarić D, Brnčić M, Brnčić SR, Badanjak M, Tripalo B, Ježek D, Novotni D, Herceg Z (2010) Texture and pasting properties of ultrasonically treated corn starch Czech. J Food Sci 28:83–93

Hernoux-Villière A, Lassi U, Hu T, Paquet A, Rinaldi L, Cravotto G, Molina-Boisseau S, Marais MF, Lévêque JM (2013) Simultaneous microwave/ultrasound-assisted hydrolysis of starch-based industrial waste into reducing sugars. ACS Sustain Chem Eng 1:995–1002

Hoover R (1995) Starch retrogradation. Food Rev Intl 11:331–346

Hu A, Lu J, Zheng J, Sun J, Yang L, Zhang X, Zhang Y, Lin Q (2013) Ultrasonically aided enzymatical effects on the properties and structure of mung bean starch. Innovative Food Sci Emerg Technol 20:146–151

Huang Q, Li L, Fu X (2007) Ultrasound effects on the structure and chemical reactivity of cornstarch granules. Starch 59:371–378

Iida Y, Tuziuti T, Yasui K, Towata A, Kozuka T (2008) Control of viscosity in starch and polysaccharide solutions with ultrasound after gelatinization. Innovative Food Sci Emerg Technol 9:140–146

Isono Y, Kumagai T, Watanabe T (1994) Ultrasonic degradation of waxy rice starch. Biosci Biotechnol Biochem 58:1799–1802

Jackson D, Chotoowen C, Waniska R, Rooney L (1988) Characterization of starch cooked in alkali by aqueous high-performance size-exclusion chromatography. Cereal Chem 65:493–496

Jadwong K, Therdthai N (2018) Effects of ultrasonic and enzymatic treatment on cooking and eating quality of Sao Hai rice. Food Appl Biosci J 6:153–165

Jambrak AR, Herceg Z, Šubarić D, Babić J, Brnčić M, Brnčić SR, Bosiljkov T, Čvek D, Tripalo B, Gelo J (2010) Ultrasound effect on physical properties of corn starch. Carbohydr Polym 79:91–100

Jiang Q, Zhang M, Xu B (2020) Application of ultrasonic technology in postharvested fruits and vegetables storage: a review. Ultrason Sonochem 69:105261

Jin J, Lin H, Yagoub AEA, Xiong S, Xu L, Udenigwe CC (2020) Effects of high power ultrasound on the enzymolysis and structures of sweet potato starch. J Sci Food Agric 100(8):3498–3506

Kang N, Zuo YJ, Hilliou L, Ashokkumar M, Hemar Y (2016) Viscosity and hydrodynamic radius relationship of high-power ultrasound depolymerised starch pastes with different amylose content. Food Hydrocoll 52:183–191

Kardos N, Luche JL (2001) Sonochemistry of carbohydrate compounds. Carbohydr Res 332:115–131

Kaur H, Gill BS (2019) Effect of high-intensity ultrasound treatment on nutritional, rheological and structural properties of starches obtained from different cereals. Int J Biol Macromol 126:367–375

Kerboua K, Hamdaoui O (2018) Numerical investigation of the effect of dual frequency sonication on stable bubble dynamics. Ultrason Sonochem 49:325–332

Knorr D, Zenker M, Heinz V, Lee DU (2004) Applications and ultrasonics in food potential of processing. Trends Food Sci Technol 15:261–266

Konwarh R, Karak N, Sawian CE, Baruah S, Mandal M (2011) Effect of sonication and aging on the templating attribute of starch for "green" silver nanoparticles and their interactions at bio-interface. Carbohydr Polym 83:1245–1252

Kringel DH, El Halal SLM, Zavareze EDR, Dias ARG (2020) Methods for the extraction of roots, tubers, pulses, pseudocereals, and other unconventional starches sources: a review. Starch 72(11–12):1900234

Lan X, Xie S, Wu J, Xie F, Liu X, Wang Z (2016) Thermal and enzymatic degradation induced ultrastructure changes in canna starch: further insights into short-range and long-range structural orders. Food Hydrocoll 58:335–342

Li C (2022) Recent progress in understanding starch gelatinization-an important property determining food quality. Carbohydr Polym 119735:119735

Li M, Li J, Zhu C (2018) Effect of ultrasound pretreatment on enzymolysis and physicochemical properties of corn starch. Int J Biol Macromol 111:848–856

Li C, Liu W, Gu Z, Fang D, Hong Y, Cheng L, Li Z (2016) Ultrasonic pretreatment improves the high-temperature liquefaction of corn starch at high concentrations. Starch 68:1–7

Lima FF, Andrade CT (2010) Effect of melt-processing and ultrasonic treatment on physical properties of high-amylose maize starch. Ultrason Sonochem 17:637–641

Liu C, Li M, Ji N, Liu J, Xiong L, Sun Q (2017) Morphology and characteristics of starch nanoparticles self-assembled via a rapid ultrasonication method for peppermint oil encapsulation. J Agric Food Chem 65(38):8363–8373

Liu J, Yu X, Liu Y (2021) Effect of ultrasound on mill starch and protein in ultrasound-assisted laboratory-scale corn wet-milling. J Cereal Sci 100:103264

Lorimer JP, Mason TJ (1987) Sonochemistry part 1. The physical aspects. Chem Soc Rev 16:239–274

Luo Z, Fu X, He X, Luo F, Gao Q, Yu S (2008) Effect of ultrasonic treatment on the physicochemical properties of maize starches differing in amylose content. Starch 60:646–653

Miles MJ, Morris VJ, Ring SG (1985) Gelation of amylose. Carbohydr Res 135:257–269

Montalbo-Lomboy M, Khanal SK, van Leeuwen J, Raj Raman D, Dunn L Jr, Grewell D (2010) Ultrasonic pretreatment of corn slurry for saccharification: a comparison of batch and continuous systems. Ultrason Sonochem 17:939–946

Nikolić S, Mojović L, Rakin M, Pejin D, Pejin J (2010) Ultrasound-assisted production of bioethanol by simultaneous saccharification and fermentation of corn meal. Food Chem 122: 216–222

Nitayavardhana S, Shrestha P, Rasmussen ML, Lamsal BP, van Leeuwen J, Khanal SK (2010) Ultrasound improved ethanol fermentation from cassava chips in cassava-based ethanol plants. Bioresour Technol 101:2741–2747

Özbek B, Ülgen KÖ (2000) The stability of enzymes after sonication. Process Biochem 35:1037–1043

Park SH, Bean SR, Wilson JD, Schober TJ (2006) Rapid isolation of sorghum and other cereal starches using sonication. Cereal Chem J 83:611–616

Park DJ, Han JA (2016) Quality controlling of brown rice by ultrasound treatment and its effect on isolated starch. Carbohydr Polym 137:30–38

Patist A, Bates D (2008) Ultrasonic innovations in the food industry: from the laboratory to commercial production. Innovative Food Sci Emerg Technol 9:147–154

Pejin DJ, Mojović LV, Pejin JD, Grujić OS, Markov SL, Nikolić SB, Marković MN (2012) Increase in bioethanol production yield from triticale by simultaneous saccharification and fermentation with application of ultrasound. J Chem Technol Biotechnol 87:170–176

Rahaman A, Kumari A, Zeng XA, Farooq MA, Siddique R, Khalifa I, Siddeeg A, Ali M, Manzoor MF, Manzoor MF (2021) Ultrasound based modification and structural-functional analysis of corn and cassava starch. Ultrason Sonochem 80:105795

Riesz P, Kondo T (1992) Free-radical formation induced by ultrasound and its biological implications. Free Radic Biol Med 13:247–270

Sang Y, Bean S, Seib PA, Pedersen J, Shi YC (2008) Structure and functional properties of sorghum starches differing in amylose content. J Agric Food Chem 56:6680–6685

Setyawati YD, Ahsan SF, Ong LK, Soetaredjo FE, Ismadji S, Ju YH (2016) Production of glutinous rice flour from broken rice via ultrasonic assisted extraction of amylose. Food Chem 203:158–164

Shi H, Hu X (2013) Preparation and structure characterization of carboxymethyl corn starch under ultrasonic irradiation. Cereal Chem 90:24–28

Singla M, Sit N (2021) Application of ultrasound in combination with other technologies in food processing: a review. Ultrason Sonochem 73:105506

Sit N, Misra S, Deka SC (2014) Yield and functional properties of taro starch as affected by ultrasound. Food Bioprocess Technol 7:1950–1958

Srichuwong S, Jane J (2007) Physicochemical properties of starch affected by molecular composition and structures: a review. Food Sci Biotechnol 16:663–674

Sujka M, Jamroz J (2013) Ultrasound-treated starch: SEM and TEM imaging, and functional behaviour. Food Hydrocoll 31:413–419

Szent-Györgyi A (1933) Chemical and biological effects of ultrasonic radiation. Nature 131:278

Tian Y, Zhu Y, Bashari M, Hu X, Xu X, Jin Z (2013) Identification and releasing characteristics of high-amylose corn starch-cinnamaldehyde inclusion complex prepared using ultrasound treatment. Carbohydr Polym 91:586–589

Wambura P, Yang W, Wang Y (2008) Power ultrasound enhanced one-step soaking and gelatinization for rough rice parboiling. Int J Food Eng 4:1

Wang D, Hou F, Ma X, Chen W, Yan L, Ding T et al (2020) Study on the mechanism of ultrasound-accelerated enzymatic hydrolysis of starch: analysis of ultrasound effect on different objects. Int J Biol Macromol 148:493–500

Wang L, Wang YJ (2004a) Application of high-intensity ultrasound and surfactants in rice starch isolation. Cereal Chem J 81:140–144

Wang L, Wang YJ (2004b) Rice starch isolation by neutral protease and high-intensity ultrasound. J Cereal Sci 39:291–296

Wang R, Wang F, Kang X, Wang J, Li M, Liu J, Strappe P, Zhou Z (2021) Ultrasonication enhanced the multi-scale structural characteristics of rice starch following short-chain fatty acids acylation. Int J Biol Macromol 190:333–342

Wang L, Wang M, Zhou Y, Wu Y, Ouyang J (2022) Influence of ultrasound and microwave treatments on the structural and thermal properties of normal maize starch and potato starch: a comparative study. Food Chem 377:131990

Wang H, Xu K, Ma Y, Liang Y, Zhang H, Chen L (2020) Impact of ultrasonication on the aggregation structure and physicochemical characteristics of sweet potato starch. Ultrason Sonochem 63:104868

Wu Y, Du X, Ge H, Lv Z (2011) Preparation of microporous starch by glucoamylase and ultrasound. Starch 63:217–225

Yang W, Kong X, Zheng Y, Sun W, Chen S, Liu D, Zhang H, Fang H, Tian J, Ye X (2019) Controlled ultrasound treatments modify the morphology and physical properties of rice starch rather than the fine structure. Ultrason Sonochem 59:104709

Yu S, Zhang Y, Ge Y, Zhang Y, Sun T, Jiao Y, Zheng XQ (2013) Effects of ultrasound processing on the thermal and retrogradation properties of nonwaxy rice starch. J Food Process Eng 36: 793–802

Zhang Z, Feng H, Niu Y, Eckhoff SR (2005) Starch recovery from degermed corn flour and hominy feed using power ultrasound. Cereal Chem J 82:447–449

Zhang Z, Niu Y, Eckhoff SR, Feng H (2005) Sonication enhanced cornstarch separation. Starch 57: 240–245

Zhang B, Xiao Y, Wu X, Luo F, Lin Q, Ding Y (2021) Changes in structural, digestive, and rheological properties of corn, potato, and pea starches as influenced by different ultrasonic treatments. Int J Biol Macromol 185:206–218

Zhang L, Zuo B, Wu P, Wang Y, Gao W (2012) Ultrasound effects on the acetylation of dioscorea starch isolated from *Dioscorea zingiberensis* C.H. Wright. Chem Eng Process Process Intensif 54:29–36

Zheng J, Li Q, Hu A, Yang L, Lu J, Zhang X, Lin Q (2013) Dual-frequency ultrasound effect on structure and properties of sweet potato starch. Starch 65:621–627

Zhu F (2015) Impact of ultrasound on structure, physicochemical properties, modifications, and applications of starch. Trends Food Sci Technol 43(1):1–17

Zhu J, Li L, Chen L, Li X (2012) Study on supramolecular structural changes of ultrasonic treated potato starch granules. Food Hydrocoll 29:116–122

Zhu B, Liu J, Gao W (2016) Process optimization of ultrasound-assisted alcoholic-alkaline treatment for granular cold water swelling starches. Ultrason Sonochem 38:579

Zuo YYJ, Hebraud P, Hemar Y, Ashokkumar M (2012) Quantification of high-power ultrasound induced damage on potato starch granules using light microscopy. Ultrason Sonochem 19: 1126–1126

Zuo JY, Knoerzer K, Mawson R, Kentish S, Ashokkumar M (2009) The pasting properties of sonicated waxy rice starch suspensions. Ultrason Sonochem 16:462–468

Chapter 11
Milling Process of Starch

Yu Tian, Xingxun Liu, Enpeng Li, and Yu Jiang

Abstract Whole cereal grains require mechanical forces to obtain cereal flour of fine particles used for subtle purposes. Although different milling methods increase the application of cereal grains, the physicochemical properties of the starch change due to the changed starch structures including the damaged starch content, starch crystalline structure, and starch molecular structure. This chapter gives detailed information about the effects of milling on starch structure and corresponding changes in the physical and chemical properties of starch and flour including pasting properties, solubility, swelling, and in vitro digestibility. Two cases of wheat and rice, which are widely planted worldwide and used in the food industry, are discussed on how damaged starch affects product quality. Accordingly, the content of damaged starch should be considered as an important parameter for industry users.

Keywords Starch · Milling · Structure · Properties

Y. Tian
College of Food Science and Engineering, Nanjing University of Finance and Economics, Nanjing, Jiangsu, China

Department of Plant and Environmental Sciences, University of Copenhagen, Frederiksberg C, Denmark

X. Liu (✉)
College of Food Science and Engineering, Nanjing University of Finance and Economics, Nanjing, Jiangsu, China

E. Li
Yangzhou University, College of Bioscience and Biotechnology, Yangzhou, China

Y. Jiang
Ardent Mills, Department of Research, Quality and Technical Services, Denver, CO, USA

Z. Sui, X. Kong (eds.), *Physical Modifications of Starch*,
https://doi.org/10.1007/978-981-99-5390-5_11

11.1 Importance of Milling in Starch Industry

Milling is a gradual mechanical process to produce flour from grain. Dry milling is commonly found in producing flour from grains such as wheat, rice, millet, buckwheat, and legume. By applying one or two mechanical forces such as impact, compression, shear, and attritions, the milling equipment can reach milling capacity from several kilograms in a lab to hundred tones in a commercial mill in 1 h, with particle from ultrafine (20 microns) to coarse (hundred microns) sizes. During the dry milling process, flour streams containing endosperm, bran, and germ may be separated, milled, and reconstituted differently to fit the needs for different product applications.

Starch is the main component in cereal grains and a major glycemic carbohydrate in the human diet. Starch is commonly derived from corn, potato, tapioca, and wet milling process is more common than dry milling to extract starch. The grains are steeped, pulped, milled, separated, centrifuged, sieved, dried, and ground to form starch powder. During the process, fiber and protein are mostly isolated earlier than starch and some acid or alkali may be applied to facilitate separation of the components.

Milling is important because different milling equipment and processes lead to products of different particle size profiles and possibly change of physical structure in starch, protein, fiber, and others. Especially for starch that has over 60% content of most grains, structure changes from milling process can largely affect the functional properties of the finished product, either flour or starch. These include pasting profile, thermal properties, swelling properties, solubility, and digestibility. This chapter covers some recent discoveries in how milling technology changes starch profiles in terms of structure and physical properties; how differently milled products impact food application, and importance of quality control in flour milling.

11.2 Milling Equipment

11.2.1 Commercial Mills System

Modern commercial flour milling is a complicated process involving many mill units and each mill unit setting is dependent on different stages or functions of milling process. While roller mill is the principal and basic grinding machine in commercial flour mills, supplementary mills, such as impact and debranning systems, would help reduce the cost due to wear out and high maintenance of roller mill. In general, the grinding process is composed of four systems: break, sizing, reducing, and retailing. Break system opens up the wheat kernel, which contains endosperm, bran, and germ. Out of break rolls that are mostly corrugated, there is a distribution of large middlings with small amount of refined flour and middling too fine to be purified before reduction into flour. Sizing system separates small bran pieces on large pieces of endosperm in order to obtain clean middling. Rolls from sizing system can be either finely corrugated or smooth. Unlike corrugated rolls, smooth rolls produce more flour or fine sizing, while corrugated rolls help detach the bran and clean sizing that contains middlings and bran particles. Both are easily separated in a purifier. Wheat germ usually is separated from sizing stream, in which the germ is attached to large particles of bran. Reduction system reduces the endosperm to flour. After purification, middlings pass through a series of smooth rolls. The rolls are spaced such that each successive reduction produces finer particles. Flour is sifted out after each reduction milling. Most fine bran is removed from the germ in this step. Tailing system recovers small portion of endosperm by reducing their size with mostly smooth rolls or corrugated rolls with low milling pressure. Impact rolls may also be found to reduce the endosperm and to break up the flakes.

11.2.2 Laboratory Mills System

Laboratory setting allows mills to operate under limited space and capacity. The mills deliver less streams and by-product than commercial mills, yet they are convenient to be adjusted by grinding rolls, sieve selection, and flow to materials. Direct impacts of mills at different settings are the degree of starch damage in the flour and variations in flour color from bran contamination. During milling process, flour yield and corresponding ash content of flour are mainly evaluated. There are more types of laboratory mills than commercial mills because milling efficiency, space design, and energy loss are not concerns. The principal forces of grinding are compression, shear, friction/abrasion, and impact. The most common grinding machines are roller mills, attrition mills, impact mills, ball mills, and disk mills. Their applications are mostly found in grain grinding for food or animal feeds.

11.2.2.1 Stone Mill

There are two stones in a stone mill: one is stationary and the other one is rotated by a pulley and a shaft. The force of compression, shear, and abrasion is used in the milling during which wheat is fed into the stationary stone. The grain is ground between the two stones as it is propelled radially to the circumference by furrow on the surface of the stone. The milled flour is discharged by the rotating plate and outlet.

11.2.2.2 Roller Mill

Roll mill is the principal grinding equipment in commercial wheat flour mills. Flour is produced by cutting or squeezing the grain through a series of rollers. The most popular ones are the four-roll or eight-roll roller mills which divide half rolls on either side of the stand. The mills always work in pairs, and the grain material is subject to shear and compressive forces due to corrugations on the roll surfaces and pressure exerted by the rolls when sending particle toward the nip. If the rolls rotate at the same speed, compression is the primary force used. If the rolls rotate at different speeds, shearing and compression are the primary forces used. If the rolls are grooved, a tearing or grinding component is introduced. In commercial mill environment, the rolls that make up a pair are 9–12 inches (23–30.5 cm) in diameter, and their ratio of length to diameter can be as great as 4:1 (Posner 2005). Each pair of rolls is counter-rotating. For improved size reduction one roll rotates faster than the other. This results in a differential in speed between the roll pair. Typical differentials range from 1.2:1 to 2:1 (Glossary: Roller mill 2018) (fast to slow). Milling force/stress on the grain varies to the rate and uniformity of flow of stock to rolls, the roll velocities, ratio of speeds between the fast and slow rolls, the gap between the rolls, the type of the roll surfaces, and hardness of the grain kernels. Advantage of roller mills includes energy efficiency: for coarse reduction of grain, a roller mill may have as high as 85 percent over a hammer mill in terms of throughput/kwh of energy (Glossary: Roller mill 2018), uniform particle size distribution, little noise and dust; disadvantage of roller mills has little or no effect on fiber, particles tend to be irregular in shape, and when required, maintenance can be expensive. Some common brand names for roller mills, such as Buhler, Brabender, and Chopin can be found in a lab or production scale.

11.2.2.3 Pin Mill

Pin mill is a type of disk grinder, also known as centrifugal impact mills. It consists of a feed inlet, a set of two disks with concentric rings of pins (one stationary at the door and one rotating connected to the motor), a narrow grinding chamber, and a product outlet. The finished product out of a pin mill can be as fine as less than 40 microns and with a uniform product size. The main force of grinding is shearing and impact. The material is impacted by a series of hardened steel pins in rotation at high speed and therefore the grinding action is controlled by the air flow, feed rate, and the speed of the motor (the differential speed of the outer two rows of pin).

In the pill mill, the pins are located symmetrically and concentrically. The product particles are shattered/impacted into fragment and progressed toward the outlet, smaller in size as they approach periphery. To avoid temperature increase, cooled inert gas is blasted during grinding. This also prevents accumulation of particles in between the pins and on the disks. The fineness of final product depends on distribution of pins, circumferential speed of the rotors, feed rate, and physical properties of the material. No screen is used in pin mill to control top particle size. The advantages of pin mill are very fine grinding at relatively low energy consumption, little floor space required, and easy cleaning ensures high hygiene standards. Materials that tend to be sticky and clog the hammer mill screen work fine in pin mills. Under cryogenic processing much lower particle sizes can be achieved. The disadvantages include low capacity, narrow range of size reduction, and internal cooling may be required.

11.2.2.4 Hammer Mill

Hammer mill works under the principle that most materials crush, shatter, or pulverize upon impact. Inside the mill, there are free-swinging hammers that are suspended from rods running parallel to the shaft and through the rotor disks. The material is crushed or shattered by a combination of repeated hammer impacts, collisions with the walls of the grinding chamber, and particle on particle impacts. Perforated metal screens or bar grates covering the discharge opening of the mill retain coarse material for further grinding while allowing properly sized materials to pass. There are both laboratory scale hammer mill as well as large high production industrial hammer mills. Fast rotor speed, small screen, large hammer give fine finished particle size and vice versa. Hammer mill is highly effective as well as nonstop continuous pulverizing process with particle size of a wide range, and most of the time it is designed to achieve medium to fine particle sizes (50–70 micron) (Hosakawa 2018). It has compact size and is dust free. However, it requires high energy compared to a roller mill (top speed for motor of the hammer can be over 100 meters per second), generates excessive dust, produces greater particle size variability (less uniform), and is considered as a hazardous operation.

11.2.2.5 Ball Mill

Ball mill is a tumbling mill using spherical balls to reduce the materials to flour inside a cylindrical rotating shell. The material size is decreased by impact and compression when the balls tumble on it in a rotation cylinder. It lost certain popularity because of concerns of high starch damage, low capacity, and need for high maintenance.

11.2.2.6 Cutting/Knife Mill

In a knife mill, there are knife blades inside the steel cylindrical cage and particle size is reduced by shearing forces. It can also be considered as type of hammer mill except it is knife instead of heavy duty hammer, yet it may share the same milling chamber as hammer mill. It is considered as a coarse grinder for milling of biomass such as grass, straw, bran, and corn stover. When material is fed into the cylinder, it is cut by the knife until its size is small enough to pass through the screen. The granulation is controlled by the spacing of the knives on the motor and the liner and the screen applied. It is more often seen as pre-grinding device for substances with a high water, oil, or fat content as well as for grinding dry, soft, and medium-hard products.

11.2.3 Other Mills

In many cases, starch production is based on wet milling. Using corn milling as an example: corn starch production usually includes cleaning, steeping, milling/separating (obtaining germ), milling/sieving (obtaining fiber and starch), centrifugation (obtaining gluten), cleaning, and dry (obtaining more starch). The purpose of corn cleaning is to remove impurities by pre-cleaner and magnet separator. Clean corn is pumped into steeping system and steeped with H_2SO_3. The purpose of steeping is to soften the corn kernels for milling purpose and restrain microorganism growing (germination and fermentation). Corn is transferred into disk mills and broken into 4–6 pieces, where most germ can be separated by germ cyclone separator. A disk mill is a type of crusher, similar as stone mill. It consists of two irons or steel disks/plates. One disk is stationary and the other one rotates (there are mills with two rotating disks, too). Grain feed between the plates is crushed and sheared. Type and spacing of plates determine fineness of the feed. Impact mill is later used for fine milling to remove bound starch from the fiber, breaking starch free from the fiber and reducing the particle size of starch grits. Slurry materials after fine grinding are transferred into pressure arc sieve and coarse starch solution is obtained from this stage. Wet starch went through dehydrating by vacuum filter machine, hydro cyclone, and drying machine.

The above mills described are commonly found in grain milling industry (Fig. 11.1). The principles and operation of the mills can be extended to other manufacturing industries, such as fine chemicals, pharmaceuticals, and food

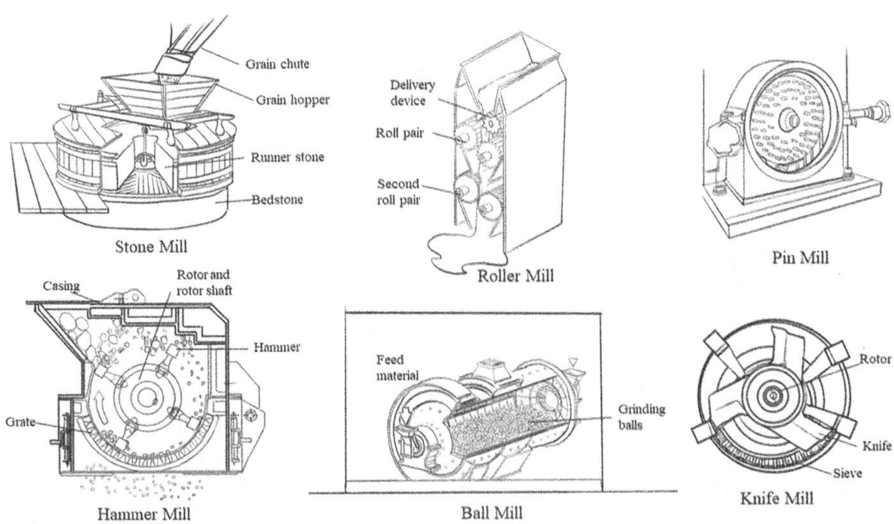

Fig. 11.1 Different types of mills

ingredients other than grain, in which some other mills may also be found, such as air classified mills and fluidized bed jet mills.

11.3 The Multi-Scale Structure Change in Milling Process

Native starch granule is a multi-scale structured polymer from nanometer to micron: amylose and amylopectin chains (~nm), crystalline and amorphous lamellar structure (9–10 nm), alternating amorphous and semi-crystalline growth rings (100–400 nm), and starch granules (<1–100 mm). Amylose and amylopectin are the two basic components in starch granule. Amylopectin (AP) is the main structural contributor to the well-ordered lamellar- and semi-crystalline structures of the starch granule. There are two models developed for its complex structure: "cluster model" and "building block backbone model". The latter revised model is characterized by three different points: (1) The long AP chains are linked to each other and form collectively concentric, not radial, amorphous backbone chain sheets. (2) Branched building blocks formed by 1,6-branch points are attached to the backbone chains and outspread along the amorphous backbone. (3) Short AP chains extend from these branched building block structures and form double helices, which align to form concentric crystalline lamellae (Bertoft 2017). Amylose is the minor, linear but slightly branched molecule, which is primarily located in the amorphous state in the starch granules (Bertoft 2017). However, amylose also possibly participates in crystallinity in high amylose starch by (1) short amylose chains co-crystallize with amylopectin side-chains within the crystalline lamellae; (2) intermediate amylose chains penetrate the crystalline and amorphous lamellae as tie-chains; (3) long amylose chains orient within amorphous lamellae (Zhong et al. 2022a, b) (Fig. 11.2).

11.3.1 Starch Granule Change: Grain and Process Conditions

Damaged starch is starch of which granule structure is broken into smaller fragments, sometimes with starch hila exposed during milling (Dhital et al. 2010; Hasjim et al. 2009). Higher degree of damage to starch granules is related to greater mechanical force and longer grinding time and increases with the severity of grinding process (Nowakowski et al. 1986; Han et al. 2002). It has been reported that milling of cereal grains with higher grain hardness, especially those with higher protein content, results in greater damage to starch granules than the milling of softer grains, due to the greater amount of mechanical energy required to break the structure of harder grains (Hasjim et al. 2009; Williams et al. 1987). The surface of severely damaged starch granules sometimes appears rougher and more porous compared with that of intact native starch granules (Evers et al. 1984; Craig and

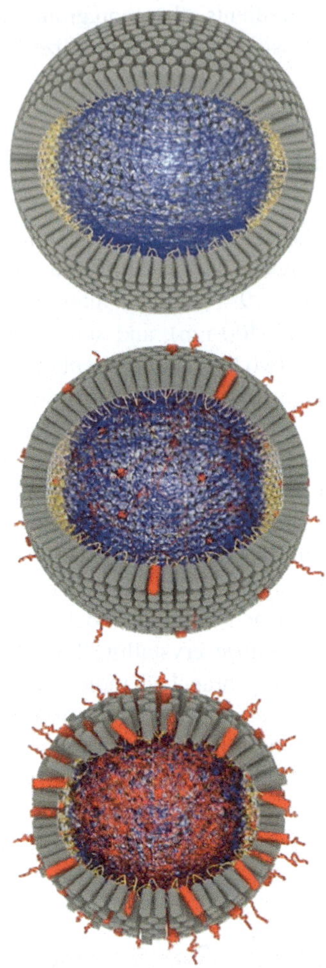

Fig. 11.2 3D schematic representations of a 9–10 nm lamella in a 100% amylopectin (top), normal (15–30% AM; middle) and high amylose starch (>50% AM; bottom) granule represented by the building block backbone model. The gray and red cylinders are the amylopectin and amylose double helices, respectively. The blue lines are amylopectin backbone chains, and the yellow lines are amylopectin connector chains of double helices and backbone chains. The red lines are amylose chains (Zhong et al. 2022a, b). (Reused with permission from *Elsevier Publications*)

Stark 1984; Tamaki et al. 1998). Normally, damaged starch content tends to increase with the decrease in flour particle size (Siliveru et al. 2017), and flour blends consisting of smaller particles result in increased percentage of damaged starch (Monnet et al. 2019). The size of the damaged starch granules, however, does not always decrease with high degree of damage to starch granules. This is due to the agglomeration of damaged starch granules, especially when they are severely damaged.

Grinding conditions including moisture content, temperature, and grinding force also affect the degree of damage to starch granules. Grinding of cereal grains in excess water (also named wet milling) commercially isolates starch granules with minimal damage (Syahariza et al. 2010; Singh et al. 1997), since water molecules can act as a plasticizer increasing elasticity and fracture toughness of the starch granules, minimizing the damage to the starch granules. Even so, the damage (< 5%)

is stilled present in both commercially and laboratory isolated starch granules (Dhital et al. 2010; Hasjim et al. 2009). In the case of dry milling, the damage mostly forms pinholes on granule surface, and channels inside the granule are attributed to hydrolysis by endogenous enzymes in grains instead of by mechanical force. Significantly low starch damage was found on the isolated starch granules which were cryogenically ground with 60% moisture (Dhital et al. 2010), probably because of the ability of water molecules to absorb and to reduce the mechanical energy experienced by the starch granules. Most milling processes are performed at atmospheric dry condition (~15% moisture content), and the temperature of grinder or mill does not exceed 100 °C, keeping the starch granules from gelatinization and heat induced damage. Compared with milling at ambient temperature, cryogenic milling and pin milling of cryogenically frozen sample produce less damage to starch granules due to the low grinding temperature (Nowakowski et al. 1986; Tran et al. 2011). In contrast, wet milling is the only technique that allows starch gelatinization during milling, yet relatively higher temperature (> 80 °C) may be needed to gelatinize starch with limited moisture content, although some starches can be gelatinized at 50 °C in excess water (Jang and Pyun 1996; Burt and Russell 1983). In fact, extrusion, a common method to gelatinize starch or to cook flour at low moisture condition using the combination of shear and heat, requires external energy to keep the heating barrels above 100 °C in addition to the heat generated by the shearing of starch granules or flour (Hasjim and Jane 2009; Liu et al. 2010).

11.3.2 Starch Crystalline Structure

The loss of starch crystalline structure by milling was widely reported for isolated starch granules (Liu et al. 2011; Dhital et al. 2010; Morrison et al. 1994; Tamaki et al. 1998; Chen et al. 2003). At prolonged grinding, the crystalline structure of starch can be completely destroyed, which is confirmed by the absence of defined peaks in the X-ray diffractogram. The loss of crystalline structure of starch granules by milling is accompanied by the loss of double helices (Dhital et al. 2010; Liu et al. 2011; Morrison et al. 1994), implying that the loss of crystallinity is more likely related to the splitting of double helices.

The loss of starch crystalline structure by milling is different from thermal starch gelatinization (Jang and Pyun 1996; Burt and Russell 1983), since most milling treatments (except wet milling) are performed at atmospheric dry condition (~15% moisture content) and the temperature of grinders or mills during grinding is usually below starch gelatinization temperature at limited moisture condition. Isolated starch granules with lower amylose contents experienced greater loss in the degree of crystallinity than those with higher amylose contents or, equivalently, lower amylopectin contents under the same grinding condition (Dhital et al. 2010). It reveals that glycosidic linkages in the double helices of amylopectin are susceptible to the cleavage by grinding force, as anhydro glucose units in the flexible and amorphous conformation are less rigid than that of those in the double helices. This is in accordance with the lower susceptibility of amylose molecules to degradation by milling than amylopectin molecules, which will be discussed in the next section.

11.3.3 Starch Lamellar Structure

Milling led to the structural disorganization of starch, resulting in a reduction in both the perfection and ordering degree of its semi-crystalline structure (Tan et al. 2015; Zhang et al. 2021). The reduced degree of milling-treated starch is influenced by raw materials and milling conditions. Starch with higher content of amylose showed higher resistance to the mechanical disruption during the milling treatment although they have similar hardness with waxy starch (Xu et al. 2018) and the greatest destruction to the amorphous lamellae for high amylose corn starch and greatest destruction to the crystalline lamellae for waxy corn starch (Tan et al. 2015).

The thickness of amorphous and crystalline lamellae of starch was increased for both dry ball-milling and wet-media milling (Li et al. 2020; Liu et al. 2021). The moisture could enter the granules more easily and cause the swelling of the crystalline and amorphous regions during wet milling (Li et al. 2020). As for dry milling, the increase in amorphous lamellae might be attributed to the noncompact packing of amorphous lamellae and the movement of double helices in the crystalline lamellae which lead to an increase in crystalline lamellae (Liu et al. 2021).

11.3.4 Starch Molecular Structure

Starch molecules comprise mainly amylopectin that has a high number of short branches, amylose that has a few long branches, and minor percentage of intermediate components in forms of α-glucan and phytoglycogen. The number of intermediate components can be abundant in some mutant starches, such as high amylose maize starch (Li et al. 2008) and sugary-1 maize starch (Wong et al. 2003).

Amylose and amylopectin molecules can be degraded during the milling of cereal grains and isolated starch granules. Morrison and Tester (Morrison et al. 1994) reported that the degree of molecular degradation increased with grinding time and they found that 41% of native amylopectin molecules in isolated wheat starch granules were converted to low molecular-weight amylopectin when the starch granules were ball-milled for 24 h. Similarly, pin milling also decreased the molecular weight of amylopectin in oat grains from 8.37×10^8 to 4.32×10^8 (Stevenson et al. 2007). Furthermore, Tran, Shelat (Tran et al. 2011) and Dhital, Shrestha (Dhital et al. 2010) observed that the amylopectin molecules in hammer-milled or cryogenically milled rice grains were both degraded to the size similar to amylose molecules. Analysis results of size exclusion chromatography (SEC) showed the cleavage should happen at the inner chains of amylopectin molecules by shearing: if the cleavage occurred at or near the outer branch-chains, molecular size of degraded amylopectin would be smaller than that of amylose as amylopectin branches are much smaller than amylose branches (degree of polymerization of 6–200 vs. 200–50,000, respectively) (Vilaplana and Gilbert 2010). However, it is opposite for waxy rice starch, which showed the degraded outer short branches (DP 6–20) during ball-milling. Besides, the molecular aggregation of amylopectin and amylose of rice starch occurred during prolonged ball-milling time.

Degradation of starch molecules by shearing occurs at α-(1 → 4) glycosidic linkages in double helical crystallites and may also happen at α-(1 → 6) glycosidic linkages/branching points in the amorphous lamellae of starch granule. Breakdown of α-(1 → 4) glycosidic linkages in double helical crystallites is indicated by disappearing of starch crystallinity after milling. Also because of the susceptibility of α-(1 → 4) glycosidic linkages, longer amylopectin branches of potato starch and rice starch were evidently cleaved during cryogenic milling, showing an apparent increased amount of shorter amylopectin branches (Dhital et al. 2010). The α-(1 → 6) glycosidic linkages breakage is evidenced by no apparent change of debranch-chain length distribution before and after milling from many isolated cereal starches, measured by SEC, high performance anion exchange chromatography (HPAEC), and fluorophore-assisted capillary electrophoresis (FACE) with enzymatic debranching treatment (Tran et al. 2011; Dhital et al. 2010; Morrison et al. 1994; Stevenson et al. 2007). This is also explained by likely very small amount of cleaved branch in the amylopectin molecules.

In general, amylose molecule is less susceptible to milling degradation compared with amylopectin molecules, because of the amorphous conformation of amylose molecules in native starch granules and/or the smaller size of amylose molecules. Molecules in amorphous conformation are more flexible than the molecules in double helical crystalline conformation, helping reduce the impact of the mechanical shear during milling. Furthermore, there may be a size range of most stabled polymer, which may be close to the size of amylose molecules, leading to least damage by shear during extrusion, milling, and SEC separation. This is proved by the fact that long amylose molecules can also be cleaved by shear during milling in hammer-milled rice grain and the finding that amylopectin can also be sheared to the size close to amylose molecules during extrusion (Lelievre 1974) and SEC separation (Vandeputte et al. 2003). Less degradation of amylopectin of non-waxy rice starch than waxy rice starch may indicate the protection ability of amylose during milling.

For finished product with similar particle size, cryogenic milling is found to cause less degree of starch damage than hammer-milled grain (Chen et al. 1999). Also milling grain into flour gives less starch damage than milling isolated starch, indicating that the non-starch components in flour, such as protein and certain polysaccharides in flours, may provide structural protection to starch granules during milling.

11.4 Change of Physical Rheological Properties during Milling

11.4.1 Gelatinization Properties

Starch gelatinization is the transition of native semi-crystalline structure in granules into amorphous structure. Differential scanning calorimetry (DSC) heats starch in the presence of water to obtain the enthalpy of gelatinization (ΔH) and the gelatinization temperature. It is a common method to assess the gelatinization property of starch. The gelatinization temperature tells heat stability of the crystallites, which is associated with the heat transfer from the surface of granules to hilum. The gelatinization temperature distribution is normally reported as onset temperature (T_o), peak temperature (T_p), and conclusion temperature (T_c). Due to the heterogeneity of starch granule size and/or shape, the range of starch gelatinization temperature (ΔT), which is the difference between T_o and T_c, can be used to indicate the heterogeneity of starch granules within a sample.

The gelatinization temperature (T_o, T_p, and T_c) of isolated starch granules is decreased after milling, especially after prolonged grinding (Morrison et al. 1994; Tamaki et al. 1998; Chen et al. 2003), without much change in the ΔT. This indicates that the thickness of the remaining crystalline lamellae and/or the length of remaining crystallites is modified by milling. The similar ΔT after milling also indicates that every starch granule is subjected to similar grinding force.

Gelatinization profile of flour is different from the isolated starch. No obvious milling effects on the T_o of flour that is milled from rice and sorghum grains can be observed, although the T_p, T_c, and ΔT are somewhat decreased with increasing severity of milling treatments (Mahasukhonthachat et al. 2010; Marshall 1992). However, for the starch granules isolated from milled rice flours, there was a slight decrease in the T_o, but not significantly altered T_p and T_c (Hasjim et al. 2013). The differences between flour and starch gelatinization profiles indicated that the gelatinization temperature of milled flour is not fully related to starch crystalline structure. Rice flour with larger particle size has a greater barrier for heat transfer. This also indicates that the damage or disruption to starch structures is more heterogenous for the starch granules in coarse milled flour than fine milled flour, and that non-starch components in flour can affect the starch gelatinization temperature.

ΔH is related to the amount of crystallites melted during heating. The reduction of ΔH of isolated starch granules is associated with the increasing severity of milling treatments, such as longer grinding time. The ΔH of starch was found to decrease with the reduction of flour particle size, which is similar to the isolated starch granules (Mahasukhonthachat et al. 2010; Hasjim et al. 2013; Marshall 1992), possibly because of the greater disruption of starch crystalline structure with increasing severity of milling treatments.

11.4.2 Pasting Properties

Pasting properties, commonly measured using Rapid Visco Analyzer (RVA) and Brabender ViscoAmylograph, demonstrates apparent viscosity of starch or flour during cooking in excess water. The standard procedure includes heating a starch or flour suspension to 95 °C, holding at 95 °C for a period of time, cooling to 50 °C, and then holding at 50 °C for a period of time. Pasting temperature is the temperature at which apparent viscosity starts to develop exponentially. Maximum viscosity during heating and holding at 95 °C is called peak viscosity. After reaching peak viscosity, the gel becomes thinner and therefore the difference between peak and trough viscosities is called breakdown. With external temperature going up, the gel then becomes thicker and the viscosity starts to pick up. The difference between final and trough viscosities is called setback indicating the degree of starch retrogradation. The peak viscosity representing the swelling of starch granules is the property of amylopectin, whereas amylose–lipid complex can inhibit granule swelling.

Milling is important for the pasting properties of starch as the pasting viscosity (peak, trough, and final viscosities) decreases with increasing severity of milling treatments, such as increasing grinding time, on isolated starch granules from normal rice, normal maize, normal potato, and high-AM maize (Chen et al. 2003; Devi et al. 2009), which should be ascribed to the fragile surface of broken milled granules that is less resistant to shear force when they get swollen during heating in water. The pasting temperature is also negatively related to the grinding time (Chen et al. 2003; Devi et al. 2009).

The pasting viscosity of flour from rice, sorghum, and maize is reduced greatly by the increasing severity in the milling of cereal grains, similar to the milling of isolated starch granules (Mahasukhonthachat et al. 2010; Hasjim et al. 2013; Becker et al. 2001). In a recent study, the pasting temperature of rice flour was found to be significantly correlated with flour particle size and the final viscosity was significantly correlated with flour particle size and degree of damage to starch granules (Hasjim et al. 2013). It was suggested that the pasting properties of rice flour are less affected by the degradation of starch molecules structure by milling, but rather they are resulted from reduction in flour particle size and the increase in damage to starch granules. A higher peak viscosity was observed from the ball-milled rice than the cryogenic milling rice, and the latter one displayed larger particle size and less damaged starch content. Larger flour particles have greater physical barrier for heat transfer and water diffusion; therefore, it takes time to paste the flour. The viscosity of cold starch paste (final viscosity) can be attributed to the presence of swollen starch granules due to high amount of damaged starch. The degraded amylose chains during ball-milling that caused a lower setback of rice starch were found. However, the specific pasting mechanism, such as the effect of molecular chain length changes on viscosity of milled flours, still needs to be researched.

The presence of non-starch components, such as protein and polysaccharide as cell-wall matrices, plays important roles in the pasting profile of flour. First it may stabilize the starch paste and prevent the rupture of swollen starch granules in flour by shearing during heating. Direct evidence is absence of peak viscosity in RVA curve for large flour particle, in which the non-starch components exists as functional whole parts. Also, a more defined peak viscosity during holding at 95 °C can be obtained under a treatment with cellulase on the flour from aged rice grains (Zhou et al. 2003). Lower pasting viscosity of rice flour is produced by the chemical and enzymatic degradation of proteins, but the pasting properties of isolated starch are not affected by the same treatments (Hamaker and Griffin 1990; Fitzgerald et al. 2003). The native protein and cell-wall structures can be degraded by severe milling treatment, therefore high possibility of weakened pasting viscosity of flour.

11.4.3 Cold-Water Solubility and Swelling

In food application, starch solubility affects the amount of solid mass loss during cooking. Milling of cereal grains (Hasjim et al. 2012; Mahasukhonthachat et al. 2010) and isolated starch granules (Stark and Yin 1986; Dhital et al. 2010; Devi et al. 2009; Tester, Properties of damaged starch granules: composition and swelling properties of maize, rice, pea and potato starch fractions in water at various temperatures, 1997) mostly lead to high cold-water starch solubility caused by damaged starch. The cold-water soluble starch molecules after milling are mostly highly branched similar to amylopectin molecules, but with smaller molecular size similar to amylose molecules (Stark and Yin 1986; Hasjim et al. 2012). The cold-water extract normally contains no or only a small amount of amylose.

Degradation of starch in molecular level may not be the precondition of increased cold-water starch solubility by milling. A recent study showed that using cryogenically milled rice flour (Hasjim et al. 2012), the cold-water starch solubility of flour at 30 °C increased with increasing degree of damage to starch granules although the degradation of starch molecules was not evident, and the cold-water soluble starch was essentially low molecular-weight amylopectin. Not all highly soluble, low molecular-weight amylopectin molecules are the products of the degradation of amylopectin molecules; some might be originally present in the native, undamaged starch granules, such as intermediate components. It is possible that increased relative surface area of starch granules due to fragmentation allows the leaching out of more molecules, and exposure of the inner part of starch granules allows the leaching out of highly soluble molecules located near the loosely packed hilum.

Swelling power is the ability of insoluble starch or flour particles to retain water molecules by forming hydrogen bond between hydroxyl groups in starch and water molecules. In general, the severe milling treatments can increase swelling of cold-water isolated starch granules. Milling causes reduction of the rigidity of amorphous growth rings in the native starch granules and allows larger space for the amorphous growth rings in the inner part of starch granules to expand to form hydrogen bonds with water. Increased surface area in starch molecule by milling also contributes to more swelling. Exception may happen when there is excess degradation of starch molecules, increasing their solubility in water and reducing the amount of swollen starch granules (Tester 1997; Tester and Morrison 1994).

Swelling of flour particles is different than that of isolated starch granules due to presence of non-starch components, such as protein and cell-wall matrices. Besides degree of damage to starch, particle size is another factor to consider because it affects how non-starch components function. For example, decreasing flour particle size increased the cold-water swelling of both hammer-milled and cryogenically milled rice flours (Hasjim et al. 2012) although the two gave different degrees of damaged starch.

11.4.4 Digestibility

The degree of damage to starch granules is commonly measured as the susceptibility of starch granules to rapid enzyme hydrolysis, normally fungal α-amylase (Gibson et al. 1992), since damaged starch granules in isolated starch granules and grain flour have greater enzyme digestibility than intact native starch granules.

Due to fragmentation, relative surface area for enzyme hydrolysis of damaged starch granules is larger than intact native starch granules. Furthermore, the exposure of the inner part of the starch granules, which is normally more susceptible to enzyme hydrolysis than the surface of the granules, is increased by breakage of starch granules, even for intact native starch granules which have smooth surface structure (lack of the pinholes and channels) and high resistance to enzyme hydrolysis, such as normal potato and high amylose maize starches (Dhital et al. 2010).

The decrease in flour particle size also increases the in vitro starch digestibility of flour from the milling of cereal grains, which can be explained by the postulation that the starch digestibility of flour is controlled by the diffusion of enzymes into the flour particles (Mahasukhonthachat et al. 2010; Al-Rabadi et al. 2009). The starch digestion rates were, however, different between barley and sorghum flours with similar particle sizes (Al-Rabadi et al. 2009) and between the sorghum flours from cryogenic milling and hammer milling with similar particle sizes (Mahasukhonthachat et al. 2010). In addition, finer particle size improved the in vivo starch digestibility of wheat grains fed to pig, and the starch digestibility of large particles was higher with the supplementation of enzymes degrading cell-wall polysaccharides (Kim et al. 2005). Therefore, the structures of non-starch components in flour, alterations of these structures by milling, and plant sources can greatly affect starch digestibility.

It is not clear which is the more important factor for starch digestibility, whether it is damage to starch granules or flour particle size. Furthermore, considering that most foods are consumed after cooking, the main nutrition implication for the digestibility of ungelatinized starch granules and uncooked flour is only animal feed. The relationship between the digestibility of milled starch/flour after cooking and its structure has not been extensively studied. Cooking process easily disrupts starch crystalline structures and grain flour structure because of starch gelatinization and the solubilization of starch molecules, proteins, and cell-wall materials. The starch or flour structure is highly swollen and can be easily ruptured even if they still exist after cooking, inducing difficulty in analyzing its size and structures. The dominant factors for starch digestibility are the molecular structures of starch and non-starch components if present, or say, whether the starch granule or flour structure disappears after cooking.

11.5 Case Study

11.5.1 Effect of Wheat Starch Damage in Bread Baking

Wheat starch makes up 67–68% of whole wheat flour content and 78–82% of the refined wheat flour. In yeast leavened bread, wheat starch plays the following roles: (1) dilutes wheat gluten and provides a surface for strong bonding with wheat gluten; (2) produces maltose for fermentation via enzyme hydrolysis; (3) provides flexibility for loaf expansion during partial gelatinization while baking; (4) sets loaf structure by providing a rigid network to prevent loaf from collapsing on cooling; (5) gives structural and textural properties to baked products; (6) holds/losses water and contributes to staling (Maningat et al. 2009). Farrand recognized early in his work that to make a loaf of good volume and texture, gluten must spread far enough to cover the total surface of the starch, and therefore the potential of using damage starch to maximize water absorption is limited by protein/gluten content (Acklin 1997).

The damaged starch granules are left with cracks and fissures and therefore are ready for water penetration. When placed in excess water, intact starch granules absorb about 30% of its weight but damaged starch takes up more water and swells more: water absorption of damaged starch on average is reported slightly differently, ranging from 2% to 4%, probably due to different degrees of damage, wheat hardness, and milling severity (Rakszegi et al. 2010). Besides high water absorption, the damaged starch is highly susceptible to enzyme attack, degrading itself into maltose and dextrins under a combination of α- and β-amylase.

Damaged starch affects many properties of flour and bread process, such as water absorption of flour, dough mixing properties, dough power to produce gas, dough handling properties (stickiness), bread loaf volume, softness of bread crumb, and color of bread crust. For example, the amount of damaged starch present in flour can significantly influence its properties and bread process. Insufficiently damaged starch can result in decreased water absorption, high gelatinization temperature, low bread volume, pale crust color, and dense bread crumb. Conversely, excessively damaged starch can cause similar negative effects on bread volume (such as bread collapse), along with higher water absorption, lower gelatinization temperature, visible blemishes (e.g., more red coloration), and a gummy, rapidly hardening crumb. The optimal level of damaged starch content is crucial for achieving superior flour and bread properties, characterized by low gelatinization temperature, high water absorption, high bread volume, and acceptable crust color. Therefore, it is important to understand and control the amount of damaged starch in the flour to achieve the desired outcomes (Barrera et al. 2007).

Great starch damage increases flour absorption and in most cases, bake absorption, too. It is known that for breadmaking, there is a range of bake absorption for certain flour to reach optimum loaf volume and flour with high starch damage is reported to impair tolerance to variations in absorption. For example, for flour with high starch damage, bread loaf volume decreased more rapidly at lower than optimum absorption and less rapidly at higher than optimum absorption (Farrand 1972).

Relatively high starch damage positively affects the water absorption capacity of flour and may lead to a better dough yield. Farrand derived a relationship between optimum water absorption, protein, starch damage, and moisture content (Farrand 1968).

$$\text{Water absorption} = 1.4P + 0.38D\text{-}[1.6\,M + 0.004D(M + P)] + 57.3 + \text{correction factor}$$

D: damage starch level in Farrand unit.

P: protein content (12% moisture basis).

M: moisture of flour.

In general, gas power was significantly higher in flour with higher starch damage (60 Farrand unit), comparing to flour of lower starch damage (34 Farrand units) (Dexter et al. 1985). Degree of this effect may vary from one situation to another, depending on when and what supplementary sugar source is added to the fermentation process. For example, in sponge and dough method, the initial sponge fermentation is mainly dominated by maltose consumption via breakdowning of damaged starch, and the final dough has a large amount of glucose and fructose produced by sugar inversion after remixing; however, for straight dough system, the fermentation is dominated by consumption of all three sugars, glucose and fructose first and then maltose (Tang et al. 1972). As a result, damaged starch promotes fermentation more readily under sponge and dough process than straight dough process.

However, excessive starch damage deteriorates bread quality. Early works by Tipple and Farrand suggest that in order to make stable dough and a loaf of good volume and texture, the gluten must spread far enough to contain the total surface of starch; failure of gas retention happens when there is insufficient gluten to cover increased surface area caused by increased amount of damaged starch (Tipple 1969; Farrand 1972). Study of flour baking with severe damaged starch levels (ranges from 33 to 79 Farrand units) showed dramatic decrease in specific volume. Meantime, too much starch damage indicates very little starch and very much dextrin left after fermentation. In this case, after being utilized for starch gelatinization, water is left for protein hydration and dextrin-water binding and leads to gumminess texture. There's only a limited amount of starch available to form a gel, which is essential for setting the loaf structure by providing a rigid network.

Optimum starch damage for breadmaking has been studied for years, mostly from the original Farrand's theory and it evolves with changing baking process and analytical tests. Below are commonly found equations for calculation of optimum starch damage.

$$\text{S.D.(in Farrand Unit)} = \text{Protein}^2/6 \tag{11.1}$$

(Farrand 1968)

$$\text{S.D.(in Farrand Unit)} = \text{Protein}^2/5.2\text{, for no-time dough} \tag{11.2}$$

(Acklin 1997)

Application of Eq. 11.2 applies to no-time dough process. For example, Ferrand's study for optimum starch damage used long fermentation process. But in fast bread processes, with short resting time, the effect of starch damage in providing substrate is minimal.

Early study by Farrand (Farrand 1972) stated that satisfactory bread could be made from pin milled flour over a range of 18–30 Farrand units using a straight dough method. The flour protein is 12.2–12.3% and was with lean formulation. A range of 4.5–8.0% (flour dry weight) was recommended for bread baking by Pyler (Pyler and Gorton 2000). According to Chopin Technologies, with appropriate alpha-amylase use and flour protein, it is end-products that determine the range of starch damage. For example, French type (Baguette): 16–20 UCD; pan bread: 19–23 UCD; biscuit/cookie: 14–16 UCD (Chopin 2018).

Controlling the content of damaged starch is developed in the milling industry. Below is a chart that serves as database for relationship between damaged starch content (in UCD unit created by Chopin) in wheat flour and baked products (Fig. 11.3).

11.5.2 Effect of Rice Starch Damage in Gluten-Free Products

The non-wheat diets prepared with rice, buckwheat, and corn are widely popular and recommended for celiac patients due to their beneficial health effects such as reduced risk of diseases and allergic reaction (Heo et al. 2013). Specially, rice has been widely used as a primary gluten-free flour base for scientific and industrial purposes. Rice proteins do not form a stable network structure, so that the viscoelastic quality of rice products depends primarily on the properties of the starch component (Xu et al. 2021; Zhong et al. 2022b).

Dry milling and wet milling are the two major processes used to produce rice flour. Dry grinding involved production of flour under dry condition using various grinding machines. In wet milling, rice is soaked in water, drained, and then being ground with extra water added (Asmeda et al. 2016). Both processes require milling

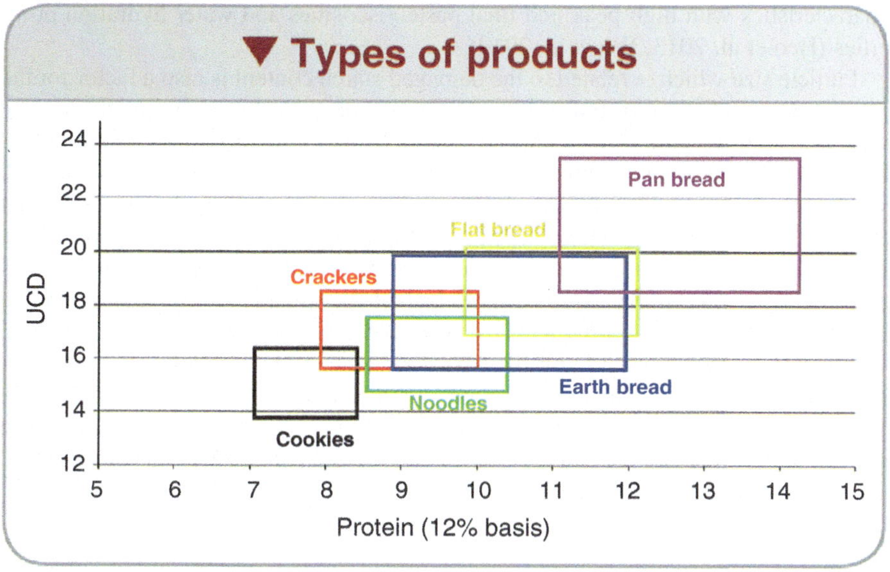

Fig. 11.3 Optimum starch damage for baked products

facilities and operations for flour drying and wastewater handling. There is also semi-dry grinding technique that consists of three steps to grind the grains, which involves soaking, drying to remove excess water, and grinding (Arendt and Zannini 2013; Tong et al. 2015). Different milling methods have been reported to affect the properties of rice flour significantly, especially the particle size of the rice flour, the content of damaged starch, and the state of the starch granules. Dry-milled rice flour, which causes a higher degree of starch damage, exhibited greater water hydration properties than wet-milled and semi-dry milled rice flour at room temperature due to the expanded surface area, favoring the smooth binding with water. However, wet-milled rice flour (less damaged starch) showed a higher value of peak viscosity (Heo et al. 2013), demonstrating its better swelling power upon starch gelatinization, which might be explained that higher damaged starch content facilitates the formation of amylose network that inhibits the granules swelling (Wang et al. 2020). And compared with cryogenic milling which causes less damage starch, semi-dry milling rice flour also exhibited higher peak viscosity, indicating that neither higher nor lower damaged starch are benefit for the peak viscosity of rice flour. Besides, the contributions of non-starch constituents to pasting property of flour should also be taken into account. The percentage of damaged starch granules also had a significant negative relationship with the pasting temperature (Asmeda et al. 2016). It seems that the rice flour produced by wet milling and semi-dry milling which caused suitable damaged starch have more possible commercial applications of rice noodles and gluten-free rice bread from the point of view of sensory and textural characteristics which showed higher quality of whiteness, chewiness, prominent pasting characteristics with high peak and final paste viscosities and water hydration properties (Heo et al. 2013; Wu et al. 2019).

Particle size which is related to the damaged starch content is also a factor for the physicochemical properties of making gluten-free products by increasing the surface area per volume unit. The damaged starch, water binding capacity, solubility, lightness, the final and setback viscosities were increased, crude protein and yellowness were decreased as the particle size of rice flour decreased (Kim and Shin 2014). Kim and Shin (Kim and Shin 2014) studied the effect of the particle size of rice flour on gluten-free rice cupcakes, finding that those prepared with moderately sized (<125 μm) rice flour particles had a better quality. It was also found that rice flour with the medium-particle size (138–165 μm) yielded rice-based vermicelli with better scores for textural profile (Athy and Bhat 1994). Gelatinization temperature was strongly negatively correlated with average particle size of the rice granules (Asmeda et al. 2016). Fine rice flours showed a higher susceptibility to enzymatic hydrolysis and extrusion process increased that effect (Martínez et al. 2014).

11.6 Quality Control in Flour Milling

The quality of starch or flour is very important for both manufacturer and end users. Traditionally, flour yield, color, and ash are considered as main parameters in the milling industry. However, starch granules in cereal grains are damaged during milling. This review provides a better understanding of the effects of milling on starch structure and corresponding changes in physical and chemical properties of starch and flour, demonstrating why the damaged starch content should be developed as an important parameter for industry user.

Different Methods were used to measure damaged starch, including extraction procedures ("Blue Value"), dye-staining procedures, NIR procedures, and enzyme digestion procedures. Of these, the enzyme digestion procedures are generally preferred for its accuracy. Starch damage assay kit from Megazyme is a method adjusted from AACC Method 76–31.01. Cereal and fungal α-amylase preparations have been employed as crude malt extracts, or crude fermentation broths. The degree of hydrolysis has traditionally been measured using non-specific reducing-sugar methods such as ferricyanide titration, or reaction with dinitrosalicylic acid (DNS). This method is accurate in determining the damaged starch content, because it is a direct method while blue value and others may be dependent on the max wavelength absorbance of amylose, which varies in plant sources and intermediate components in starch.

11.7 Conclusions

This chapter describes common milling process of starch and flour from grains, the multi-scale change ranging from starch granules to molecular structure during grain or starch milling, and the milling effects on physical and rheological properties of finished products. Using wheat flour and rice milling as examples, this chapter also advocates starch damage as specification for quality control in flour milling industry.

Milling of cereal grains causes damage to starch granules, or even disruption of starch crystalline structure and degradation of starch molecules. Amylose is mostly in amorphous conformation in native starch granules. It is flexible and thus less affected by milling, compared to amylopectin double helices with rigid crystallinities. The grinding of isolated starch granules, however, does not replicate the milling of cereal grains, because the effects on starch structure are more heterogeneous in flour than isolated starch granules. Non-starch components in flour particles may protect starch from grinding forces. Among all milling techniques, cryogenic milling of cereal grains produces the least damage to starch granules and least degradation of starch molecules. Starch gelatinization properties are highly related to starch crystalline structure; however, non-starch components in flour particles can affect starch gelatinization profiles. The pasting properties of flour are highly governed by flour particle size and by the degree of damage to starch granules.

The starch solubility of flour is more strongly related to the damage to starch granules than flour particle size and the degradation of starch molecules, indicating that molecular degradation is not the precondition of increased starch solubility by milling. The cold-water swelling of flour shows high correlations with the damage to starch granules and flour particle size and weak correlations with starch molecular structures. The starch digestibility of raw flour is greatly affected by flour particle size and damage to starch granules; however, it is not well understood if the same applies to the digestibility of gelatinized starch in cooked flour or flour applications.

References

Acklin T (1997) Technical profile: starch damage and British baking practices. World Grain, New York

Al-Rabadi G, Gilbert R, Gidley M (2009) Effect of particle size on kinetics of starch digestion in milled barley and sorghum grains by porcine alpha-amylase. J Cereal Sci 50(2):198–204

Arendt EK, Zannini E (2013) Cereal grains for the food and beverage industries. Elsevier, Amsterdam

Asmeda R, Noorlaila A, Norziah M (2016) Relationships of damaged starch granules and particle size distribution with pasting and thermal profiles of milled MR263 rice flour. Food Chem 191: 45–51

Athy JHV, Bhat K (1994) Effect of particle size on viscoamylographic behaviour of rice flour and vermicelli quality. J Texture Stud 25(4):469–476

Barrera GN, Pérez GT, Ribotta PD, León AE (2007) Influence of damaged starch on cookie and bread-making quality. Eur Food Res Technol 225:1–7

Becker A, Hill S, Mitchell J (2001) Milling—a further parameter affecting the rapid visco analyser (RVA) profile. Cereal Chem 78(2):166–172

Bertoft E (2017) Understanding starch structure: recent progress. Agronomy 7(3):56

Burt D, Russell P (1983) Gelatinization of low water content wheat starch—water mixtures. A combined study by differential scanning calorimetry and light microscopy. Starch 35(10): 354–360

Chen J, Lii C, Lu S (2003) Effect of particle size on physicochemical and morphological analyses on damaged Rice starches. J Food Drug Anal 11(4):283–289

Chen J, Lu S, Lii C (1999) Effects of milling on the physicochemical characteristics of waxy rice in Taiwan. Cereal Chem 76(5):796–799

Chopin (2018). http://www.chopin.fr/media/applications/na23_estimating_the_optimum_ds_level_gb.pdf

Craig S, Stark J (1984) Molecular properties of physically-damaged sorghum starch granules. J Cereal Sci 2(3):203–211

Devi A et al (2009) Physical properties of cryomilled rice starch. J Cereal Sci 49(2):278–284

Dexter J, Preston K et al (1985) Relationship of flour starch damage and flour protein to the quality of Brazilian-style hearth bread and remix pan bread produced from hard red spring wheat. Cereal Foods World 30:511–514

Dhital S, Shrestha A, Gidley M (2010) Effects of cryo-milling on starches: functionality and digestibility. Food Hydrocoll 24(2–3):152–163

Evers A, Baker G, Stevens D (1984) Production and measurement of starch damage in flour. Part 2. Damage produced by unconventional methods, 1984. Starch 36(10):350–355

Farrand A (1968) Starch damage and alpha-amylase as bases for mathematical models. Relating to flour water-absorption. Cereal Chem 46:103–116

Farrand A (1972) Controlled levels of starch damage in a commercial United Kingdom bread flour and effects on absorption, sedimentation value, and loaf quality. Cereal Chem 49(4):479–483

Fitzgerald M et al (2003) Viscosity of rice flour: a rheological and biological study. J Agric Food Chem 51(8):2295–2299

Gibson TS, Al Qalla H, McCleary BV (1992) An improved enzymic method for the measurement of starch damage in wheat flour. J Cereal Sci 15:15–27

Glossary: Roller mill (2018). http://www.feedmachinery.com/glossary/equipment/roller_mills/

Hamaker B, Griffin V (1990) Changing the viscoelastic properties of cooked rice through protein disruption. Cereal Chem 67(3):261–264

Han X-Z et al (2002) Consequence of starch damage on rheological properties of maize starch pastes. Cereal Chem 79(6):897–901

Hasjim J, Jane J (2009) Production of resistant starch by extrusion cooking of acid-modified Normal-maize starch. J Food Sci 74(7):C556–C562

Hasjim J, Li E, Dhital S (2012) Milling of rice grains: the roles of starch structures in the solubility and swelling properties of rice flour. Starch 64(8):631–645

Hasjim J, Li E, Dhital S (2013) Milling of rice grains: effects of starch/flour structures on gelatinization and pasting properties. Carbohydr Polym 92(1):682–690

Hasjim J et al (2009) Kernel composition, starch structure, and enzyme digestibility of opaque-2 maize and quality protein maize. J Agric Food Chem 57(5):2049–2055

Heo S, Lee SM, Shim J-H, Yoo S-H, Lee S (2013) Effect of dry-and wet-milled rice flours on the quality attributes of gluten-free dough and noodles. J Food Eng 116(1):213–217

Hosakawa (2018) Hosokawa micron powder systems. Mikro-Pulverizer-Hammer-Screen-Mill. http://mikropulverizer.com/wp-content/uploads/2016/04/Mikro-Pulverizer-Hammer-Screen-Mill.pdf

Jang J, Pyun Y (1996) Effect of moisture content on the melting of wheat starch. Starch 48(2):48–51

Kim J, Mullan B, Pluske J (2005) A comparison of waxy versus non-waxy wheats in diets for weaner pigs: effects of particle size, enzyme supplementation, and collection day on total tract apparent digestibility and pig performance. Anim Feed Sci Technol 120(1–2):51–65

Kim J-M, Shin M (2014) Effects of particle size distributions of rice flour on the quality of gluten-free rice cupcakes. LWT 59(1):526–532

Lelievre J (1974) Starch damage. Starch 26(3):85–88

Li X, Fan M, Huang Q, Zhao S, Xiong S, Zhang B, Yin T (2020) Effect of wet-media milling on the physicochemical properties of tapioca starch and their relationship with the texture of myofibrillar protein gel. Food Hydrocoll 109:106082

Li L et al (2008) Characterization of maize amylose-extender (ae) mutant starches. Part I: relationship between resistant starch contents and molecular structures. Carbohydr Polym 74(3): 396–404

Liu W, Halley P, Gilbert R (2010) Mechanism of degradation of starch, a highly branched polymer, during extrusion. Macromolecules 43(6):2855–2864

Liu C, Jiang Y, Liu J, Li K, Li J (2021) Insights into the multiscale structure and pasting properties of ball-milled waxy maize and waxy rice starches. Int J Biol Macromol 168:205–214

Liu T et al (2011) The effect of ball milling treatment on structure and porosity of maize starch granule. Innov Food Sci Emerg Technol 12(4):586–593

Mahasukhonthachat K, Sopade P, Gidley M (2010) Kinetics of starch digestion in sorghum as affected by particle size. J Food Eng 96(1):18–28

Maningat C, Seib P, Bassi S, Woo K (2009) Wheat starch: production, properties, modification and uses. In: BeMiller J, Whistler R (eds) Starch chemistry and technology, 3rd edn. Academic Press, Cambridge, MA, pp 481–482

Marshall W (1992) Effect of degree of milling of brown rice and particle size of milled rice on starch gelatinization. Cereal Chem 69(6):632–632

Martínez MM, Calviño A, Rosell CM, Gómez M (2014) Effect of different extrusion treatments and particle size distribution on the physicochemical properties of rice flour. Food Bioprocess Technol 7(9):2657–2665

Monnet AF, Eurieult A, Berland S, Almeida G, Jeuffroy MH, Michon C (2019) Damaged starch in pea versus wheat flours: fragmentation behavior and contribution of fine and coarse fractions. Cereal Chem 96(3):465–477

Morrison W, Tester R, Gidley M (1994) Properties of damaged starch granules. II. Crystallinity, molecular order and gelatinisation of ball-milled starches. J Cereal Sci 19(3):209–217

Nowakowski D, Sosulski F, Hoover R (1986) The effect of pin and attrition milling on starch damage in hard wheat flours. Starch 38(8):253–258

Posner ES (2005) Wheat flour milling. In: Wheat: chemistry and technology, 4th edn. Elsevier, Amsterdam, pp 119–152

Pyler E, Gorton L (2000) Baking science and technology, vol 2, 4th edn. Sosland Publishing Co, Kansas City

Rakszegi M et al (2010) Effect of milling on the starch properties of winter wheat genotypes. Starch 62(2):115–122

Siliveru K, Ambrose RK, Vadlani PV (2017) Significance of composition and particle size on the shear flow properties of wheat flour. J Sci Food Agric 97(8):2300–2306

Singh S et al (1997) Comparison of laboratory and pilot-plant corn wet-milling procedures. Cereal Chem 74(1):40–48

Stark J, Yin X (1986) The effect of physical damage on large and small barley starch granules. Starch 38(11):369–374

Stevenson D, Jane J, Inglett G (2007) Structure and physicochemical properties of starches from sieve fractions of oat flour compared with whole and pin-milled flour. Cereal Chem 84(6): 533–539

Syahariza Z, Li E, Hasjim J (2010) Extraction and dissolution of starch from rice and sorghum grains for accurate structural analysis. Carbohydr Polym 82(1):14–20

Tamaki S et al (1998) Structural change of maize starch granules by ball-mill treatment. Starch 50(8):342–348

Tan X, Zhang B, Chen L, Li X, Li L, Xie F (2015) Effect of planetary ball-milling on multi-scale structures and pasting properties of waxy and high-amylose cornstarches. Innovative Food Sci Emerg Technol 30:198–207

Tang R, Robinson R, Hurley W (1972) Quantitative changes in various sugar concentrations during breadmaking. Bakers Digest 46:48

Tester R (1997) Properties of damaged starch granules: composition and swelling properties of maize, rice, pea and potato starch fractions in water at various temperatures. Food Hydrocoll 11(3):293–301

Tester R, Morrison W (1994) Properties of damaged starch granules. V. Composition and swelling of fractions of wheat starch in water at various temperatures. J Cereal Sci 20(2):175–181

Tipple K (1969) The relation of starch damage to the baking performance of flour. Baker's Digest 44:28–33

Tong L-T, Gao X, Lin L, Liu Y, Zhong K, Liu L, Zhou X, Wang L, Zhou S (2015) Effects of semidry flour milling on the quality attributes of rice flour and rice noodles in China. J Cereal Sci 62:45–49

Tran T et al (2011) Milling of Rice grains. The degradation on three structural levels of starch in Rice flour can be independently controlled during grinding. J Agric Food Chem 59(8): 3964–3973

Vandeputte G et al (2003) Rice starches. II. Structural aspects provide insight into swelling and pasting properties. J Cereal Sci 38(1):53–59

Vilaplana F, Gilbert R (2010) Two-dimensional size/branch length distributions of a branched polymer. Macromolecules 43(17):7321–7329

Wang Q, Li L, Zheng X (2020) A review of milling damaged starch: generation, measurement, functionality and its effect on starch-based food systems. Food Chem 315:126267

Williams P et al (1987) Measuring wheat hardness by revolutions per minute reduction. Cereal Chem 64(6):422–427

Wong K-S et al (2003) Structures and properties of amylopectin and phytoglycogen in the endosperm of sugary-1 mutants of rice. J Cereal Sci 37(2):139–149

Wu T, Wang L, Li Y, Qian H, Liu L, Tong L, Zhou X, Wang L, Zhou S (2019) Effect of milling methods on the properties of rice flour and gluten-free rice bread. LWT 108:137–144

Xu J, Li Z, Zhong Y, Zhou Q, Lv Q, Chen L, Blennow A, Liu X (2021) The effects of molecular fine structure on rice starch granule gelatinization dynamics as investigated by in situ small-angle X-ray scattering. Food Hydrocoll 121:107014

Xu B, Mense A, Ambrose K, Graybosch R, Shi YC (2018) Milling performance of waxy wheat and wild-type wheat using two laboratory milling methods. Cereal Chem 95(5):708–719

Zhang B, Yuan Z, Qiao D, Zhao S, Lin Q, Xie F (2021) Wet ball milling of Indica Rice starch effectively modifies its multilevel structures and pasting behavior. ACS Food Sci Technol 1(4): 636–643

Zhong Y, Qu J, Li Z, Tian Y, Zhu F, Blennow A, Liu X (2022b) Rice starch multi-level structure and functional relationships. Carbohydr Polym 275:118777

Zhong Y, Tai L, Blennow A, Ding L, Herburger K, Qu J, Xin A, Guo D, Hebelstrup KH, Liu X (2022a) High-amylose starch: structure, functionality and applications. Crit Rev Food Sci Nutr:1–23

Zhou Z et al (2003) Effect of rice storage on pasting properties of rice flour. Food Res Int 36(6): 625–634

Chapter 12
Dry Heat Modification of Starch

Qingjie Sun

Abstract This chapter introduces the preparation methods of dry heating modified starch, including dry heating treatment (DHT) combined with or without food additives, DHT with other physical treatments such as annealing, ultrasonication, and microwave radiation. DHT plays a significant role in improving the physicochemical properties of starch. The influencing factors of DHT, such as starch types, treatment temperature, treatment times, additives, were summarized. In addition, the effects of DHT on the physicochemical properties of starch were described in detail. The DHT does not destroy the integrity of starch granules, but some dents and cracks appear on the surface of starch granules. After DHT, the crystalline structure of starch does not change, but it usually leads to the partial dissociation of the double helix structure in the crystalline region, resulting in the reduction of short-range molecular order. At the same time, freeze-thaw stability and rheological properties of dry heated starch could be improved to a certain extent. DHT modified starch is widely used in food stabilizer, baked food, functional food, 3D printing of food materials, and other fields because of its green and clean characteristics.

Keywords Starch · Dry heating · Modification · Properties

12.1 Introduction

Starch is a kind of biopolymer that exists widely in nature. It is easy to find in all kinds of plant sources, such as wheat, rice, corn, potato, and cassava. The molecular structure plays a vital role in the physicochemical properties of starch, including the proportion of amylose and amylopectin, molecular weight, and the distribution of chain length (Zou et al. 2019). Starch has various applications in food and several industrial commodities (Zou et al. 2019). However, because of the poor stability of native starch in the traditional heating and shearing process, especially in the presence of acid, the utilization of starch is limited. In addition, native starch is

---

Q. Sun (✉)
College of Food Science and Engineering, Qingdao Agricultural University, Qingdao, China

Z. Sui, X. Kong (eds.), *Physical Modifications of Starch*,
https://doi.org/10.1007/978-981-99-5390-5_12

easy to regenerate after gelatinization. Therefore, native starch is often modified to improve its properties using chemical methods, physical methods, and enzyme methods. In general, chemically modified starch is the most common modification method that can be used in the manufacture of various foods, such as those requiring mechanical processing resistance and heat resistance (Zia-ud-Din et al. 2015).

However, in the past few decades, consumer concerns about application of chemically modified starch in food are limited by the chemical substances remaining in food. As a substitute for chemically modified starch, based on consumer demand, physically modified starch has occupied an important position in starch modification. Furthermore, the application of physically modified starch in food processing can represent a "clean label," because no chemical agents are added to these starches to keep their natural characteristics. In contrast, those starches modified by chemical agents are called "modified food starch."

Dry heat method is a simple, safe, and environment-friendly physically modification method of starch (Oh et al. 2018a, b). In the process of preparing modified starch by dry heat process, water content in starch should be reduced to less than 10% to inhibit starch gelatinization, and the whole system is usually heated to 110–150 °C. Although the starch chain may rearrange in the starch granules during the heat treatment, the morphology and crystallinity of starch can be kept unchanged. The dry heated starch is considered to be functionally equivalent to chemically crosslinked starch.

12.2 Method of Dry Heat Modification of Starch

Dry heat treatment (DHT) is a method to improve the properties of starch by heat treatment at the temperature of 110–150 °C at low moisture content (≤10%). This physical modification method is similar to the chemical crosslinking method in improving the properties of starch. Compared with chemical modification method, this method has the advantages of simple production operation, no pollution, and high yield. It is a promising green method.

Several studies have investigated the effects of DHT on different kinds of starch, such as rice (Li et al. 2008), cassava (Chandanasree et al. 2016), and corn (Ji et al. 2016). In the case of continuous DHT, starch can quickly reach the degree of modification and the balance point of various properties. However, when the modification of starch reaches the equilibrium state, it will have a better effect if it is cooled to room temperature and then heated several times. Gou et al. (2019) reported on repeated DHT, where starch samples were heated at 130 °C for 3 h, then cooled at 25 °C for 60 min (1 cycle), and the process was recovered 6 times. The results showed that the original A-type crystallinity of sweet potato starch remained unchanged after DHT. The gelatinization temperature of the native starch is 69.85 °C, and the gelatinization temperature of the two dry heat methods is higher. The highest gelatinization temperature is 71.47 °C for the sample after continuous DHT for 18 h. The peak temperature (77.88 °C) of modified starch is also higher than that of the native starch (76.12 °C). While the native starch ΔH is 12.81 J/g, in the starch treated by the two dry heat methods, after repeated DHT for 1 h, ΔH reached the maximum value of 15.29 J/g, and then decreased as the number of repetitions increased, while ΔH of starch modified by continuous dry heating should always be less than that of native starch. Compared with the native starch, the swelling power of the starch treated by the two dry heat methods is lower. In addition, after the same time treatment, the paste viscosity and digestibility of starch samples treated with continuous DHT were lower than those treated with repeated DHT. The resistant starch content of the starch treated by the two dry heat methods was less than that of the native starch. Moreover, compared with repeated DHT, continuous DHT had a great effect on the surface structure of starch granules. In general, it can be concluded that repeated DHT has a great impact on the structure, physicochemical, and digestive properties of sweet potato starch compared with continuous DHT.

12.3 Combined Methods of Dry Heat Modification

12.3.1 Annealing

Annealing is an important physical method to enhance the thermal stability of starch through simple hydrothermal treatment (Guo et al. 2015). Annealing is a method to treat starch with high moisture content (> 40%) and moderate temperature lower than gelatinization temperature. Unfortunately, the rigidity of starch chain in natural granules greatly limits the annealing efficiency of starch modification. Importantly, DHT may improve the flexibility of starch chains, because starch is degraded after high temperature treatment. If the starch treated with dry heat is annealed, the starch with higher flexibility/deformation is expected to be better reorganized in the annealing process. Therefore, the multi-scale structure of starch can be reorganized, so as to reduce the enzymatic digestion of starch. Synergistic modification may be beneficial to the rearrangement of starch during annealing, so as to significantly reduce the digestibility of starch.

Chi et al. (2019) reported starch (water content less than 10%) was heated at 130 °C for 2 h and then cooled to room temperature and then annealed in a 50 °C water bath. Compared with single dry heat modified starch, the gelatinization temperature of starch treated by dry heating combined annealing was 68.64 °C, which was higher than the gelatinization temperature of starch treated by single annealing (67.82 °C). The ordered structure of starch after dry heat and annealing treatment is increased compared to that of original starch. The ordered structure of potato starch after dry heat and annealing treatment is 6.86 nm, while that of original starch is 6.62 nm. The results showed that the synergistic modification changes the thickness of starch layer and increases the dual helix sequence. In addition, the enzyme digestibility of starch treated by dry heating combined with annealing was the lowest. Compared with native starch, the slow digested starch of corn starch treated by dry heating combined annealing increased by about 7.90%, and the resistant starch of potato starch increased by about 5.04%.

12.3.2 Ultrasonic Treatment

The combination of DHT and ultrasonic treatment is conducive to the expansion of starch particles, which improves starch peak viscosity. Gao et al. (2021) described combination of DHT and ultrasonic treatment can improve the freeze-thaw stability of potato starch, and the retrogradation trend became weaker (412 cP), smaller than 720 cP of the native starch. According to the report, ultrasound may not destroy the crystal structure and characteristic diffraction peak intensity of corn starch.

12.3.3 Microwave Radiation

Microwave radiation technology is a modification technology, which is a nonionizing physical technology. Gao et al. (2021) described combination of DHT and microwave treatment decreased the freeze-thaw stability of potato starch, and the retrogradation trend is stronger with setback value of 739 cP, more than 720 cP of the original starch.

12.3.4 Ozone Treatment

As a starch modification technology, ozone treatment has been investigated for different starch sources. Lima et al. (2020) placed cassava starch suspension in a glass reactor and treated with ozone for 30 min, then dry heated. Another method is to treat starch with dry heating followed by ozone treatment. After DHT followed ozone treatment, starch granules had cracks on the surface, while this phenomenon was not observed in the starch modified by ozone treatment followed DHT. The hydrogel formed by DHT and ozone modified starch showed more resistant to rupture than single DHT modified starch and native starch hydrogels. The dual modified starch hydrogel was about 6.6 times stronger than the native starch gel (based on the force acting on the gel). Conversely, ozone and DHT modified starch formed a weak gel. Generally speaking, ozone treatment is a process that contributes more to the oxidation properties (oxidation and depolymerization degree) of starch.

12.3.5 Cold Plasma

As a no heat treatment technology, cold plasma (CP) attracts great interest in starch modification (Han et al. 2020). Ge et al. (2021) reported the effects of DHT (130 °C, 1, 3, and 9 h) and cold plasma (40 V, 1, 5, and 10 min) on the physicochemical properties of red bean starch. The results showed that the morphology of DHT or cold plasma changed little, and the diffraction pattern of starch remained basically unchanged. Under the dual modification of the two treatments, various properties changed to a certain extent, such as the content of amylose, crystallinity, molecular weight, long chain of amylopectin, enthalpy change value, swelling, and so on. T_o, T_P, T_C, and ΔH of native starch were 50.78, 55.01, 65.54 °C, and 13.52 J/g, respectively. After dual treatments, T_o (55.94–57.16 °C), T_P (58.29–59.55 °C), and Tc (66.03–66.27 °C) of modified starch increased, and the Δ H (13.18–12.03 J/g) was decreased. The increase in gelatinization temperatures may be due to the enhancement of the interaction between amylose and amylopectin and the formation of more stable crystals. DHT treatment may induce the rearrangement of starch molecules and raise the interplay between starch chains, which leads to the reduction of the content of the double helix and the improvement of the integrity of the dual helix. Compared with single treatment, the dual modification has deeper changes. The apparent amylose content of natural adzuki bean starch was 24.33%. The amylose content of DHT (23.47–16.25%) and CP (23.22–16.93%) starch was lower than that of native starch. In particular, it was found that the amylose content of dual modified starch was between 18.90 and 10.38%, which was lower than that of native and single treated starch. Starch treated by dry heating and plasma is a kind of green starch, which can improve the structure of starch and improve some properties of starch.

12.4 Factors Influencing Dry Heat Treatment

12.4.1 Types of Starch

Normal and waxy wheat starches were investigated when subjected to DHT. After DHT, waxy wheat starch exhibited higher solubility, swelling power, and crystallinity than normal wheat starch. Moreover, waxy wheat starch showed decreased gelatinization enthalpy while normal wheat starch had increased ones after DHT. Therefore, wheat starches of different amylose content exhibited varied structural and physicochemical characteristics before and after DHT (Zhang et al. 2021a, b).

In contrary to other heat treatment modifications, treatment at 180 °C, 10 min for rice and 170 °C, 20 min for barnyard millet showed a higher increment in RS content. As a method of DHT, convective microwave heating had increased the amount of RS in rice and barnyard millet by 22% and 14%, respectively. The formation of RS in rice and barnyard millet differs in terms of DHT conditions (Kanagaraj et al. 2019).

DHT transformed the X-ray pattern of potato starches from B to B + A and the solubility and swelling power were enhanced in varying degrees. Potato starches showed a decrease in the pasting temperature while an increase in peak viscosity after treatment (Liu et al. 2019). While applying DHT to proso millet starch, all treated samples were of the classic "A" type. DHT exhibited a decrease in both pasting temperature and peak viscosity while an increase in the final viscosity to proso millet starch. Compared with native sample, treated proso millet starch exhibited more compact gel structures and plumper particles, which was useful for modifying its pasting and structural properties (Sun et al. 2014). Consequently, studies mentioned above demonstrated that DHT could make a difference in the improvement of starch qualities from different sources.

12.4.2 Temperatures and Times of Treatment

Different DHT conditions could affect the physical and digestible properties of starch. For example, high amylose rice starch was subjected to different temperatures (110, 130, and 150 °C) and periods (0, 1, 2, and 4 h). With the increasing in temperature and time, the gelatinization temperature and enthalpy change decreased, which indicated the change in semi-crystalline region. Moreover, DHT modified rice starch above 130 °C have increased gel strength and pasting viscosity whereas exhibited decreases as the heating time increased. DHT starch gel had lowest predicted glycemic index when treated at 130 °C for 1 h. Results demonstrated that the in vitro starch digestibility was negatively correlated with the treating temperature. Therefore, DHT could enhance the physical and nutritional properties of high amylose rice starch (Im et al. 2018).

DHT was performed in cassava starch under condition at 130 °C for 2 and 4 h to evaluate cassava starch properties, for the application of hydrogel in 3D printing. Different gelatinization conditions (65, 75, 85, and 95 °C) were conducted after DHT process. With the increase in DHT time, DHT starches showed higher carbonyl content and bigger granule size whereas lower water absorption index and molecular size. Furthermore, DHT-4 h exhibited the lowest peak apparent viscosity and the strongest gels under all the evaluated conditions which resulted in a better printability. Therefore, DHT might expand the potential of cassava starch for 3D printing (Maniglia et al. 2019).

Cooling intervals could also reduce changes in starch properties during DHT (Zou et al. 2020). Repeated dry heating had influenced maize starch significantly, for instance, decreasing crystallinity and thermal stability, weakening short-range order, and impairing paste stability. Compared with continuous dry heating, repeated dry heating could affect starch characteristics more due to the cooling procedure (Zhang et al. 2021a, b).

Waxy corn starches were subjected to repeated and continuous DHT to investigate their effects on physicochemical properties. Repeated DHT samples illustrated lower solubility and swelling power, poorer short-range molecular order, lower crystallinity and transmittance than continuous DHT ones. During repeated DHT, cooling procedures might inhibit the arrangement but accelerate the destruction of starch granules. Therefore, repeated DHT waxy corn starches could be applied to noodle manufacture due to their lower solubility and play roles in flavor encapsulation for sustainable release (Zou et al. 2019).

12.4.3 Additives

Adding reactive compounds to starch could elevate the effect of dry heating. The chemical reaction between starch and additives could affect the rearrangement of starch granules thermally (Ganesh et al. 2020). Dry heating combined with glucose could induce a synergistic effect, thus altering the pasting properties and gel texture of starch. Addition of glucose to waxy corn starch enhanced the gel hardening during dry heating which indicated the reinforced starch network. Therefore, glucose could allow chemical linkages in starch chains by acting as an intermediate. This kind of combination effectively improved the pasting and gelatinizing characteristics of various starches (Lee et al. 2021).

Starch modified by DHT combined with anionic food gums could remarkably change the pasting properties of starch. Moreover, waxy maize starch tended to be influenced by dry heating with anionic gum compared with potato starch. It was found that DHT enhanced the functional properties of corn starch, potato starch, and pea starch obviously. After adding CMC, the paste viscosity of potato starch was increased, while corn and pea starch were decreased (Sun et al. 2013).

Under mild alkaline conditions, phytic acid combined with DHT could phosphorylate waxy maize and waxy rice starches effectively by primarily forming mono-starch monophosphate. The phosphorylated waxy starches showed increased breakdown and peak viscosity while decreased pasting temperature. Phosphorylation with phytic acid enhanced the paste clarity and swelling power but reduced the melting temperature and enthalpy of samples. To obtain improved physicochemical properties, the kind of modification could be applied to various starch samples of different botanical sources (Park and Lim 2019).

Normal corn starch was subjected to DHT at 130 °C for 1, 2, and 4 h cooperating with different concentrations of ethanol and acid to produce starch nanoparticles. Starch granules were easily fragmented into smaller particles with the ethanol concentration decreasing which could be due to preferentially hydrolyzed starch chains in amorphous regions. In various ethanol concentrations, lower ones ($< 30\%$) caused damage in long- and short-range crystalline structures of starch granules. Though 30% of ethanol concentration destroyed the long-range crystalline order, the short-range order remained intact and uniform nanoparticles with a mean diameter of 46.4 nm were generated. However, higher ethanol concentrations (50% and 95%)

could not produce homogenous nanoparticles. This DHT produced starch nanoparticles that could readily disperse in water without physical fragmentation, which might promote the industrial production of starch nanoparticles (Miskeen et al. 2019).

Tapioca starch was studied by DHT at various heating temperatures (100, 120, and 140 °C for 5 h) and citric acid concentrations (10, 30, and 50 g/100 g dry starch). Results showed that increased citric acid concentrations and temperatures led to the increment of degree of esterification. Esterification conditions determined the formation of RS and starch digestibility. Moderate levels of citric acid under high heating temperatures could produce the maximum RS content. Therefore, compared with native starch, citric acid esterification of tapioca starch played role in the increment in RS content and slow digestion rate (Srikaeo et al. 2019).

12.5 Effects of Dry Heat Modification on Physicochemical Properties of Starch

12.5.1 Thermal Properties

A multivariate differential scanning calorimeter (DSC) is usually applied to measure the thermal properties of starch during gelatinization. The measured parameters included onset temperature (To), peak temperature (Tp), conclusion temperature (Tc), and gelatinization enthalpy change (ΔH). The transition temperature (To, Tp, Tc) of starch refers to the degree of perfection of the double helix (Cooke and Gidley 1992). To and Tc represent weak and strong crystalline melting temperatures, respectively (Xu et al. 2018). The effect of DHT on the thermal properties of starch is not always consistent, and this inconsistency may be attributed to the starch source and reaction conditions (Table 12.1).

DHT increased the To, Tp, and Tc of mung bean starch and decreased the ΔH. This is because during the DHT, starch granules undergo partial gelatinization, which leads to the rearrangement of the intermolecular molecules of the granules. This process results in increased interactions between starch chains, resulting in a more stable granular structure (Liang et al. 2021). Similarly, Gou et al. (2019) discussed the effect of repeated DHT and continuous DHT on the thermal properties of sweet potato starch. They found that DHT increased the To, Tp, and Tc of sweet potato starch and decreased the ΔH of sweet potato starch. With the increase in processing time and cycle number, To, Tp, and Tc increased continuously, and ΔH decreased continuously. This suggests that DHT can stabilize the weak crystallization in starch granules. The increase in the gelatinization transition temperature indicates that the double helix of the starch after DHT is more perfect, and a higher temperature is required to break it compared to the native starch. ΔH represents the content of the double helix. After DHT treatment, the decrease in ΔH was attributed to partial damage of starch and gelatinization of starch granules. In another study,

Table 12.1 Effects of different dry heat treatment (DHT) conditions on thermal properties of different starches

Starch	Modified conditions	T_O, T_P, and T_C (°C)	ΔH (J/g)	References
Maize starch	140–200 °C, <10%, 2 h	The effect of DHT on T_c was not significant, but T_o decreased by 2.9–10.6% T_p decreased by 1.0–7.1%	ΔHdecreased by 16.4–45.9%	Lei et al. (2020)
Waxy corn starch	140 °C, <10%, 4 h, Repeat 5 cycles	T_o decreased by 0.3–2.5% T_p decreased by 0.2–0.7 T_c decreased by 1.1–2.5%	ΔH decreased from 8.1 J/g of native starch to 5.15 J/g	Zou et al. (2019)
	140 °C, <10%, 4–20 h	T_o decreased by 0.3–1.8% T_p decreased by 0.2–0.7 T_c decreased by 1.1–2.0%	ΔH first increased from 8.1 J/g of native starch to 9.08 J/g. after prolonging the processing time, ΔH was reduced to 6.19 J/g	
Normal maize starch	130 °C, <10%, 2 h	T_o decreased by 2.2% T_p decreased by 2.0 T_c decreased by 1.7%	ΔH decreased by 14.4%	Chi et al. (2019)
Potato starch		T_o decreased by 6.3% T_p decreased by 5.5% T_c decreased by 1.0%	ΔH decreased by 13.1%	
Maize starch	140 °C, <10%, 4 h, Repeat 5 cycles	T_o decreased by 1.2–3.1% T_p decreased by 1.7–1.8% T_c decreased by 1.4%	ΔH decreased by 17.5–18.1%	Zou et al. (2020)
	140 °C, <10%, 4–20 h	T_o decreased by 1.2–2.7% T_p decreased by 0.9–1.7% T_c decreased by 0.4–1.0%	ΔH decreased by 10.8–17.5%	
Waxy potato starch	110 °C, <10%, 0.5–2.5 h	T_o decreased by 5.8–8.0% T_p decreased by 4.4–7.1% T_c decreased by 3.6–6.8%	ΔH decreased by 12.5–32.8%	Liu et al. (2019)
High amylose rice starch	150 °C, <10%, 1–4 h	T_o decreased by 9.1–9.8% T_p decreased by 5.4–17.5%	ΔH decreased by 60.4%	Oh et al. (2018a, b)
Normalwheat starch	130 °C, <10%, 3 h, Repeat 6 cycles	T_p decreased by 0.01–1.4% T_c decreased by 0.7–1.6%	ΔHincreased by 4.6–15.7%	Zhang et al. (2021a, b)
	130 °C, <10%, 3–18 h	T_p decreased by 0.01–1.4% T_c decreased by 0.8–1.6%	ΔHincreased by 3.2–15.7%	
Sweet potato starch	130 °C, <10%,	T_o increased by 0.04–1.3%	ΔH decreased with increasing treatment cycle	Gou et al. (2019)

(continued)

Table 12.1 (continued)

Starch	Modified conditions	T_O, T_P, and T_C (°C)	ΔH (J/g)	References
	3 h, Repeat 6 cycles	T_p increased by 0.2–1.1% T_c decreased by 0.1–0.8%	(from 12.81 J/g to 10.49 J/g)	
	130 °C, <10%, 3–18 h	T_o increased by 0.04–2.3% T_p increased by 0.2–2.3% T_c increased by 0.01–2.2%	ΔH decreased with increasing treatment time (from 12.81 J/g to 10.12 J/g)	
Mung bean Starch	130 °C, <10%, 3 h, Repeat 6 cycles	T_o increased by 2.1–5.2% T_p increased by 0.2–4.4% T_c increased by 0. 5–7.4%	ΔHincreased by 1.2 ~ 33.5%	Liang et al. (2021)
	130 °C, <10%, 3–18 h	T_o increased by 2.1–8.2% T_p increased by 1.7–5.8% T_c increased by 0.5–15.9%	ΔHincreased by 1.2–32.6%	
Millet starch	130 °C, 8%, 2–4 h	T_o increased by 0.7–1.2% T_p increased by 2.1–2.4% T_c decreased by 0.9%	ΔHdecreased by 14.3–17.8%	Sun et al. (2014)
Quinoa starch	140 °C, <10%, 4 h, Repeat 5 cycles	T_o increased by 9.3–28.3% T_p increased by 5.7–13.6% T_c was slightly higher than native starch	ΔH first decreased from 0.257 J/g of native starch to 0.168 J/g, and then with the increase in cycle times, ΔH was higher than that of native starch	Zhou et al. (2021)
	130 °C, <10%, 4–20 h	T_o increased by 9.8–28.3% T_p increased by 7.2–13.6% T_c was slightly higher than native starch		

T_o onset temperature, T_p peak temperature, T_c conclusion temperature, ΔH gelatinization enthalpy

DHT increased To, Tp, and Tc of quinoa starch, while ΔH showed a trend of decreasing first and then increasing (Zhou et al. 2021). The increase in gelatinization transition temperature indicated that DHT improved the thermal stability of quinoa starch. The first decrease in ΔH was due to the partial gelatinization of quinoa starch during the DHT process, resulting in a decrease in the content of the double-helical structure. The subsequent increase in ΔH was due to the recrystallization of starch granules after DHT modification, which is the increase in the structural stability of starch granules.

Differently, Zou et al. (2019) studied the effects of repeated DHT and continuous DHT on the thermal properties of maize starch. They found that DHT reduced the To, Tp, Tc, and ΔH of maize starch. This indicates that DHT can destroy the integrity of the molecular structure of starch. With the increase in treatment time and cycle number, To gradually decreased. When the samples were treated for 12 h or 3 cycles, the Tp, Tc, and ΔH values showed a gradually increasing trend. This is because the starch chains rearranged before 3 cycles or 12 h, resulting in high crystallinity. The Tp, Tc, and ΔH values decreased when the treatment time was extended to 20 h and 5 cycles. The subsequent decrease was due to DHT destroying the structure of starch molecules, leading to a decrease in crystallinity. In addition, the reason for the decrease in ΔH was the partial gelatinization of starch granules during DHT (Zou et al. 2020). Similarly, DHT resulted in decrease in To, Tp, Tc, and ΔH values in potato starch. Especially, the Tp of potato starch decreased from 66.95 °C to 63.27 °C after DHT, and the ΔH decreased from 20.53 J/g to 17.84 J/g. This suggested that DHT disrupted the ordered structure of starch (Chi et al. 2019). In addition, Zhang et al. found that DHT decreased the To, Tp, and Tc of wheat starch, which indicated that DHT disrupted the molecular structural integrity of wheat starch. In addition, the ΔH value in the RDHT wheat starch samples was higher than that in the continuous DHT starch samples. This might be because the cooling process of repeated DHT promoted the reorganization of starch chains and enhanced the stability of the double helix structure (Zhang et al. 2021a, b).

12.5.2 Pasting Properties

Pasting parameters of starch measured by a rapid viscosity analyzer include peak viscosity (PV), trough viscosity (TV), breakdown (BD), final viscosity (FV), and setback (SB). When starch granules are heated in excess water, they absorb water and swell. At this time, amylopectin and amylose molecules are leached from the swollen starch granules, leading to an increase in the viscosity of the system. PV is related to the expansion of starch granules at peak temperature (Ahmed et al. 2016). The rupture of the expanded starch granules at the end of the heating period is related to the TV. The SB value and FV characterize the stability of the starch paste upon cooling and are related to the aging of the starch (Zou et al. 2019). The BD value represents the degree to which the granules expand under the influence of heat and shear. Maniglia et al. studied the effect of DHT at different times (2 h, 4 h) on the pasting properties of wheat starch. They found that DHT significantly reduced the PV of wheat starch, and the viscosity decreased further with increasing treatment time. This might be due to the weakening of starch granules caused by DHT, which made the modified wheat starch less able to maintain granule integrity at peak temperatures. In addition, with the extension of DHT time, TV and FV further decreased, which indicated that DHT modified wheat starch granules were more fragile and easily broken (Maniglia et al. 2020a, b). Similarly, Zou et al. (2020) find that DHT reduced the overall viscosity of maize starch. With the increase in cycle times and treatment time, the PV of maize starch decreased gradually. Similarly, the PV of millet starch after DHT for 2 h and 4 h also showed a downward trend (Sun et al. 2014). In the study, the repeated DHT samples had a higher viscosity profile than the continuous DHT samples. This suggested that strong intramolecular/intermolecular interactions were formed in the repeated DHT samples (Zou et al. 2020). The difference was that DHT increased the PV of waxy potato starch (Liu et al. 2019). Similar phenomena were also observed in DHT modified rice starch (Qin et al. 2016). This might be due to the enhanced interaction between starch molecules during DHT treatment. Therefore, under thermal conditions, the shear resistance and shear stability of starch granules were increased. In general, the differences in pasting properties are related to the structural characteristics of starch granules and processing conditions.

12.5.3 Morphological Characteristics Analysis

DHT can maintain the integrity of the surface of starch granules, but some dents and cracks appear on the surface of the granules. The effect of DHT temperature on the structure of wheat flour was investigated by González et al. They found that the granules structure of wheat flour treated at 50 °C did not change. However, at 100 °C and above, the surface of starch granules appeared dents. This indicated that DHT caused starch granules to break and destroyed the integrity of starch granules. In addition, they argued that the dents on the granule surface might be related to the initial gelatinization, thermal fracture, and melting of starch granules (Gonzalez et al. 2021). Similarly, Ge et al. (2021) observed cracks or a small number of pores on the surface of red adzuki bean starch granules after DHT. In addition, cracks also appeared on the surface of water chestnut starch granules after DHT. This might be related to the leaching of amylose and the accelerated intermolecular movement at high temperatures (Gul et al. 2014). Similarly, some dents and wrinkles appeared on the granule surface after DHT for both common and waxy wheat starches (Zhang et al. 2021a, b). Zou et al. (2019) demonstrated that the surface of the waxy corn starch granules also had dents by DHT. Furthermore, Miao et al. (2021) found that after heating tigernut tuber starch at 130 °C for 2 h, the granule surface had no significant change compared with native starch. However, after heating at 130 °C for 4 h, the granular structure of starch was destroyed. This could be explained by amylose leaching and heating effects.

12.5.4 Crystalline Structure

The crystalline structure of starch can be classified into A, B, and C types on the basis of the arrangement of branched chains and the formation of the double helix structure. Zhou et al. (2021) discussed the effect of continuous DHT and repeated DHT on the crystalline structure of quinoa starch. The native quinoa starch is A-type starch, and the main diffraction peaks are at 15, 17, 18, and 23°. After DHT, the diffraction peaks did not change significantly, indicating that the crystalline structure of quinoa starch did not change after DHT. This is consistent with previous studies that the diffraction peaks of proso millet starch and waxy rice starch were almost unchanged after DHT. This indicated that the crystal types of proso millet starch and waxy rice starch also did not change (Li et al. 2013; Sun et al. 2014). These results suggested that DHT mainly changed the amorphous region of starch. After DHT, the relative crystallinity of quinoa starch samples was higher than that of native starch. This could be explained by the rearrangement of starch molecules during the DHT process, leading to a relatively perfect crystal structure (Zhou et al. 2021). In addition, Zhang et al. concluded that DHT increased the crystallinity of wheat starch (Zhang et al. 2021a, b). This might be due to the rearrangement of the disrupted double helix structure, which further enhanced the interaction between starch chains, resulting in the formation of more perfect crystals. In contrast, the relative crystallinity of rice flour (Zhou et al. 2019), red adzuki bean starch (Ge et al. 2021), and maize starch (Lei et al. 2020) decreased after DHT. The decrease in crystallinity was mainly due to the increase in amorphous regions or the degradation of crystalline regions. In general, crystallinity is influenced by various factors, such as crystal size and number, double helix orientation, and degree of interaction. The decrease in relative crystallinity of starch may be related to the degradation of the starch crystalline region during the DHT process, partial gelatinization of starch granules, or changes in crystalline orientation (Zou et al. 2019).

12.5.5 *Fourier Transformation Infrared Spectroscopy (FTIR) Analysis*

Infrared spectroscopy is used to describe the short-range molecular structure of starch. Generally, the changes in the characteristics of starch crystalline regions and amorphous regions can be reflected by the absorbance ratio in the 1048 cm^{-1}/1022 cm^{-1} band. In addition, this ratio is also used to express the number of ordered domains (double helix and single helix) and amorphous domains (linear chain structure) of amylopectin. Ge et al. studied the effect of DHT on the starch structure of red adzuki beans. They did not observe new absorption peaks in the FTIR spectra (4000 ~ 400 cm^{-1}) of all samples. This suggested that DHT did not alter the chemical groups of native starch. Notably, the 1048 cm^{-1}/1022 cm^{-1} ratio of the starch samples decreased after DHT. This was attributed to the high temperature breaking the original hydrogen bonds of starch. At the same time, this also led to the dissociation of the double helix structure in the crystal region. In addition, the decrease in molecular weight and increase in the proportion of short chains caused by DHT also led to structural changes and increased crystal disorder (Ge et al. 2021). Similarly, the native waxy corn starch and DHT modified starch had similar absorption peaks, which indicated that no new chemical groups were formed in the waxy corn starch after DHT. In contrast, the 1048 cm^{-1}/1022 cm^{-1} ratio of waxy maize starch first increased and then decreased during DHT (Zou et al. 2019). The decrease in this ratio might be related to the dissociation of the double helix structure in the crystal region, while the increase in this ratio might be related to the association between starch chain polymers (Liang et al. 2021). Therefore, DHT may first strengthen the binding between starch chain polymers and then decompose the double helix structure of the crystalline region. In a recent study, Zhou et al. (2021) found that neither repeated DHT nor continuous DHT modified quinoa starch exhibited new absorption peaks. These results indicated that DHT did not destroy the original chemical groups of native starch, nor does it generate new chemical groups.

12.5.6 Chrominance

In the DHT of whole barley, the L value of starch decreased and the a* and b* increased. The results showed that the value of L* decreased from 93.34 to 73.20, indicating that the color of barley flour became darker with the increase in dry heat modification temperature. In addition, the parameter a* increases remarkably from 1.05 to 8.35. This parameter correlates the redness characteristic of flour, that is, under the influence of DHT, flour appears browning phenomenon to a certain extent. In similar, the parameter b* was increased from 5.63 to 25.98. This indicates that the DHT can increase the greenness characteristic. Boyd et al. (2017) also reported the influence of DHT on whole barley. Overall, the chromaticity results showed that dry heating produced browning in wheat flour, probably due to the Maillard reaction and caramelization reaction caused by the baking of flour components (Chung et al. 2012). For oat and millet flour, the color change caused by the dry heat process was more pronounced. With the exception of barley flour, these two starches are the brightest (L*). Because the Maillard and caramelization reactions cause non-enzymatic Browning of food, the color changes of these two types of flour are more visually noticeable under the conditions of drying and heating than those of naturally darker flours. No similar phenomenon was observed for barley flour during treatment because it contained more black grains from the bran. Under the influence of tannin and polyphenol oxidase activity, the total color difference of sorghum after two kinds of heat treatment was more obvious. At the beginning of the heating process, a slightly higher temperature causes Browning of the enzyme, while further heating inactivates the enzyme (Dlamini et al. 2007).

12.5.7 Freeze-Thaw Stability

All kinds of starch-based foods often freeze and thaw during storage and transportation. Therefore, starches used in the production of these foods must be stable against freezing and thawing. When starch gels are repeatedly frozen and thawed, interchain interactions extrude and release residual water in the gel matrix (Zhang et al. 2019). For example, when the cycle time increased to 5, the dehydration rate of the native tigernut starch gel increased from 34.62% to 41.04%, suggesting that the tigernut starch gel was unstable during repeated freezing and thawing. After DHT, the viscosity increases obviously with the extension of heat treatment time. In general, DHT starch gels tend to have higher dehydration values because heat treatment increases the interaction between starch molecules, resulting in lower water holding capacity (Zhang et al. 2021a, b). Ultrasonic treatment and DHT can improve the freeze-thaw stability of potato starch, so that the degradation trend of potato starch is weak. Dry heat, microwave and ultrasonic treatments had significant effects on the gelatinization enthalpy of potato starch with or without pectin. The gelatinization temperature range of potato starch-pectin treated with dry heating was wide, and the gelatinization enthalpy change was the lowest.

12.5.8 Rheological Properties

Dry heating can increase the apparent viscosity of starch paste. The n value (flow property index) of starch decreased continuously with the prolongation of time after DHT. This may be due to the destruction of glycosidic bonds and hydrogen bonds in starch particles by continuous heating, which leads to the failure of the original dense network structure of starch to be maintained and collapse along with the disintegration of the chain, finally leading to the collapse of the gel (Oh et al. 2018a, b). After the dual treatment of $CaCl_2$ and dry heating, the molecular structure of starch was seriously disintegrated, the crystallinity of original starch particles was completely lost, and the disappearance of endothermic peak in DSC thermography could well confirm the above speculation. Thus, the dual treated potato starch had a very low swelling power, which prevented the 6% (w/v) potato starch suspension from forming a paste. The results showed that the cooked suspension of potato starch after dual treatment with $CaCl_2$ and dry heating showed very weak viscoelastic (1 h DHT) or non-viscoelastic (2 h and 4 h DHT). For steady flow state of starch gels, obviously, the apparent viscosity η of both native and DHT starches decreases with the increase in shear rate (γ), suggesting all samples displayed the typical shear thinning behavior. To further parameterize the flow property, the power law model ($\eta = K\times \gamma^{n-1}$) is fitted to the viscosity flow curves of starch gels, where K is consistency index and n is the flow behavior index. This research compares the differences between DHT samples. Generally, with heating temperature increasing, K value declines, indicating the smaller intermolecular resistance to flow and the thinner liquid of starch (Dangi et al. 2019). The n value of all investigated samples demonstrated the non-Newtonian flow and pseudoplastic behavior (Górecki et al. 2018).

12.6 Applications

12.6.1 Clean Label Starch

Dry heating modified starch has been widely used in food field because of its green and pollution-free characteristics. At the same time, as a substitute for chemically modified starch, starch manufacturers have paid more attention to physical modified starch based on consumer demand. In addition, physically modified starch is often used in processed foods due to its green and clean properties, such as providing "Clean labels." Since these starches are not chemically modified, rather than chemical additives, they are generally considered as natural ingredients and can be referred to as "starch" in food labels. Chemically modified starch is labeled as "modified food starch" (Keppler et al. 2018).

12.6.2 Starch-Based Particulate Stabilizers

The combination of dry heating with hydrophobic modification is considered as a promising method for the preparation of starch-based particle stabilizer. The physical stability of pickling emulsion determines the shelf life of pickling emulsion products in the food industry. Some studies first used octenyl succinic anhydride to esterify rice starch and modified rice starch by electric dry heat or baking dry heat. The properties of starch prepared by compound modification were better than that by single modification.

12.6.3 Bakery Products

Dry heat modification as a green technology is widely used in flour and starch products to enhance the water absorption and swelling behavior of starch granules. It has been studied as an alternative to chlorination of flour in many European countries and in places such as Australia for large-scale production of food ingredients with similar functions (Dudu et al. 2019). These methods apply to wheat flour as well as other flours. The modification of flour properties by dry heat modification was first reported by Manels in 1934. The most common use of DHT is related to starch pasting properties. In addition to modifying starch viscosity properties, dry heat modification can also be used to modify proteins, alter their original structures or give them new structures to improve nutrient availability, and inactivate toxic heat-resistant compounds and other enzyme inhibitors (He et al. 2020). The changes of other components such as protein and lipid in flour after DHT are also important factors for the improvement of flour quality. It is generally believed that proteins and lipids on the surface of starch particles play a role in preventing water absorption, which will weaken the expansion of starch in flour processing (Qiu et al. 2015). The effect of dry heat modification on protein structure is also reflected by protein denaturation and oxidation of sulfhydryl groups to disulfide bonds. When dry heated, the folded structure of the three-dimensional protein of the proteotype is destroyed, exposing the lipid core wrapped by hydrophobic amino acids. The exposed lipids greatly enhance the functional properties of flour by interacting with proteins and starch rather than by their chemical modification (Sun et al. 2014). At the same time, water is more likely to migrate to the surface of starch particles during starch gelatinization. In addition, a more hydrophobic surface increases the contact between starch and bubbles (Yano et al. 2017). This effect is particularly significant in improving air absorption in baked products and making them more resistant to collapse (Keppler et al. 2018). In terms of applications, dry heat modified flour is used as a baking amendment in bread baking, partially replacing wheat flour (2–5%), while 50–100% substitution is considered in cake making. Keppler et al. (2016) observed that dry heat modification improved starch expansion properties at high temperatures, increased gluten networks, and enhanced interactions between flour polymers in high-proportion cakes (i.e., sponge cakes). Dudu et al. (2019) observed that the addition of DHT tapioca to wheat flour slightly increased the water absorption capacity of flour, softening degree, and dough development time. Bread made of DHT cassava flour has enhanced resistance to deterioration and slightly higher specific volume. Maniglia et al. observed that modified starch formed from DHT modified cassava and wheat starch was able to form stronger gels with lower dehydration rates and higher prospects for 3D printing applications as "inks" (Maniglia et al. 2020a, b).

12.6.4 3D Printing

The printability of wheat starch-based hydrogel (ink) was studied by DHT. The size of starch granule increased slightly by dry heat modification. However, it did not change the shape of the starch particles or their surface condition. DHT does not alter starch functional groups, indicating that oxidation was negligible. However, a slight depolymerization of starch was observed, especially in medium-sized molecules, as well as a decrease in the crystalline zone in starch particles. The modified starch hydrogels showed lower apparent viscosity and stronger static structure strength during gelatinization. The hardness of hydrogel increases with the extension of dry heat modification time, and the dehydration is also serious. The promoting effect of dry heat modification on wheat starch (especially after 4 h treatment) makes hydrogels have better printability and reproducibility in 3D printing. Dry heat modification is a simple and feasible way to improve the properties of wheat starch and extend the texture possibilities of printed samples based on wheat starch hydrogels.

12.6.5 Functional Foods

Dry heat modified starch can be used as a functional food to broaden the application of starch products in the food field. Dry heating sorghum hybrid full meal is a source of resistant starch and phenolic compounds. In a previous study, it decreased inflammatory markers, such as the transcription factor P65-NFκB, by inhibiting Akt phosphorylation in adult Wistar rats fed a high-fat and high-fructose diet. In addition, the consumption of the sorghum meal decreased the oxidative stress, inhibited the production of nitric oxide, promoted the secretion of antioxidant enzymes, and significantly increased the total antioxidant capacity of plants. Thus, dry heated whole sorghum meal is an excellent compositional alternative grain that effectively inhibits inflammation and oxidative stress and greatly reduces the harmful effects of a high-fat and high-sugar diet. In the DHT, convective microwave heating had the best effect on high RS content in rice and millet, which increased by 22% and 14%, respectively. Through orthogonal experiments, it is found that different temperature–time combinations can obtain higher RS content. The effects of different DHT on RS formation in rice and millet were reported. Rice varieties with high amylose and heat treatment above 100 °C have the potential to significantly increase RS content in grains, which can be used to develop functional foods such as low GI as meal replacement staples for specific populations. The combined treatment of dry heating and annealing has a potential synergistic recombination effect on starch structure, and the final result is to reduce the molecular weight and digestibility of starch (Chi et al. 2019). The glucose-lowering effect of DHT on rice starch was proved by comparing the starch digestibility in vitro and glucose response in vivo. DHT with xanthan gum can significantly reduce the content of RDS to 38.9%, increase the content of RS to 28.1%, and starch digestibility HI is the lowest. These results confirmed that DHT and xanthan altered the chain arrangement within the starch molecule, thereby altering the packing arrangement of the starch gel, consistent with the glucose response in vivo. In terms of starch digestion in mice, DHT with xanthan gum starch mixture also effectively delayed starch digestion within 30 min, with the lowest GI value. Correlation analysis showed that starch fractions (RDS, SDS and RS) in vitro correlated well with GI values in vivo.

References

Ahmed J, Thomas L, Taher A, Joseph A (2016) Impact of high pressure treatment on functional, rheological, pasting, and structural properties of lentil starch dispersions. Carbohydr Polym 152: 639–647

Boyd L, Storsley J, Ames N (2017) Effect of heat treatments on starch pasting, particle size, and color of whole-grain barley. Cereal Chem J 94(2):325–332

Chandanasree D et al (2016) Effect of hydrocolloids and dry heat modification on physicochemical, thermal, pasting and morphological characteristics of cassava (Manihot esculenta) starch. Food Hydrocoll 52:175–182

Chi C, Li X, Lu P, Miao S, Zhang Y, Chen L (2019) Dry heating and annealing treatment synergistically modulate starch structure and digestibility. Int J Biol Macromol 137:554–561

Chung SY, Han SH, Lee SW, Rhee C (2012) Effect of Maillard reaction products prepared from glucose-glycine model systems on starch digestibility. Starch-Stärke 64(8):657–664

Cooke D, Gidley MJ (1992) Loss of crystalline and molecular order during starch gelatinisation origin of the enthalpic transition. Carbohydr Res 227:103–112

Dangi N, Yadav BS, Yadav RB (2019) Pasting, rheological, thermal and gel textural properties of pearl millet starch as modified by guar gum and its acid hydrolysate. Int J Biol Macromol 139: 387–396

Dlamini NR, Taylor JRN, Rooney LW (2007) The effect of sorghum type and processing on the antioxidant properties of African sorghum-based foods. Food Chem 105(4):1412–1419

Dudu OE, Li L, Oyedeji AB, Oyeyinka SA, Ma Y (2019) Structural and functional characteristics of optimised dry-heat-moisture treated cassava flour and starch. Int J Biol Macromol 133:1219–1227

Ganesh A, Singh B, Duttagupta A, Kalita D, Zhong Y, Blennow A, Singh H (2020) Preparation of starch citrates using solvent free reaction and comparison with aqueous and ethanol mediated reactions. Starch-Stärke 72(5–6):1900260

Gao H et al (2021) Effect of dry heat, microwave and ultrasonic treatments on physicochemical properties of potato starch with or without pectin. Trop J Pharm Res 18(7):1365–1373

Ge X, Shen H, Su C, Zhang B, Zhang Q, Jiang H, Li W (2021) The improving effects of cold plasma on multi-scale structure, physicochemical and digestive properties of dry heated red adzuki bean starch. Food Chem 349:129–159

González M, Vernon-Carter EJ, Alvarez-Ramirez J, Carrera-Tarela Y (2021) Effects of dry heat treatment temperature on the structure of wheat flour and starch in vitro digestibility of bread. Int J Biol Macromol 166:1439–1447

Górecki A, Błaszczak W, Lewandowicz J, Thanh-Blicharz J, Penkacik K (2018) Influence of high pressure or autoclaving-cooling cycles and Pullulanase treatment on buckwheat starch properties and resistant starch formation. Pol J Food Nutr Sci 68(3):235–242

Gou M et al (2019) Effects of repeated and continuous dry heat treatments on properties of sweet potato starch. Int J Biol Macromol 129:869–877

Gul K, Riar CS, Bala A, Sibian MS (2014) Effect of ionic gums and dry heating on physicochemical, morphological, thermal and pasting properties of water chestnut starch. LWT Food Sci Technol 59(1):348–355

Guo X et al (2015) Changes in physicochemical properties and in vitro digestibility of common buckwheat starch by heat-moisture treatment and annealing. Carbohydr Polym 132:237–244

Han Z, Shi R, Sun DW (2020) Effects of novel physical processing techniques on the multi-structures of starch. Trends Food Sci Technol 97:126–135

He C, Zheng J, Liu F, Woo MW, Xiong H, Zhao Q (2020) Fabrication and characterization of oat flour processed by different methods. J Cereal Sci 96:103123

Im K, Young IB et al (2018) Effect of dry heat treatment on physical property and in vitro starch digestibility of high amylose rice starch. Int J Biol Macromol 108:568–575

Ji Y et al (2016) Impact of dry heating on physicochemical properties of corn starch and lysine mixture. Int J Biol Macromol 91:872–876

Kanagaraj SP, Ponnambalam D, Antony U (2019) Effect of dry heat treatment on the development of resistant starch in rice (Oryza sativa) and barnyard millet (Echinochloa furmentacea). J Food Process Preserv 43(7):e13965

Keppler S, Bakalis S, Leadley CE, Fryer PJ (2016) A systematic study of the residence time of flour in a vibrating apparatus used for thermal processing. Innovative Food Sci Emerg Technol 33: 462–471

Keppler S, Bakalis S, Leadley CE, Sahi SS, Fryer PJ (2018) Evaluation of dry heat treatment of soft wheat flour for the production of high ratio cakes. Food Res Int 107:360–370

Lee SJ, Zhang C, Lim ST, Park EY (2021) Effect of combination of dry heating and glucose addition on pasting and gelling behavior of starches. Int J Biol Macromol 183:1302–1308

Lei N, Chai S, Xu M, Ji J, Mao H, Yan S, Gao Y, Li H, Wang J, Sun B (2020) Effect of dry heating treatment on multi-levels of structure and physicochemical properties of maize starch: a thermodynamic study. Int J Biol Macromol 147:109–116

Li Y, Shoemaker CF, Ma J, Shen X, Zhong F (2008) Paste viscosity of rice starches of different amylose content and carboxymethylcellulose formed by dry heating and the physical properties of their films. Food Chem 109(3):616–623

Li Y, Zhang H, Shoemaker CF, Xu Z, Zhu S, Zhong F (2013) Effect of dry heat treatment with xanthan on waxy rice starch. Carbohydr Polym 92(2):1647–1652

Liang S, Su C, Saleh ASM, Wu H, Zhang B, Ge X, Li W (2021) Repeated and continuous dry heat treatments induce changes in physicochemical and digestive properties of mung bean starch. J Food Proc Preserv 45(3):e15281

Lima DC et al (2020) Dual-process of starch modification: combining ozone and dry heating treatments to modify cassava starch structure and functionality. Int J Biol Macromol 167:894–905

Liu K, Hao Y, Chen Y, Gao Q (2019) Effects of dry heat treatment on the structure and physicochemical properties of waxy potato starch. Int J Biol Macromol 132:1044–1050

Maniglia BC, Lima DC, Junior M, Le-Bail P, Augusto P (2019) Preparation of cassava starch hydrogels for application in 3D printing using dry heating treatment (DHT): a prospective study on the effects of DHT and gelatinization conditions. Food Res Int 128:108803

Maniglia BC, Lima DC, da Matta Junior M, Oge A, Le-Bail P, Augusto PED, Le-Bail A (2020a) Dry heating treatment: a potential tool to improve the wheat starch properties for 3D food printing application. Food Res Int 137:109731

Maniglia BC, Lima DC, Matta Junior MD, Le-Bail P, Le-Bail A, Augusto PED (2020b) Preparation of cassava starch hydrogels for application in 3D printing using dry heating treatment (DHT): a prospective study on the effects of DHT and gelatinization conditions. Food Res Int 128:108803

Miao W-B, Ning Y-Y, Huang H-R, Liu H-M, Cai X-S, Wang X-D (2021) Effect of dry heat modification and the addition of Chinese quince seed gum on the physicochemical properties and structure of tigernut tuber starch. Arab J Chem 14(11):103407

Miskeen S, Park EY, Kim JY (2019) Controlled fragmentation of starch into nanoparticles using a dry heating treatment under mildly acidic conditions. Int J Biol Macromol 123:810–816

Oh IK, Bae IY, Lee HG (2018a) Complexation of high amylose rice starch and hydrocolloid through dry heat treatment: physical property and in vitro starch digestibility. J Cereal Sci 79: 341–347

Oh IK, Bae IY, Lee HG (2018b) Effect of dry heat treatment on physical property and in vitro starch digestibility of high amylose rice starch. Int J Biol Macromol 108:568–575

Park EY, Lim ST (2019) Characterization of waxy starches phosphorylated using phytic acid. Carbohydr Polym 225:115225

Qin Y, Liu C, Jiang S, Cao J, Xiong L, Sun Q (2016) Functional properties of glutinous rice flour by dry-heating treatment. PLoS One 11(8):e0160371

Qiu S, Yadav MP, Chen H, Liu Y, Tatsumi E, Yin L (2015) Effects of corn fiber gum (CFG) on the pasting and thermal behaviors of maize starch. Carbohydr Polym 115:246–252

Srikaeo K, Hao PT, Lerdluksamee C (2019) Effects of heating temperatures and acid concentrations on physicochemical properties and starch digestibility of citric acid esterified tapioca starches. Starch-Stärke 71(1–2):1800065

Sun Q, Si F, Xiong L, Chu L (2013) Effect of dry heating with ionic gums on physicochemical properties of starch. Food Chem 136(3–4):1421–1425

Sun Q, Gong M, Li Y, Xiong L (2014) Effect of dry heat treatment on the physicochemical properties and structure of proso millet flour and starch. Carbohydr Polym 110:128–134

Xu M, Saleh ASM, Liu Y, Jing L, Zhao K, Wu H, Zhang G, Yang SO, Li W (2018) The changes in structural, physicochemical, and digestive properties of red adzuki bean starch after repeated and continuous annealing treatments. Starch-Stärke 70(9–10):1700322

Yano H, Fukui A, Kajiwara K, Kobayashi I, Yoza K-I, Satake A, Villeneuve M (2017) Development of gluten-free rice bread: pickering stabilization as a possible batter-swelling mechanism. LWT Food Sci Technol 79:632–639

Zhang C, Lim S-T, Chung H-J (2019) Physical modification of potato starch using mild heating and freezing with minor addition of gums. Food Hydrocoll 94:294–303

Zhang B, Zhang Q, Wu H, Su C, Ge X, Shen H, Han L, Yu X, Li W (2021a) The influence of repeated versus continuous dry-heating on the performance of wheat starch with different amylose content. LWT Food Sci Technol 136:110380

Zhang H et al (2021b) Thermal, pasting, and rheological properties of potato starch dual-treated with $CaCl_2$ and dry heat. LWT Food Sci Technol 146:111467

Zhou W, Song J, Zhang B, Zhao L, Hu Z, Wang K (2019) The impacts of particle size on starch structural characteristics and oil-binding ability of rice flour subjected to dry heating treatment. Carbohydr Polym 223:115053

Zhou Y, Cui L, You X, Cui Y (2021) Effects of repeated and continuous dry heat treatments on the physicochemical and structural properties of quinoa starch. Food Hydrocoll 113:106532

Zia-ud-Din et al (2015) Physical and chemical modification of starches-a review. Crit Rev Food Sci Nutr 57(12):2691–2705

Zou J, Xu M, Tian J, Li B (2019) Impact of continuous and repeated dry heating treatments on the physicochemical and structural properties of waxy corn starch. Int J Biol Macromol 135:379–385

Zou J, Xu M, Tang W, Wen L, Yang B (2020) Modification of structural, physicochemical and digestive properties of normal maize starch by thermal treatment. Food Chem 309:125733

Chapter 13
Modifications of Starch by Pulsed Electric Fields

Zhong Han, Ying Li, and Xin-An Zeng

Abstract Pulsed electric field (PEF) is a non-thermal processing technology, which plays an important role in the macromolecules in food components, especially in improving the structure, viscosity, and gelatinization characteristics of starch. This chapter reviews the physicochemical properties, granular and molecular structure of PEF modified starch. The effect of PEF as a synergistic method on the starch esterification process and the mechanism of PEF-assisted starch esterification modification are discussed. PEF technology has the potential to be used in starch modification to create a series of functions and provide a feasible environmentally friendly and economical method for the assisted chemical modification of biological macromolecules.

Keywords Starch · Structure · Amylose · Amylopectin

13.1 Introduction

Pulsed electric field (PEF) technology applies short pulses of high electric fields with a very short duration of micro- to milli-seconds and field strength of ~1 to 80 kV/cm (Zhu 2018). The short duration is to avoid excessive heat and unwanted electrolytic reactions. The samples are placed between a set of electrodes. The treatment time is calculated by multiplying the pulse duration with the number of pulses applied. The electric field can be in the form of bipolar, square wave, exponential decaying, or oscillatory pulses (Hong et al. 2016a, b). A typical experimental system of PEF for starch modifications is illustrated (Fig. 13.1). In the studies on PEF modifications of starch, the temperature of the starch suspension is controlled by a water bath system

Z. Han (✉) · Y. Li
School of Food Science and Engineering, South China University of Technology, Guangzhou, China
e-mail: fezhonghan@scut.edu.cn

X.-A. Zeng
Department of Food Science, Foshan University, Foshan, Guangdong, China

Z. Sui, X. Kong (eds.), *Physical Modifications of Starch*,
https://doi.org/10.1007/978-981-99-5390-5_13

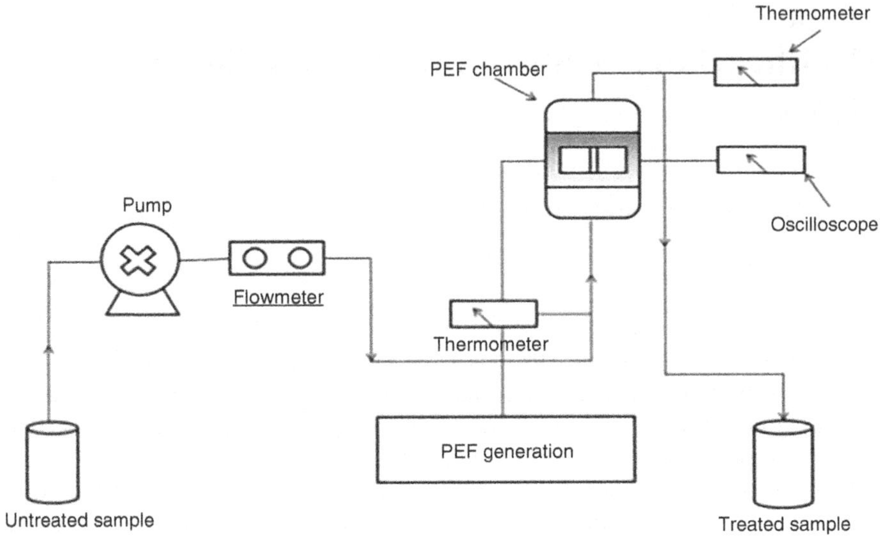

Fig. 13.1 Schematic diagram of the continuous PEF system used in this study (Hong et al. 2016b) (Reused with permission from Elsevier Publications)

to be under 40–45 °C (Han et al. 2012b). Therefore, PEF systems with an efficient temperature controlling unit remain to be developed.

13.2 Effect of PEF on Granular and Chemical Structure of Starch

13.2.1 Effect of PEF on the Surface Morphology of Starch Granules

The effect of PEF on the granular structure of starches from different botanical sources has been reported. Zhong Han et al. (Han et al. 2009a, b) studied the effect of PEF on the structure and properties of corn and potato starch. The surface morphology of native starch granules was smooth. After PEF treatment at 30 $kV \cdot cm^{-1}$, the shape and size of starch did not change. After being treated at 40 $kV \cdot cm^{-1}$, for corn starch, some pits emerged and some small particles grouped together to form a big one. For potato starch, the granule shape remained while some pits emerged. Under the condition of high field strength (50 $kV \cdot cm^{-1}$), the starch was significantly deformed. Furthermore, some fragments were produced and congregated to form larger particles (Han et al. 2009a, b). The aggregation and surface disruption may be due to surface gelatinization of starch caused by PEF. PEF may locally create "high" temperature due to the specific electric conductivity of starch granules (Fig. 13.2).

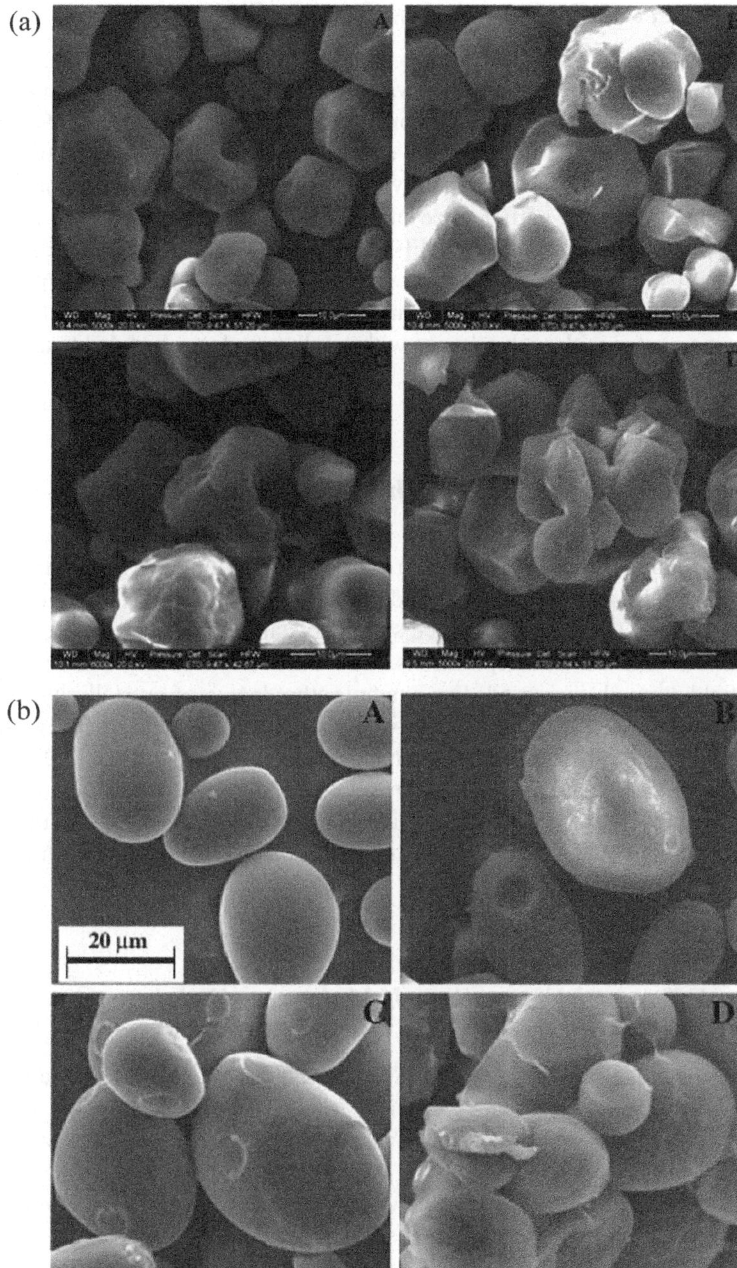

Fig. 13.2 SEM micrographs of corn (**a**) and potato (**b**) starches treated at different electric field strengths: (*A*) native; (*B*) 30 kV cm^{-1}; (*C*) 40 kV cm^{-1}; and (*D*) 50 kV cm^{-1} (Han et al. 2009a, b) (Reused with permission from Elsevier Publications)

13.2.2 Effect of PEF on Starch Particle Size

The volume average particle size of waxy corn starch (WS), normal corn starch (NS), corn starch with 50% amylose content (G50), corn starch with 80% amylose content (G80) are 16 μm, 14 μm, 11 μm, and 10 μm, respectively. The particle size of starch decreases with the increase in amylose content. After PEF treatment at 50KV/cm, the volume average particle size of WS, NS, G50, G80 increased to 23 μm, 22 μm, 16 μm, 13 μm. Further, after PEF treatment, with the increase in PEF strength and treatment time, the starch particle size increased. As the content of amylose increases, the effect of PEF on its particle size decreases. The modified starch granules aggregated with each other or with small granules, which led to an increase in granule size.

13.2.3 Effect of PEF on Starch Crystal Structure

PEF had no influence on the polymorph type of starch, but it did reduce the relative degree of crystallinity, according to wide-angle X-ray diffraction study (Han et al. 2009a, b, 2012b). The degree to which starch crystallinity decreases depends on the type of starch and the PEF conditions. The starch crystallinity decreased as the electric field strength increased. For example, the degree of crystallinity of potato starch reduced dramatically from 27% to 3% at a high field strength (50 kV/cm), and the degree of crystallinity of cassava starch decreased significantly from 24% to 7%. In contrast, the PEF treatment had little effect on the starch's double helical structure, which is related to crystallinity formation (Han et al. 2012b). Small angle X-ray scattering (SAXS) analysis suggested that PEF decreased the lamellar thickness of starch (Zeng et al. 2016).

13.2.4 Effect of PEF on Starch Molecular Weight

The weight average molecular weight (Mw) and number average molecular weight (Mn) of corn starch are 1.018×10^8 and 4.683×10^7, respectively. After PEF treatment, under the condition of 30 kV/cm, as the PEF treatment time increased to 424 μs, 848 μs, 1272 μs, the Mw of corn starch decreased to 8.855×10^7, 8.459×10^7, 8.175×10^7, respectively. Under the condition of 40 kV/cm, as the PEF treatment time increased to 424 μs, 848 μs, 1272 μs, the Mw of corn starch decreased to 5.712×10^7, 5.001×10^7, 4.472×10^7, respectively. Under the condition of 50 kV/cm, as the PEF treatment time increased to 424 μs, 848 μs, 1272 μs, the Mw of corn starch decreased to 3.511×10^7, 2.806×10^7, 1.306×10^7, respectively (Han et al. 2012a). Mn decreases with the increase in PEF intensity and treatment time. After high field strength PEF (50 kV/cm) treatment, Mn of corn starch decreases from 4.683×10^7 of native starch to 4.6×10^6. PEF of 50 kV/cm significantly reduced the molecular weight of maize starch. However, PEF slightly decreased the molecular weight and z-average radius of gyration of waxy rice starch from 10.38×10^7 to 8.9×10^7 (g/mol) and from 337 to 302 nm, respectively (Zeng et al. 2016). The difference may be due to the experimental conditions (e.g., different electrical conductivity and treatment time) as well as the starch type (waxy rice vs. maize) and concentrations (8% vs. 10%). After PEF treatment, the surface layer of corn starch is destroyed, which means that the amylopectin entangled on the surface layer of the starch is broken, resulting in a decrease in the molecular weight of corn starch.

13.3 Effect of PEF on Physicochemical Properties of Starch

13.3.1 Swelling Degree and Solubility

The swelling degree and solubility of starch are important properties. The swelling degree and solubility of corn starch increase with the increase in temperature and reach the maximum at about 90 °C. The swelling degree and solubility of corn starch are waxy corn starch > normal corn starch > G50 (amylose content in starch is 50%) > G80 (amylose content in starch is 80%). With the increase in amylose content, the swelling degree and solubility of corn starch decrease. After PEF treatment, with the increase in electric field intensity and treatment time, both the swelling degree and solubility of starch decreased. With the increase in amylose content, the influence of PEF on its swelling degree and solubility weakened. The swelling degree of starch is related to the intermolecular force. The lower the swelling degree, the stronger the intermolecular force. The PEF treatment destroys the crystalline regions in the starch molecules that are not tight enough to make the undamaged crystalline regions more stable and uniform.

13.3.2 Viscosity Properties

Pasting temperature (PT) is the temperature at which an increase in viscosity (onset) is observed. After PEF treatment with high electric field strength (e.g., 50 kV/cm), starch tends to decrease the viscosity during pasting event. For example, treating corn starch with PEF at 50 kV/cm reduced the peak viscosity from 335 BU (untreated starch) to 250 BU during pasting (Han et al. 2009a). For potato starch, when the electric field strength was increased from 30 $kV{\cdot}cm^{-1}$ to 40 $kV{\cdot}cm^{-1}$ and 50 $kV{\cdot}cm^{-1}$, the difference of peak viscosity between native and treated sample was correspondingly increased from 190 BU to 256 BU and 320 BU, respectively. Meanwhile, the breakdown viscosity was also decreased with increasing electric field strength. When the electric field strength was increased from 30 $kV{\cdot}cm^{-1}$ to

40 kV·cm^{-1} and 50 kV·cm^{-1}, the difference in breakdown viscosity between native potato starch and treated sample was correspondingly increased from 221 BU to 267 BU and 318 BU, respectively.

13.3.3 Thermal Properties

Thermal analysis by DSC (differential scanning calorimetry) showed that PEF tends to decrease the ΔH (enthalpy change) and gelatinization temperatures as well as the gelatinization temperature range (Han et al. 2009a, b, 2012b). For example, treating corn starch with PEF at 50 kV/cm reduced the peak temperature of gelatinization from 73.5 °C (untreated starch) to 71.5 °C as measured by DSC (Han et al. 2009a). The reduced gelatinization temperature range suggests that PEF simultaneously disrupted the crystalline region with similar structure and differentially destroyed those with different crystalline structure. Thermogravimetric analysis revealed that PEF decreased the decomposition temperature of maize starch. PEF treatment tends to cause partial gelatinization of starch and reduce the gelatinization temperatures and ΔH and viscosity of starch during pasting event.

13.3.4 In Vitro Digestive Properties

PEF decreased the resistant starch (RS) and slowly digestible starch (SDS) contents of waxy rice starch while increasing that of rapidly digestible starch (RDS) (Wu et al. 2019; Zeng et al. 2016). For example, treating waxy starch with PEF of 50 kV/cm decreased the content of resistant starch from 22% to 14%. Table 13.1 shows the amounts of rapidly digestible starch (RDS), slowly digestible starch (SDS) and resistant starch (RS) in native and PEF treated starches. All PEF treated starches showed higher RDS levels and lower SDS and RS levels than that of native starch. Furthermore, with the increase in PEF intensity, RDS levels increased, while SDS and RS levels decreased.

PEF treatment increased the sensibility of starch toward digestive enzyme. The disruption of starch granule packaging could be responsible for the increased RDS

Table 13.1 The amount of RDS, SDS, and RS of native and modified starches (Zeng et al. 2016) (Reused with permission from Elsevier Publications)

Starch sample	RDS (%)	SDS (%)	RS (%)
Native starch	32.4 ± 1.4^{d}	45.5 ± 2.0^{a}	22.1 ± 1.7^{a}
30 kV/cm	37.4 ± 0.9^{c}	42.0 ± 0.8^{ab}	20.7 ± 1.1^{ab}
40 kV/cm	43.1 ± 1.1^{b}	39.4 ± 1.0^{b}	17.6 ± 1.2^{bc}
50 kV/cm	50.4 ± 2.4^{a}	35.2 ± 1.3^{c}	14.4 ± 0.6^{c}

Values followed by the different superscripts in the same column are significantly different ($P < 0.05$)

and decreased SDS and RS levels, which facilitates the entry of the enzyme into the granule interior.

13.4 Effect of PEF on Modification of Starch Esterification

13.4.1 Effect on Degree of Substitution (DS) and Reaction Efficiency (RE)

Starch is composed of amylopectin and amylose. The position where PEF acts on starch is related to the ratio of amylose to amylopectin. Therefore, the change in starch chain structure has different effects on the degree of substitution (DS) of esterified starch. Waxy corn starch, normal corn starch, corn starch with 50% amylose content, and corn starch with 80% amylose content are named W, N, G50, G80, respectively. Figure 13.3 shows the effect of different amylose contents on DS of acetylated starches by conventional and PEF-assisted methods. The DS of acetylated starch increased with the increase in amylose, and the DS of G50 and G80 improved significantly. After traditional esterification of W, N, G50, and G80, the DS was 0.0646, 0.0650, 0.0694, 0.0744, respectively. After the four starches by

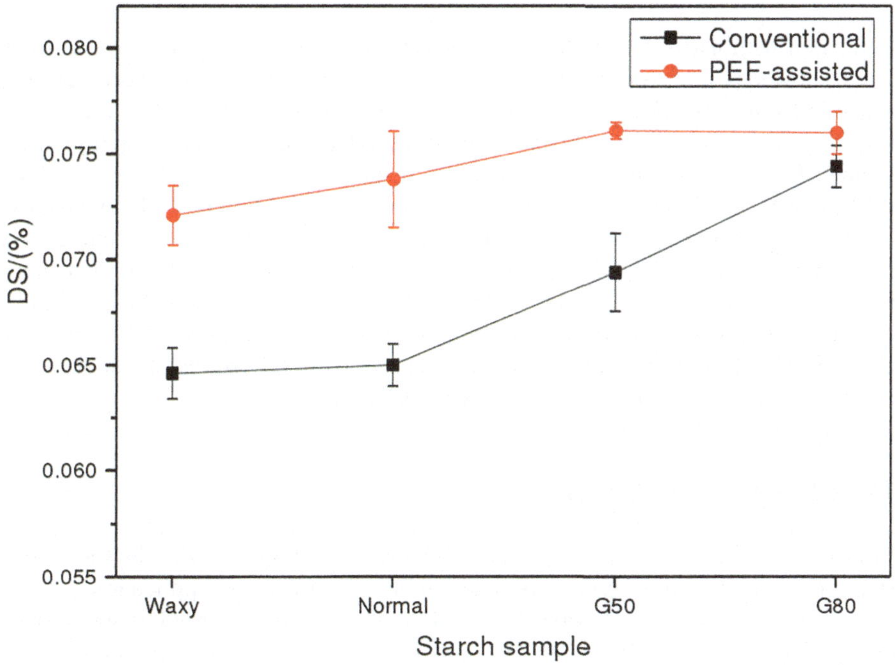

Fig. 13.3 Effect of different amylose contents on DS of starches esterified by conventional and PEF-assisted methods

PEF-assisted, the DS was 0.0721, 0.0738, 0.0761, and 0.0760, respectively. A similar trend was observed in PEF-assisted octenyl succinic anhydride (OSA) esterified potato starch. The DS and RE of PEF-assisted OSA starches increased from 0.0102 to 0.0190 and from 43.3% to 82.4%, respectively, compared to the no-PEF control (B.-R. Chen et al. 2021). PEF assistance significantly improves the acetylation effect of high-amylopectin starch, while it has no obvious effect on the acetylation effect of high-amylose starch. The study of (Wang et al. 2013) showed that the esterification reaction mostly occurs on the surface of starch granules. The surface of starch is mainly composed of amylose, so starch with higher amylose content is more prone to esterification substitution reaction. Furthermore, the application of PEF destroys the crystalline structure of starch, causing more hydroxyl groups of amylopectin to be released, acetyl substitution reaction is more likely to occur. PEF can increase the DS value of esterified starch in a short time, which not only provides a new method for the industrial production of esterified starch, but also reduces reaction costs and environmental pollution while improving reaction efficiency.

13.4.2 The Effect of PEF on the Structure of Esterified Starch

13.4.2.1 Fourier Translation Infrared Spectroscopy

Compared with native starch, the FT-IR spectrum of PEF synergistically esterified starch showed new absorption peaks at 1733 cm^{-1} (C=O stretching of acetyl group) and 1560 cm^{-1} (carboxyl stretching of acetyl group). It shows that there is a stretching vibration of the carboxyl group of acetylated starch in the starch granule sample. Figure 13.4 shows FT-IR spectra of native (a) and pulsed electric field (PEF)-assisted octenyl succinylated potato starches at different PEF intensities. Compared to native starch, the FT-IR spectra of esterified starch presented two characteristic octenyl succinic group peaks. One peak at 1728 cm^{-1} was characteristic of ester carbonyl group stretching (C=O). Another band at 1567 cm^{-1} was caused by the asymmetric stretching vibration of carboxyl groups (COONa−) (Zhou et al. 2009). These phenomena indicated that the hydroxyl groups of starch molecules had been successfully connected with OSA groups.

13.4.2.2 Crystalline Structure

The crystalline structure of starch has an important relationship with the molecular structure and formation conditions and ultimately affects its application performance. The crystallinity of waxy corn starch, native corn starch, G50 starch, and G80 starch is 24.3%, 20.1%, 13.7%, and 12.8%, respectively. This tendency is largely the result of the increasing amylose content which reduces the formation of double helices by intermolecular hydrogen bonds in amylopectin segments leading

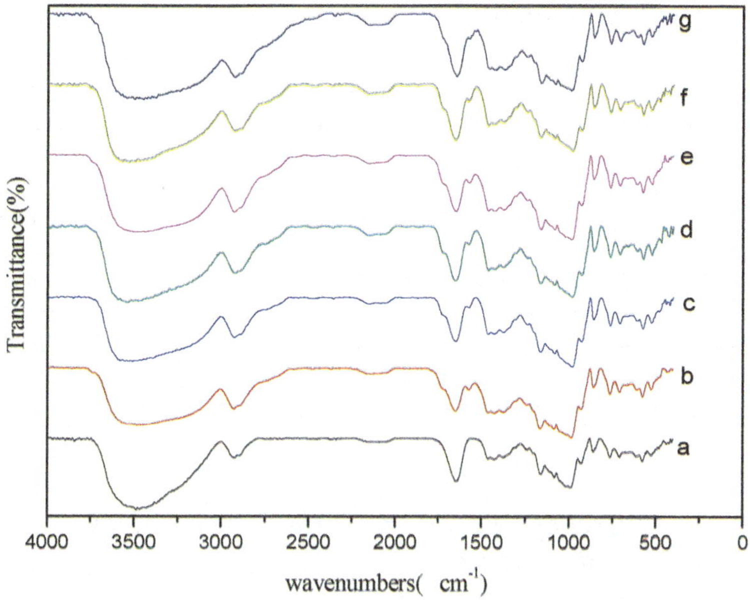

Fig. 13.4 FT-IR spectra of native (a) and pulsed electric field (PEF)-assisted octenyl succinylated potato starches at different PEF intensities. (b) Control (0 kV/cm), (c) 2 kV/cm, (d) 3 kV/cm, (e) 4 kV/cm, (f) 5 kV/cm, (g) 6 kV/cm. (B.-R. Chen et al. 2021) (Reused with permission from Elsevier Publications)

to more unordered conformation (Diop et al. 2011; Noda et al. 2009). The crystallinity of native corn starch (20.4%), G50 starch (12.9%), and G80 starch (12.6%) modified by PEF synergistic esterification is slightly higher than that of native corn starch (19.7%), G50 starch (11.7%), and G80 starch (12.3%) modified by traditional esterification. The crystallinity of starch modified by esterification is reduced. This can be explained by the replacement of hydroxyl groups by acetyl groups, thus decreasing the inter- and intramolecular hydrogen bonds and resulting in the destruction of the original ordered crystal structure (Halal et al. 2015). Compared with the traditional esterification reaction, an increase in crystallinity appeared in native corn starch, G50 and G80 starches after PEF synergistic esterification treatment, indicating the reconstruction of the molecular structure and more double helices formation by intermolecular hydrogen bonds. Furthermore, after the waxy corn starch is esterified by PEF, its crystal structure changes from the A-type crystal structure to the B-type crystal structure. This demonstrates that the molecular order and alternative repeating stacks of the amorphous and crystalline lamella in starch esters with low amylose content can be easily destroyed by the treatment of conventional as well as dual esterification (Brewer et al. 2012).

13.4.3 Amylose Content

Previous studies have shown that esterification decreased the amylose content of potato starch, and the amylose content of PEF-assisted OSA starches was lower than that of the no-PEF control. The effect was more obvious for starch under higher PEF intensity conditions. The amylose content decreased from 28.4% to 21.3% with increasing electric intensity from 2 kV/cm to 6 kV/cm, while the amylose content of the control sample obtained by the conventional method was 28.5% (B.-R. Chen et al. 2021). A previous study revealed the effect of OSA esterification on amylose chains might be due to the incorporation of a long hydrophobic chain, which leads to reduce the absorption of iodine (Lopez-Silva et al. 2020). As the PEF intensity increased, there were more octenyl succinic groups (higher DS) grafted onto the starch molecular chains, which further prevented iodine absorption.

13.4.4 The Effect of PEF on the Surface Morphology of Esterified Starch

After esterification, defects and bumps can be observed on the granular surface without obviously deformed shape. Figure 13.5 shows the SEM images of native starch and esterified starch. Starch granules of G50CE, G50PE, G80CE, and G80PE displayed heavier deformation and roughness than waxy and normal starch granules. As PEF was applied during the esterification reaction, more damage and deformation occurred to the granules as seen in WPE, NPE, G50PE, and G80PE. It can be speculated that starch granules with higher amylose content (≥50%) were more sensitive to esterification and detectable deformation was found in esterified starch granules. This may be attributed to the amorphous area which is easy to be attacked by acetyl groups and higher amylose content means larger area of amorphous lamella. The higher amylose content of starch granules contained, the more hydrogen of hydroxyl groups in starch granules was substituted during acetylation. In addition, esterification increased the particle size of potato starch. The increase in the mean diameter could be due to the incorporation of OS groups and increased disorder in the granule structure. This caused the granules to swell to a greater extent.

13.4.5 The Effect of PEF on the Physical and Chemical Properties of Esterified Starch

13.4.5.1 Pasting Properties

Pasting temperature is an extremely important attribute of the functionality of starch-based products. Pasting temperature (PT) is the temperature at which an increase in

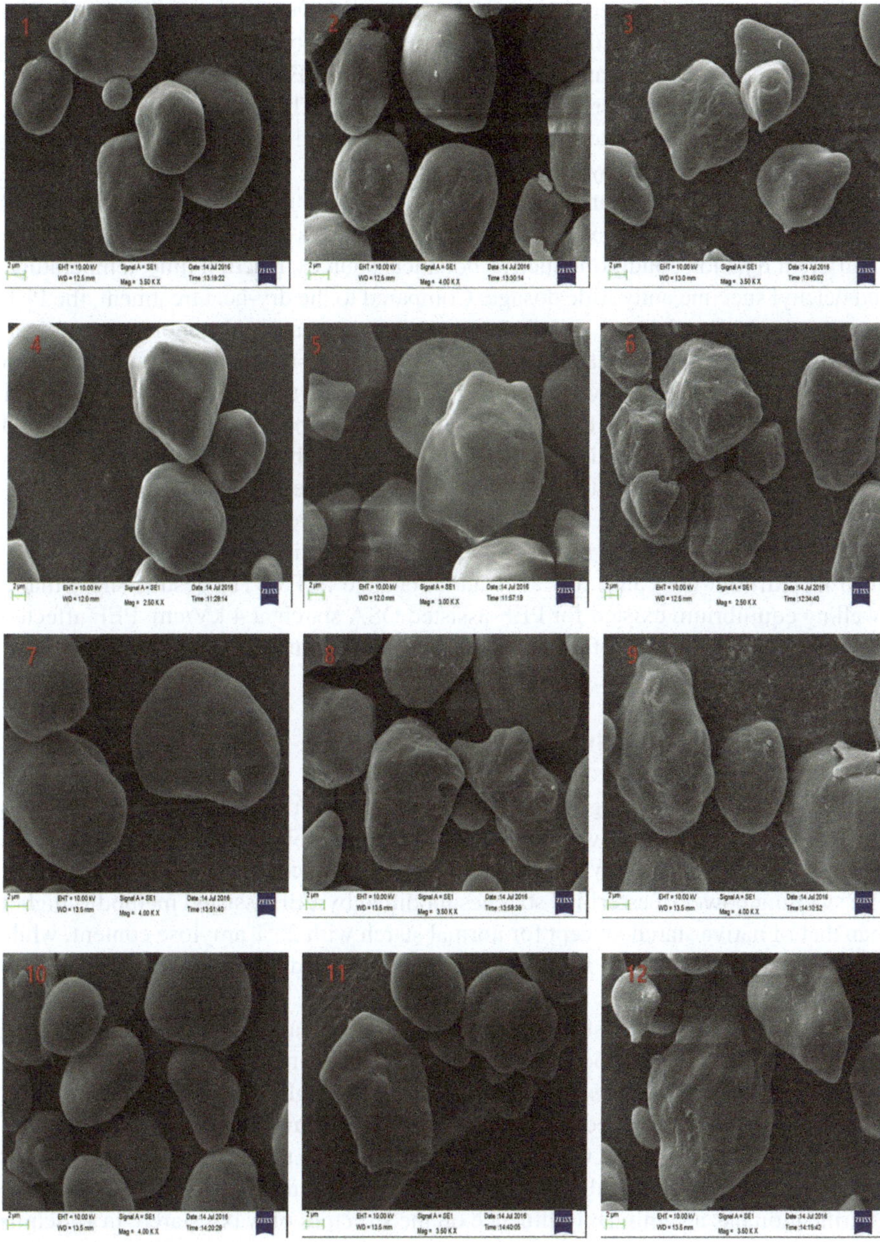

Fig. 13.5 SEM images of native and esterified starches. NO. 1–12 represent NW, WCE, WPE, NN, NCE, NPW, NG50, G50CE, G50PE, NG80, G80CE, G80PE, respectively. NW, NN, G50, and G80 correspond to native waxy starch, native normal starch, starch with 50% amylose, and starch with 80% amylose. WCE and WPE represent waxy starch esterified by conventional method and PEF-assisted dual acetylation methods (Hong et al. 2018a) (Reused with permission from Elsevier Publications)

viscosity (onset) is observed. Compared with the native potato starch, the pasting temperature of PEF-assisted OSA starch (PEF intensity of 5 kV/cm) reduced by 15.1 °C. Conversely, the pasting temperature of non-PEF-assisted octenyl succinic anhydride (OSA) starch decreased by only 3 °C (B.-R. Chen et al. 2021). The pasting temperatures of PEF-assisted OSA starches decreased significantly compared to those starches esterified by other synthesis process, such as dry heating post-treatment (130 °C, 2 h) (Sandhu et al. 2015) and hydrothermal pre-treatment (60 ° C, 3 h) (X. Chen et al. 2014). The variation of PT values in the literature may be due to different reaction conditions such as botanical sources, reaction time, temperature, and octenyl succinic anhydride dosage. Compared to the dry-heat treatment, the PEF treatment as a non-thermal processing technique appears to be more economical. Low pasting temperature starch can be used as a performance additive, and it can promote the sustainable development of paper and textile industries. This kind of starch can also be used in the production of food, especially instant noodles and potato chips, with the added advantages of reduced edible oil consumption.

Peak viscosity (PV) is the water absorption and swelling ability of starch granules. The PV of PEF-assisted OSA starches gradually reduced as the PEF intensity increased from 0 kV/cm to 4 kV/cm. However, further increasing the PEF intensity (to 6 kV/cm) did not cause any substantial change in PV. The results show that a swelling equilibrium existed for PEF-assisted OSA starch at 4 kV/cm. PEF affected the swelling performance of granules, leading to a decline in viscosity.

13.4.5.2 Solubility and Swelling Power

The solubility and swelling power of native and esterified starches with various amylose contents are shown in Table 13.2. The solubility of native starches increased from 2.18% (NW) to 6.38% with amylose content increased (G80CE). The swelling power of esterified starches modified by PEF-assisted method is higher than that of native starch, except for normal starch with 23% amylose content, while lower than esterified starch by the conventional method with significant difference ($p < 0.05$) as seen in Table 13.2.

The introduction of substituting groups in amylose and amylopectin chains had bulk hindrance, which helped retain water molecules, resulting in higher swelling power of esterified starches than native starches (Singh et al. 2004; Zhong et al. 2006). Starch esters showed a decrease in swelling power by the modification of PEF-assisted at level of $p < 0.05$, which could be due to the PEF-induced damage of the granular structure, resulting in a weakened water retention ability. Moreover, esterification has a significant influence on the structure of waxy starch, in which its crystalline area was prone to be destroyed.

Substituted groups were most likely occurred in the interior crystalline domains composed by amylopectin main chains, causing structural damage and inducing the higher solubility of waxy starch esters, particularly in the PEF-assisted dual modification (Miao et al. 2014).

Table 13.2 Solubility and swelling power of esterified starches differing in amylose content as modified by conventional and dual methods (Hong et al. 2018a)

Sample	Solubility/(%)	Swelling power/(g/g)
NW	2.18 ± 0.58	6.74 ± 0.03
WCE	11.25 ± 0.82	8.76 ± 0.87
WPE	18.99 ± 0.23	7.61 ± 0.11
NN	4.31 ± 0.37	12.53 ± 0.08
NCE	5.89 ± 0.02	13.47 ± 0.15
NPE	9.05 ± 0.64	10.74 ± 0.86
NG50	5.24 ± 0.09	5.48 ± 0.01
G50CE	12.65 ± 0.54	10.67 ± 0.26
G50PE	19.20 ± 0.09	9.50 ± 0.38
NG80	6.38 ± 0.21	6.03 ± 0.05
G80CE	12.39 ± 0.76	7.22 ± 0.04
G80PE	17.04 ± 1.47	7.00 ± 0.01

Results were means ± standard deviation. Data of same variety in the same column with different letters (a–c) are significantly different ($p < 0.05$). NW, NN, G50, and G80 correspond to native waxy starch, native normal starch, starch with 50% amylose, and starch with 80% amylose. WCE and WPE represent waxy starch esterified by conventional method and PEF-assisted dual acetylation methods

13.4.5.3 Freeze-Thaw Stability

Freeze-thaw stability, expressed as percent syneresis, is a significant indicator for assessing the ability of starch gels to withstand undesirable physical changes during repeated freezing and thawing (Meng et al. 2014). During the first 3 cycles, esterified starches have lower syneresis and better freeze-thaw stability than native starches, especially for starches with higher amylopectin content modified by conventional method. This is due to the increased steric hindrance induced by more amylopectin starch branches, which helps prevent water molecules from separating from the gel network.

13.4.5.4 In Vitro Digestibility

Rapidly digested starch (RDS) is defined as starch that can be digested quickly in the first 20 min of digestion in the small intestine, while slowly digested starch (SDS) is defined as starch digested in 20–120 min, and resistant starch (RS) is defined as starch that cannot be digested within 120 min (Englyst et al. 1992). The amylose content in native maize starch increases, it is more resistant to enzyme digestion by α-amylase and amyloglucosidase (Brewer et al. 2012). Both conventional and PEF-assisted maize starch esterification can change glycemic digestibility and decrease the RDS fraction while increasing the SDS and RS fractions. Moreover, the higher amylose content of a starch, the lower the digestibility improvement due to acetylation. It is notable that native maize starch shows reacts strongly to decreased digestibility after esterification, especially with PEF-assisted (Hong et al. 2018b). After PEF-assisted esterification, more noticeable bulges and roughness appeared on the surface, as well as serious deformation of particle shape, as compared to conventionally esterified starches, making granules more sensitive to enzyme hydrolysis, as indicated by a higher RDS.

13.4.6 The Mechanism of PEF-Assisted Starch Esterification Reaction

Figure 13.6 shows the synergistic mechanism of PEF in the esterification reaction.

1. The PEF system generates a high potential difference between the two electrodes in the esterification reaction processing chamber. The reaction chamber is composed of two electrode plates. After voltage is applied to both ends of the processing chamber, a high potential difference is generated between the electrode plates. When the PEF used is a bipolar square wave, the direction of the electric field strength between the two electrodes is constantly changing, which can change the direction of movement of the charged particles in the esterification reaction. Therefore, during the esterification reaction, the high potential difference produced by PEF can accelerate the movement rate of the charged particles in the reaction system. Conversely, the bipolar pulse wave can change the direction of the field strength between the two plates, promote the effective collision of charged particles between the reaction systems, and promote the formation of esterification products. The high potential difference and the change of the direction of the electric field can not only accelerate the mobility of the reaction particles, but also change the direction of their movement, which leads to an increase in effective collisions and improves the efficiency of the esterification reaction.
2. When PEF synergizes the esterification reaction, a large amount of energy is input to the reaction system, and PEF provides energy for the energy barrier transition of the esterification reaction, thereby accelerating the progress of the esterification reaction.
3. PEF can destroy the structure of starch granules and facilitate the entry of esterification agent. Previous studies have shown that during the PEF treatment process, after 100–300 V/cm PEF treatment, pores are formed on the surface of the organism (C. Chen et al. 2006; Parniakov et al. 2015). Under the action of PEF, the ions present in the solution move and gather on the surface of starch granules to form a macroscopic space charge, that is, space polarization charge. With the continuous increase in the PEF voltage, an instant high-voltage discharge is generated on the outer layer of the starch granule, which ruptures the outer layer of the starch granule, and the amylopectin is broken and precipitated, which leads to the destruction of the starch granule. Therefore, PEF treatment can destroy the structure of starch granules and provide a convenient way for ester group substitution.

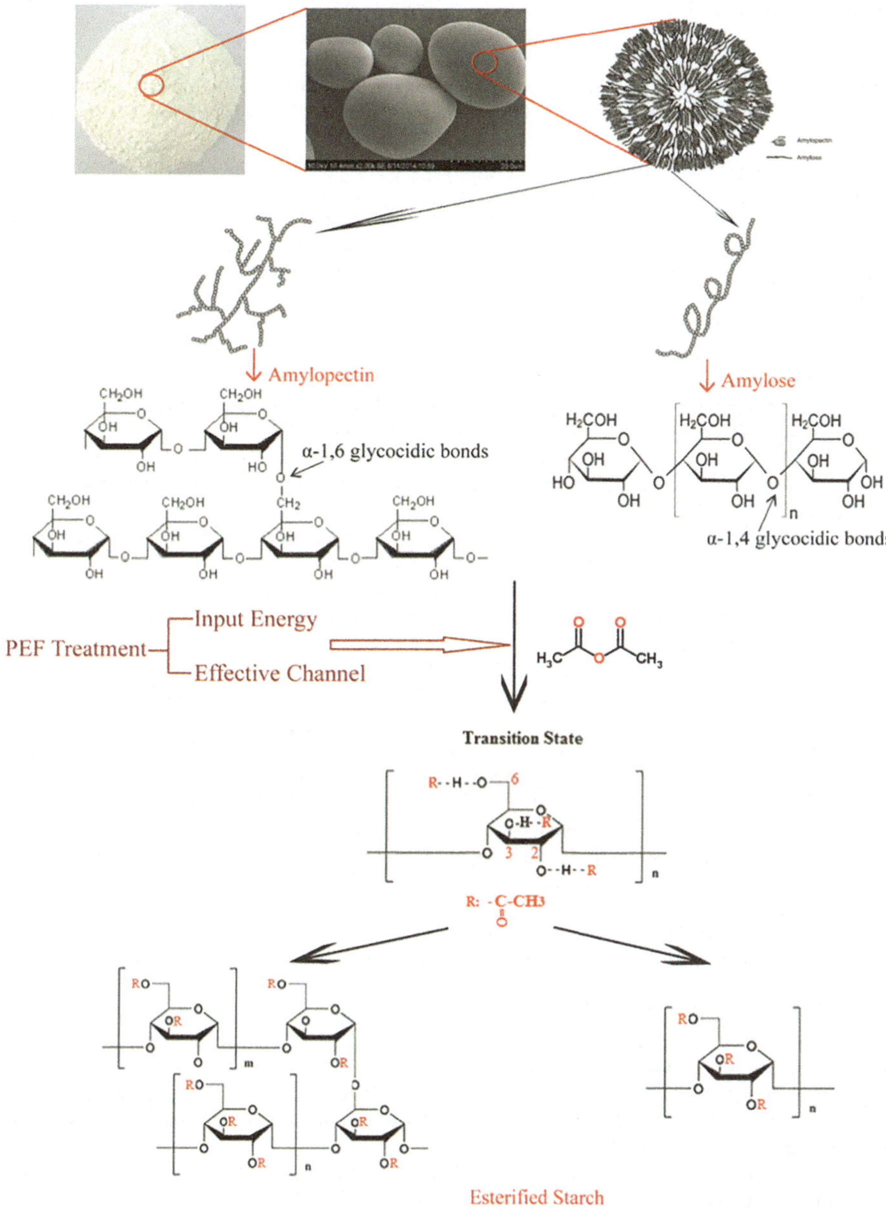

Fig. 13.6 Diagram of the action mechanism of pulsed electric field assisted starch esterification reaction

References

Brewer LR, Cai L, Shi Y-C (2012) Mechanism and enzymatic contribution to in vitro test method of digestion for maize starches differing in amylose content. J Agric Food Chem 60(17): 4379–4387

Chen C, Smye SW, Robinson MP, Evans JA (2006) Membrane electroporation theories: a review. Med Biol Eng Comput 44(1):5–14

Chen X, He X, Huang Q (2014) Effects of hydrothermal pretreatment on subsequent octenylsuccinic anhydride (OSA) modification of cornstarch. Carbohydr Polym 101:493–498

Chen B-R, Wen Q-H, Zeng X-A, Abdul R, Roobab U, Xu F-Y (2021) Pulsed electric field assisted modification of octenyl succinylated potato starch and its influence on pasting properties. Carbohydr Polym 254:117294

Diop CIK, Li HL, Xie BJ, Shi J (2011) Effects of acetic acid/acetic anhydride ratios on the properties of corn starch acetates. Food Chem 126(4):1662–1669

Englyst HN, Kingman SM, Cummings JH (1992) Classification and measurement of nutritionally important starch fractions. Eur J Clin Nutr 46(Suppl 2):S33–S50

Halal SLME, Colussi R, Pinto VZ, Bartz J, Radunz M, Carreño NLV, Dias ARG, Zavareze E, d. R. (2015) Structure, morphology and functionality of acetylated and oxidised barley starches. Food Chem 168:247–256

Han Z, Zeng X-A, Zhang B-S, Yu S-J (2009a) Effects of pulsed electric fields (PEF) treatment on the properties of corn starch. J Food Eng 93(3):318–323

Han Z, Zeng XA, Yu SJ, Zhang BS, Chen XD (2009b) Effects of pulsed electric fields (PEF) treatment on physicochemical properties of potato starch. Innovative Food Sci Emerg Technol 10(4):481–485

Han Z, Yu Q, Zeng XA, Luo DH, Yu SJ, Zhang BS, Chen XD (2012a) Studies on the microstructure and thermal properties of pulsed electric fields (PEF)-treated maize starch. Int J Food Eng 8(1)

Han Z, Zeng XA, Fu N, Yu SJ, Chen XD, Kennedy JF (2012b) Effects of pulsed electric field treatments on some properties of tapioca starch. Carbohydr Polym 89(4):1012–1017

Hong J, Chen R, Zeng X-A, Han Z (2016a) Effect of pulsed electric fields assisted acetylation on morphological, structural and functional characteristics of potato starch. Food Chem 192:15–24

Hong J, Zeng X-A, Buckow R, Han Z, Wang M-S (2016b) Nanostructure, morphology and functionality of cassava starch after pulsed electric fields assisted acetylation. Food Hydrocoll 54:139–150

Hong J, Zeng X-A, Buckow R, Han Z (2018a) Structural, thermodynamic and digestible properties of maize starches esterified by conventional and dual methods: differentiation of amylose contents. Food Hydrocoll 83:419–429

Hong J, Zeng X-A, Han Z, Brennan CS (2018b) Effect of pulsed electric fields treatment on the nanostructure of esterified potato starch and their potential glycemic digestibility. Innovative Food Sci Emerg Technol 45:438–446

Lopez-Silva M, Bello-Perez LA, Castillo-Rodriguez VM, Agama-Acevedo E, Alvarez-Ramirez J (2020) In vitro digestibility characteristics of octenyl succinic acid (OSA) modified starch with different amylose content. Food Chem 304:125434

Meng Y-C, Sun M-H, Fang S, Chen J, Li Y-H (2014) Effect of sucrose fatty acid esters on pasting, rheological properties and freeze–thaw stability of rice flour. Food Hydrocoll 40:64–70

Miao M, Li R, Jiang B, Cui SW, Zhang T, Jin Z (2014) Structure and physicochemical properties of octenyl succinic esters of sugary maize soluble starch and waxy maize starch. Food Chem 151: 154–160

Noda T, Isono N, Krivandin AV, Shatalova OV, Błaszczak W, Yuryev VP (2009) Origin of defects in assembled supramolecular structures of sweet potato starches with different amylopectin chain-length distribution. Carbohydr Polym 76(3):400–409

Parniakov O, Barba FJ, Grimi N, Marchal L, Jubeau S, Lebovka N, Vorobiev E (2015) Pulsed electric field and pH assisted selective extraction of intracellular components from microalgae nannochloropsis. Algal Res 8:128–134

Sandhu KS, Sharma L, Kaur M (2015) Effect of granule size on physicochemical, morphological, thermal and pasting properties of native and 2-octenyl-1-ylsuccinylated potato starch prepared by dry heating under different pH conditions. LWT Food Sci Technol 61(1):224–230

Singh N, Chawla D, Singh J (2004) Influence of acetic anhydride on physicochemical, morphological and thermal properties of corn and potato starch. Food Chem 86(4):601–608

Wang C, He X, Huang Q, Fu X, Luo F, Li L (2013) Distribution of Octenylsuccinic substituents in modified a and B polymorph starch granules. J Agric Food Chem 61(51):12492–12498

Wu C, Wu Q-Y, Wu M, Jiang W, Qian J-Y, Rao S-Q, Zhang L, Li Q, Zhang C (2019) Effect of pulsed electric field on properties and multi-scale structure of japonica rice starch. LWT 116: 108515

Zeng F, Gao Q-Y, Han Z, Zeng X-A, Yu S-J (2016) Structural properties and digestibility of pulsed electric field treated waxy rice starch. Food Chem 194:1313–1319

Zhong F, Yokoyama W, Wang Q, Shoemaker CF (2006) Rice starch, amylopectin, and amylose: molecular weight and solubility in dimethyl sulfoxide-based solvents. J Agric Food Chem 54(6):2320–2326

Zhou J, Ren L, Tong J, Xie L, Liu Z (2009) Surface esterification of corn starch films: reaction with dodecenyl succinic anhydride. Carbohydr Polym 78(4):888–893

Zhu F (2018) Modifications of starch by electric field based techniques. Trends Food Sci Technol 75:158–169

Chapter 14
Chemical Changes During Physical Treatments

Chuangchuang Zhang, Solomon Abate Mekonnen, and Zhongquan Sui

Abstract Physical treatments generally induce changes only in the packing arrangements of starch polymer molecules within granules. However, some physical treatments induce changes in the chemical structures of starch polymer molecules due to the formation of various free radicals during modification. These possible chemical changes mainly include transglucosidation, functional group introduction, cross-linking, and degradation into different radiolytic products such as aldehydes (formaldehyde, acetaldehyde, acetone, and dihydroxyacetone), acids (formic, glucuronic acid, and pyruvic acid), dextrin, maltotetraose, maltotriose, maltose, and glucose, having significant impacts on the physicochemical and functional properties of starch. This chapter attempts to present detailed information about chemical changes and their mechanism during physical treatments such as gamma irradiation, ultraviolet (UV) irradiation, cold plasma, and pulsed electric field (PEF).

Keywords Chemical changes · Gamma irradiation · Ultraviolet irradiation · Cold plasma · Pulsed electric field

14.1 Introduction

Physical treatments generally induce changes only in the packing arrangements of starch polymer molecules within granules, and such structural changes can have significant impacts on the physicochemical and functional properties of starch. The chemical structures of starch polysaccharide molecules are altered mainly due to the formation of various free radicals during some physical treatments such as gamma irradiation, ultraviolet (UV) irradiation, cold plasma, and pulsed electric field (PEF),

C. Zhang · Z. Sui (✉)
Department of Food Science & Technology, School of Agriculture and Biology, Shanghai Jiao Tong University, Shanghai, China
e-mail: zsui@sjtu.edu.cn

S. A. Mekonnen
Department of Food Science and Nutrition Research, Ethiopian Institute of Agricultural Research, Addis Ababa, Ethiopia

Z. Sui, X. Kong (eds.), *Physical Modifications of Starch*,
https://doi.org/10.1007/978-981-99-5390-5_14

although these treatments do not use chemical reagents (BeMiller and Huber 2015). It should be noted that these chemical changes such as cross-linking, degradation, transglucosidation, and functional group introduction can also significantly alter the starch properties (e.g., gelatinization, pasting, gelling, and digestion properties).

The various free radicals produced by gamma radiation resulted in transglucosidation, cross-linking, and degradation into different radiolytic products such as aldehydes (formaldehyde, acetaldehyde, acetone, and dihydroxyacetone), acids (formic, glucuronic acid, and pyruvic acid), dextrin, maltose, and glucose. The macromolecule degradation may be in a certain order: amylopectin degradation mainly at α-1,6 linkage first and then simultaneous degradation of free long chains and true amylose during gamma radiation. It is worth noting that gamma radiation degradation randomly occurred at both α-1,4 and α-1,6 linkage compared to thermal degradation mainly at α-1,6 linkage.

The free radical formation during UV irradiation cleaved glycosidic bond to shorten the amylose chain and debranch the amylopectin chain, resulting in less energy being required for gelatinization. Oxygen-induced cross-linking reactions involved amylose or small molecular fractions with greater reactivity rather than amylopectin. The UV-induced depolymerization occurred in both amylopectin and amylose molecules under both aerobic and anaerobic conditions, whereas cross-linking mainly affected the amylose fractions.

Besides the primary free radicals, some secondary free radicals were formed by interactions of plasma reactive species and biopolymer surfaces. These free radicals can induce the cross-linking and depolymerization of starch chains as well as plasma etching, resulting in various surface modifications of biodegradable polymers. In addition, functional group introduction on the starch molecules can also result in starch modification due to interaction between the chemically active species and starch. It is concluded that cross-linking occurred at lower voltage and shorter period of treatment plasma, whereas depolymerization occurred at higher voltages and longer treatment time (Kalaivendan et al. 2022; Thirumdas et al. 2017a).

PEF, as one of the nonthermal and physical methods, has the potential to expand the range of starch functionalities in a cleaner and more efficient manner (Zhu 2018). Short pulses of high electric fields with a very short duration of micro- to milliseconds applied by pulsed electric field (PEF) can avoid excessive heat and unwanted electrolytic reactions and induce molecular scissions of starch (Maniglia et al. 2021).

14.2 Chemical Changes During Gamma Irradiation

The ionizing radiation (gamma, electron beam, and UV light) treatment for starches is considered to be nonthermal, cost-effective, fast, simple, and safe due to the absence of catalyst, minimal sample preparation, and no penetration of toxic substances (Sunder et al. 2022). In addition, all radiolytic end products are similar after treatment, indicating that the degradation changes of starch are reproducible and quantitative (Bhat and Karim 2009). Based on the above advantages, the ionizing radiation is an environmentally friendly alternative to common techniques of chemical and physical modifications. Gamma radiation, one of the most common physical methods used for starch modification in the food industry (Sudheesh et al. 2019a), was used in more than 40 countries and produced more than 60 different irradiated foods available in the market worldwide (Dar et al. 2018). The high-energy gamma rays emitted from ^{60}Co and ^{137}Cs can penetrate the target product up to several meters. All industries employed ^{60}Co due to technical difficulty, although both ^{60}Co and ^{137}Cs were permitted by FDA. FAO/WHO/IAEA joint expert committee proposed that the irradiated foods up to 10 kGy could be fit for human consumption (Sudheesh et al. 2019a). According to FDA, a low dose of gamma radiation (less than 2 kGy) can be used for improving the quality of food, a medium dose of gamma radiation (2–10 kGy) can be used to increase the shelf-life of food products, and a higher dose of gamma radiation (more than 10 kGy) can be used for sterilization of food products (Sunder et al. 2022). The various free radicals produced by radiation resulted in the cross-linking, grafting, and progressive reduction in molecular size of amylose and amylopectin by random cleavage of the glycosidic chains (radiolytic products: dextrin, maltose, and glucose). Starch chain length distributions showed no significant changes during thermal treatment (heat-moisture treatment and microwave heating), while the decrease in molecular size measured by size-exclusion chromatography (SEC) was observed in nondebranched starches (Chen et al. 2021; Liu et al. 2019), indicating that thermal degradation likely occurred at or near branch points of the amylopectin molecules. Thus, it is worth noting that gamma radiation degradation randomly occurred at both α-1,4 and α-1,6 linkage compared to thermal degradation mainly at α-1,6 linkage.

14.2.1 Carboxyl Content, Acidity, and pH

Free radicals generated by gamma irradiation react with starch molecules, result in the breakdown of the starch chains (amylose and amylopectin units), and produce different radiolytic products such as aldehydes (formaldehyde, acetaldehyde, acetone, and dihydroxyacetone) and acids (formic, glucuronic acid and pyruvic acid) (BeMiller et al. 2015; Zhu 2016). The radiolytic products formed during starch irradiation do not change with starch source (Bashir and Aggarwal 2019). The carboxyl content and acidity increased with increasing irradiation dose (Polesi et al. 2016; Sudheesh et al. 2019a; Verma et al. 2018).

14.2.2 Amylopectin

The cleavages of glycosidic bonds and cross-linking of the hydrolyzed fragments were observed in previous studies (Bao et al. 2005; Polesi et al. 2016). A significant decrease in the ratio of the first peak (amylopectin) intensity to the second peak (amylose) intensity measured by gel permeation chromatography (GPC) suggested that amylopectin degradation occurred after gamma irradiation (Bao et al. 2005; Rombo et al. 2004). The amylopectin degradation could be attributed to free radical depolymerization of starch chains. Degradation occurred more readily in waxy maize starch than amylose-containing (25–70%) maize starch, indicating that the amylopectin was more susceptible to degradation. The increase in branch chains DP > 25 and decrease in branch chain DP ≤ 24 were observed at 9 kGy in rice starch (Bao et al. 2005). Similarly, the increase in branch chains DP > 24 and decrease in branch chain DP ≤ 12 were observed, from 2 kGy dose in rice starch (Polesi et al. 2016). These observations suggested that cross-linking induced by free radicals might occur between branch chains. A slow dose rate (0.4 kGy/h) may be more favorable for forming cross-linking, because the starch is exposed to radiation for a longer period of time to absorb the same dose (Chung and Liu 2009). An increased proportion of β(1–3)- and β(1–4)-bonded starch molecules was observed with increasing irradiation dose in bean and maize flours (Rombo et al. 2004), which may be attributed to transglucosidation (Ghali et al. 1979).

The different effects of gamma irradiation on chain length distribution probed by high-performance anion-exchange chromatography (HPAEC) may be attributed to the difference in dose, starch type, and detection method (Zhu 2016). The proportion of amylopectin chains (DP 6–12 and ≥ 37) and average chain length increased with increasing irradiation dose in potato and bean starches, while the proportion of DP 13–24 decreased after irradiation (Chung and Liu 2010). The proportion of amylopectin chain (DP 6–12) increased with increasing irradiation dose in corn starches, while the proportions of amylopectin chains (DP 13–24 and ≥ 37) and average chain length decreased after irradiation (H. J. Chung, et al. 2009). The different trends in the proportion of longer chains between A-type starch (corn starch) and B-type starch (potato) or C-type starch (bean) may be attributed to the difference in proportion of undetectable long branch chains (DP > 65). Very long branch chains of amylopectin (DP > 65) were not detected, though they exist in starch. Shorter branch chain length with a detectable range of DP was produced due to the degradation of very long branch chains after irradiation, contributing to the increased proportions of DP ≥ 37.

The mechanism of amylopectin degradation is not clear due to very limited information on chain length distribution and molecular size distribution after irradiation. For example, the preferably cleaved and cross-linked sites in chains as well as the structural nature of these newly formed chains need to be further revealed. It is also interesting to study the effects of gamma irradiation on internal chain structure (in the form of φ-limit dextrin or φ,β-limit dextrin) of amylopectin and lamellar structure measured by small-angle X-ray scattering.

14.2.3 Amylose

The effect of gamma irradiation on apparent amylose content is most determined by colorimetric methods (based on the amylose-iodine complex formation) and depends on the radiation parameters and starch sources (Table 14.1). Besides dosage and dose rate, the radiation parameters usually include sample types (e.g., whole grains, whole flour, and starch powder) and water content (Table 14.1). For example, the apparent amylose content of corn starches with different amylose content increased first and then decreased with increasing irradiation dose (5–50 kGy) (Lee et al. 2013). A significant difference among these starches was observed in change extent of apparent amylose content with increasing dose, although these corn starches with different amylose content showed the same change trend (Lee et al. 2013). A similar result was also observed in corn starch with different amylose contents (Chung et al. 2015). With the increasing radiation dose, apparent amylose content increased first (1–10 kGy) and then decreased (10–50 kGy) in waxy corn starches, while it decreased subtly first (1–10 kGy) and then decreased significantly (10–50 kGy) in both normal and high amylose corn starches (Chung et al. 2015). Thus, gamma irradiation affected differently starch with different amylose content. The continuous and significant decrease in apparent amylose content was observed using buckwheat (5–20 kGy), oats (5–20 kGy), and chickpea (4–12 kGy) starch with the increasing radiation dose (Bashir and Haripriya 2016; Dar et al. 2018). The different trends of change in apparent amylose content with the increasing radiation dose between corn starch and other starch (buckwheat, oats, and chickpea) indicated that gamma irradiation affected differently starch with different sources.

The different change trends (increase or decrease or no significant change) of apparent amylose content at low irradiation dose may be attributed to differences in the sensitivity of starch samples to gamma radiation. Amylose was less susceptible to irradiation than amylopectin. It is possible that amylose had smaller size, different locations (the peripheral and interior parts of granules, or amorphous lamellae and crystalline lamellae), and different roles in the construction of starch granule structure. In addition, the free radicals generated by gamma radiation were more likely to break the α-1,6 linkage of amylopectin compared to the α-1,4 linkage of amylose.

On the one hand, amylopectin degradation may generate linear amylose-like chains, which explained the observed increase in amylose content at low irradiation dose. A minimum degree of polymerization (DP) 15 is required to form a helix which will be complex with iodine (Othman et al. 2015). The long amylopectin chains (B_2 and B_3 chains: DP > 36) can form an iodine complex and develop blue color, although amylopectin is not easy to form the helical complex because short and branched chains interfere or generate obstacles to complexation with iodine. It is possible that the free long chains from amylopectin obtained by breaking α-1,6 linkage can better bind iodine due to the elimination of obstacles at low irradiation dose. The long chains may be more easily cleaved from amylopectin than short chains due to different functions in starch structure. The long chains function as the interconnecting chains and are mainly confined to the amorphous lamellae, while

Table 14.1 Chemical change of starches under gamma irradiation

Sources	Amylose (%)	Dosage/ dose rate	Water content (%)	Amylose content	Other changes	References
Rice	26.43	5–50 kGy, 2 kGy/h	5	↓	pH: ↓	Lam et al. (2021)
Kithul	38.55	0.5–10kGy, 2 kGy/h	10.73	↓	pH: ↓ Carboxyl: ↑	Sudheesh et al. (2019a)
Pea, vetch	28.52–31.08	5 kGy, 5 Gy/h	12	↓		Majeed et al. (2018)
Buckwheat, potato	26.84–27.01	5–50 kGy, 2.5 kGy/h	8.72	↓	pH: ↓ Carboxyl: ↑	Verma et al. (2018)
Oat	22.41–27.33	0–20 kGy, 0.12 kGy/h	12	↓		Mukhtar et al. (2017)
Wheat	25.33	0.5–10 kGy	11.16	↑		Bashir et al. (2017)
Lentil	31.16	5 kGy	12	↓		Majeed et al. (2017)
Rice	18.13–20.62	1–10 kGy, 0.4 kGy/h	12.5	↓	pH: ↓ Carboxyl: ↑	Gul et al. (2016)
Chickpea	32.91	4–12 kGy, 2 kGy/h	12.4	↓	pH: ↓	Bashir, et al. (2016)
Elephant foot yam	28.3	5–25 kGy, 2 kGy/h	10.93	↓	pH: ↓ Carboxyl: ↑	Reddy et al. (2015)
Potato	30.39	5–30 kGy, 3.6 kGy/h	12.45	↓		Sujka et al. (2015)
Arrowhead	30.73	5–15 kGy, 5 kGy/h	8.3	↓	pH: ↓	Wani et al. (2015)
Sago	27.6	6–25 kGy, 8.3 kGy/h	13.14	Inconsistent change trends		Othman et al. (2015)
Maize	4.1–73.1	1–50 kGy, 1 kGy/h		Inconsistent change trends		Chung et al. (2015)
Banana	24.5	5–25 kGy, 2 kGy/h		↓	pH: ↓ Carboxyl: ↑	Koteswara Reddy et al. (2015)

(continued)

Table 14.1 (continued)

Sources	Amylose (%)	Dosage/ dose rate	Water content (%)	Amylose content	Other changes	References
Potato	32–32.6	5–20 kGy, 2 kGy/h	10	↓	pH: ↓ Carboxyl: ↑	Gani et al. (2014)
Rice	15.4–16	5–20 kGy, 2 kGy/h	12	↓	pH: ↓ Carboxyl: ↑	Ashwar et al. (2014)
Maize	57.67	30–60 kGy, 1.2 kGy/h		↑		Ocloo et al. (2014)
Horse chestnut	27.53	5–15 kGy, 5 kGy/h	11.67	↓	pH: ↓ Carboxyl: ↑	Wani et al. (2014)
Broad bean	51.69	5–15 kGy, 5 kGy/h	7.91	↓	pH: ↓ Carboxyl: ↑	Sofi et al. (2013)
Kidney bean	35.7–41.2	5–20 kGy, 2 kGy/h	10	↓	pH: ↓ Carboxyl: ↑	Lee et al. (2013)

short chains form the double helices of the crystalline lamellae according to the building block backbone model (Bertoft 2017). On the other hand, the decrease in apparent amylose content originated from the further breakage of glycoside bonds at chain ending in free long chains and/or true amylose at low irradiation dose. Thus, both macromolecule degradation may be in a certain order: amylopectin degradation mainly at α-1,6 linkage first and then simultaneous degradation of free long chains and true amylose.

The changes in apparent amylose content need to be further explained by high-performance size-exclusion chromatography (HPSEC) and high-performance anion-exchange chromatography (HPAEC), although the changes were revealed by colorimetric methods during radiation in most studies. The effect of gamma irradiation on true amylose content of starch is not clear since the long chains of amylopectin before and after degradation may be complex with iodine.

14.3 Chemical Changes During Ultraviolet (UV) Irradiation

The UV irradiation results in cross-linking, oxidative photodegradation (in air), depolymerization, and dextrinization via free radical formation, although it does not involve the added reagents (BeMiller et al. 2015). The radiation parameters (e.g., wavelength, intensity, and time) and starch sources can affect irradiation process.

Two distinct mechanisms of photodegradation have been proposed. One type, a photo-oxidation, included a light-sensitized hydrolysis and further oxidation and resulted in the complete conversion of amylose to CO_2 via the intermediate products (e.g., formaldehyde and formic acid) in the presence of O_2 at $\lambda < 366$ nm, while it needed to be initiated using a powerful sensitizer at $\lambda \geq 366$ nm (Peat et al. 1948). Such oxidation occurs primarily on the hydroxyl groups at the C-2 and C-3 positions, producing two aldehyde groups. The fresh a-glycol groups were released by hydrolyzing the oxidized amylose and then were further oxidated. The second form of photodegradation occurred in the absence of O_2 (Whelan and Peat 1949).

14.3.1 Depolymerization

The decrease in trough viscosity and final viscosity of UV-irradiated cassava and corn starch may be attributed to a decrease in swelling due to partial depolymerization of amylopectin and/or amylose in amorphous regions (Bertolini et al. 2000). The decrease in peak viscosity, enthalpy of gelatinization (ΔH), and gelatinization temperature and the increase in setback values were observed in white yam (*Dioscorea* sp.) starch (Hornung et al. 2018). The free radical formation during UV irradiation cleaved glycosidic bond to shorten the amylose chain and debranch the amylopectin chain, resulting in less energy being required for gelatinization.

The decrease in ΔH and paste viscosity of corn starch implied the photodegradation, which was further confirmed by a shift of peak mass of each region (amylopectin, intermediate, and amylose fractions) to a higher elution volume (based on HPSEC) under nitrogen (Fiedorowicz et al. 1999). The huge drop in molecular weight (M_w) was found for all regions during the first 5 h irradiation under nitrogen, while the subtle change in amylopectin weight fractions and the huge increase in intermediate weight fractions were observed during the first 5 h irradiation under nitrogen. The huge drop in amylopectin M_w and fractions was found during irradiation from 5 to 15 h, but no significant changes in the distribution (weight fraction vs. molar mass) were observed between 15 and 25 h under nitrogen. The number of free radicals in UV-irradiated starch reached the highest level during the first 5 h irradiation and then slowly decreased, which illustrated dominant effects of irradiation time on the above distribution. The starch irradiated under air for 5 h showed lower paste viscosity and ΔH compared to the one irradiated under nitrogen for 5 h, suggesting the sensitivity of starch to photodegradation in the presence of air (Fiedorowicz et al. 1999).

14.3.2 Cross-Linking

The increase in ΔH and paste viscosity after 15 h irradiation indicated possible cross-linkage formations among starch chains under air. Amylopectin showed a continued degradation during prolongated irradiation from 5 to 15 h, while amylose fractions showed an opposite shift to higher M_w regions, indicating that oxygen-induced cross-linking reactions involved amylose or small molecular fractions with greater reactivity rather than amylopectin. Thus, the UV-induced depolymerization occurred in both amylopectin and amylose molecules of native corn starch under both aerobic and anaerobic conditions whereas cross-linking mainly affected the amylose fractions (Fiedorowicz et al. 1999).

14.4 Chemical Changes During Cold Plasma Treatment

Cold plasma treatment of starch has recently been demonstrated as a novel method for physical modification in addition to microbial decontamination (BeMiller et al. 2015; Sarangapani et al. 2018). The modification is characterized by unique activity, nonthermal technology, low-cost and energy-processing aid, and diversity of generation and delivery. The cold plasma, the fourth state of matter, is a fully or partially ionized gas with a net neutral charge, including electrons, ions, free radicals, photons, UV radiation, reactive species, and molecules in their ground or excited states (Pankaj et al. 2018). The mechanism of modification mainly depends on type of feed gas, humidity, power, voltage, surrounding phase, and treatment time (Misra et al. 2016). Atomic oxygen and hydroxyl radicals are the most reactive due to their versatility in covalent bonding with many different compounds. Cold plasma generated using oxygen, carbon dioxide, and nitrogen gas mixtures included O^-, O_2^-, O^+, N^+, N_2^-, NO^+, and CO_2^+ (Thirumdas et al. 2017a). Cold plasma generated using nitrogen and oxygen gas induces energetic electron collision, resulting in reaction chain forming nitrogen oxides (Misra et al. 2016). Besides the primary free radicals, some secondary free radicals are formed by interactions of plasma reactive species and biopolymer surfaces. These free radicals can induce the cross-linking and depolymerization of starch chains as well as plasma etching, resulting in various surface modifications of biodegradable polymers. Starches with properties somewhat like those obtained by chemical modifications were obtained when subjected to atmospheric plasma treatment, suggesting that the reactive species can cross-link or depolymerize starch (Pal et al. 2016). In addition, functional group introduction on the starch molecules can also result in starch modification due to interaction between the chemically active species and starch. Several gases can be applied to form functional groups or free radicals on the surface of starch granules. For example, carboxylic acid, peroxides, and hydroxyl groups were introduced by oxygen plasma treatment. CO_2 gas plasma introduced hydroxyls, ketones, aldehydes, and esters, while nitrogen and ammonia plasmas introduced primary, secondary, and tertiary amines (Desmet et al. 2009). The noble gas plasma does not introduce any functional groups (Siow et al. 2006), but produces free radicals which can react with oxygen from the atmosphere. This indicates that plasma treatment does not result in a unique functionality, and therefore it is not a selective technique. It is worth noting that the aging effects like postplasma rearrangement and postplasma reactions should also be considered. It is concluded that cross-linking occurred at lower voltage and shorter period of treatment plasma, whereas depolymerization occurred at higher voltages and longer treatment time (Kalaivendan et al. 2022; Thirumdas et al. 2017a).

14.4.1 Depolymerization

Depolymerization of amylose and amylopectin into smaller fragments results from bombardment of high energetic ions of cold plasma and occurs even at low energy levels, producing abundant end products such as maltose, maltotriose, and maltotetrose. The reactive oxygen species (ROS) and reactive nitrogen species (RNS) are mainly responsible for the backbone scission of polysaccharides (Duan and Kasper 2011). The extent of depolymerization depended on the nature of the starch, plasma type, and exposure time (Lii et al. 2002b). Gas plasma can be divided into two groups according to its ability to depolymerization: (1) less active (hydrogen and air plasma) and (2) more active (ammonia and oxygen plasma). The potato starch was more easily degraded than corn, rice, and other waxy starches (Lii et al. 2002b). Amyloses were less susceptible to depolymerization than amylopectin (Sarangapani et al. 2018). Corn starch with high amylose content, due to formation of oxidation products with high molecular weight, showed great resistance to plasma action (Lii et al. 2002a). Moisture content plays an important role in depolymerization mechanism. The water molecules breakdown by the secondary electrons lead to formation of hydroxyl radicals and free electrons, and the water molecules absorbing radiation energy form the solvated hydroxyl ions and electrons (Thirumdas et al. 2017a).

The decrease in average M_w measured by GPC attached to multi-angle laser light scattering detector (MALS) was observed in potato starch after oxygen glow plasma modification (Zhang et al. 2015). The decrease in M_w was observed with increasing plasma power and treatment time (Wongsagonsup et al. 2014) and suggested molecular degradation due to strong oxidizing capacity of reactive species of plasma. The decrease in the short-range order may be attributed to the cleavages of the glycosidic bonds (Zhou et al. 2019). The amylose content and iodine binding capacity of kithul starch decreased after plasma modification due to plasma degrading amylose into glucose units (Sudheesh et al. 2019b). Similarly, a decrease in amylose content was also observed in aria, corn, and rice starch after plasma modification (Banura et al. 2018; Carvalho et al. 2021).

14.4.2 Cross-Linking

The free radicals and energetic electrons formed during plasma generation induced cross-linking or grafting between the polymeric chains of starch through interaction between active species and polymeric molecules (Thirumdas et al. 2017a). The free radicals resulted in formation of active sites for further reactions (Gomathi et al. 2008). A new C-O-C linkage between two chains was formed with removal of water molecule after cleavage between the reducing ends of two polymeric chain (C-OH) (Zou et al. 2004). The C-2 site was the most susceptible for the cross-linking among the 3 carbons present in pyranose (Zou et al. 2004).

An increase in amylose M_w was observed in rice starch after dielectric barrier discharge (DBD) plasma treatment (Sun et al. 2022). The increase in amylose content after DBD treatment suggested that the higher treatment voltage (20 kV) induced polymerization of the aria starch chains (Carvalho et al. 2021). The increase in relative peak area at 924 cm^{-1} related to C-O-C linkage as shown by Fourier transform infrared (FTIR) spectroscopy (Bie et al. 2016; Carvalho et al. 2021; Wongsagonsup et al. 2014), and therefore the cross-linking resulted in the increase in glycosidic linkage. Cross-linking was observed in non-granular waxy maize and rice starches as well as granular waxy potato and rice starches according to FTIR-attenuated total reflectance (FTIR-ATR) analysis (Okyere et al. 2022). The cross-linking can also be confirmed by the ^1H NMR measurement in addition to FTIR. The decrease in relative intensity of OH groups is due to cross-linking. The degree of rice starch hydrolysis decreased from 91% to 87% after plasma modification with low-pressure plasma systems, because the cross-linking of starch chains limited the rate of enzymatic hydrolysis (Thirumdas et al. 2017b). Cross-linking mainly depends on energy supplied due to the high bond energy of O-H bond (Wongsagonsup et al. 2014), and the percentage of cross-linking can be extended with increasing treatment time (Khorram et al. 2015).

14.4.3 Change in pH

A reduction in pH of starch slurry was observed after cold plasma treatment (Bußler et al. 2015; Thirumdas et al. 2017b), indicating that the oxidation of starch granules resulted in the formation of chemical groups with acidic characters such as carboxyl, carbonyl, and peroxides groups. The pH reduction can be explained by the increase in peak intensity at 1710 cm^{-1} corresponding to carboxylic acid with C = O stretch (Deeyai et al. 2013). X-ray photoelectron spectroscopy (XPS) analysis showed an increase in the O=C–OH bonds of corn starch due to the oxidation of some bonds such as C–O and O–C–O bonds by plasma reactive species (Bie et al. 2016).

14.5 Chemical Changes During Pulsed Electric Field (PEF) Treatment

PEF, as one of the nonthermal and physical methods, has potential to expand the range of starch functionalities in a cleaner and more efficient manner (Zhu 2018). Short pulses of high electric fields with a very short duration of micro-to milli-seconds applied by PEF can avoid excessive heat and unwanted electrolytic reactions. PEF significantly reduced the M_w of maize starch (Zhong et al. 2012), but did not cause significant difference in M_w of waxy rice and japonica rice starch (Wu et al. 2019; Zeng et al. 2016). Molecular weight distribution measured by GPC changed for wheat starch, potato starch, and pea starch after PEF treatment (Li et al. 2019). The increased molecular weight distribution index suggested that the degradation and aggregation effects may happen at the same time (Li et al. 2019). A slight reduction of the intermediate-sized molecules was observed in wheat starch, while the larger and intermediate-sized molecules of cassava starch significantly decreased after PEF treatment (Maniglia et al. 2021), indicating that molecular scissions of starch were promoted by PEF.

References

Ashwar BA, Shah A, Gani A, Rather SA, Wani SM, Wani IA, Masoodi FA, Gani A (2014) Effect of gamma irradiation on the physicochemical properties of alkali-extracted rice starch. Radiat Phys Chem 99:37–44

Banura S, Thirumdas R, Kaur A, Deshmukh RR, Annapure US (2018) Modification of starch using low pressure radio frequency air plasma. LWT 89:719–724

Bao J, Ao Z, Jane J-L (2005) Characterization of physical properties of flour and starch obtained from gamma-irradiated white rice. Starch-Stärke 57(10):480–487

Bashir K, Aggarwal M (2019) Physicochemical, structural and functional properties of native and irradiated starch: a review. J Food Sci Technol 56(2):513–523

Bashir K, Swer TL, Prakash KS, Aggarwal M (2017) Physico-chemical and functional properties of gamma irradiated whole wheat flour and starch. LWT 76:131–139

Bashir M, Haripriya S (2016) Physicochemical and structural evaluation of alkali extracted chickpea starch as affected by γ-irradiation. Int J Biol Macromol 89:279–286

BeMiller JN, Huber KC (2015) Physical modification of food starch functionalities. Annu Rev Food Sci Technol 6(6):19–69

Bertoft E (2017) Understanding starch structure: recent progress. Agronomy 7(3):56

Bertolini AC, Mestres C, Colonna P (2000) Rheological properties of acidified and UV-irradiated starches. Starch-Stärke 52(10):340–344

Bhat R, Karim AA (2009) Impact of radiation processing on starch. Compr Rev Food Sci Food Saf 8(2):44–58

Bie P, Pu H, Zhang B, Su J, Chen L, Li X (2016) Structural characteristics and rheological properties of plasma-treated starch. Innovative Food Sci Emerg Technol 34:196–204

Bußler S, Steins V, Ehlbeck J, Schlüter O (2015) Impact of thermal treatment versus cold atmospheric plasma processing on the techno-functional protein properties from Pisum sativum 'Salamanca'. J Food Eng 167:166–174

Carvalho APMG, Barros DR, da Silva LS, Sanches EA et al (2021) Dielectric barrier atmospheric cold plasma applied to the modification of Ariá (Goeppertia allouia) starch: effect of plasma generation voltage. Int J Biol Macromol 182:1618–1627

Chen X, Liu Y, Xu Z, Zhang C, Liu X, Sui Z, Corke H (2021) Microwave irradiation alters the rheological properties and molecular structure of hull-less barley starch. Food Hydrocoll 120: 106821

Chung HJ, Liu Q (2009) Effect of gamma irradiation on molecular structure and physicochemical properties of corn starch. J Food Sci 74(5):C353–C361

Chung H-J, Liu Q (2010) Molecular structure and physicochemical properties of potato and bean starches as affected by gamma-irradiation. Int J Biol Macromol 47(2):214–222

Chung K-H, Othman Z, Lee J-S (2015) Gamma irradiation of corn starches with different amylose-to-amylopectin ratio. J Food Sci Technol 52(10):6218–6229

Dar MZ, Deepika K, Jan K, Swer TL, Kumar P, Verma R, Verma K, Prakash KS, Jan S, Bashir K (2018) Modification of structure and physicochemical properties of buckwheat and oat starch by γ-irradiation. Int J Biol Macromol 108:1348–1356

Deeyai P, Suphantharika M, Wongsagonsup R, Dangtip S (2013) Characterization of modified tapioca starch in atmospheric argon plasma under diverse humidity by FTIR spectroscopy. Chin Phys Lett 30(1):018103

Desmet T, Morent R, De Geyter N, Leys C, Schacht E, Dubruel P (2009) Nonthermal plasma technology as a versatile strategy for polymeric biomaterials surface modification: a review. Biomacromolecules 10(9):2351–2378

Duan J, Kasper DL (2011) Oxidative depolymerization of polysaccharides by reactive oxygen/ nitrogen species. Glycobiology 21(4):401–409

Fiedorowicz M, Tomasik P, You S, Lim S-T (1999) Molecular distribution and pasting properties of UV-irradiated corn starches. Starch-Stärke 51(4):126–131

Gani A, Nazia S, Rather SA, Wani SM, Shah A, Bashir M, Masoodi FA, Gani A (2014) Effect of γ-irradiation on granule structure and physicochemical properties of starch extracted from two types of potatoes grown in Jammu & Kashmir, India. LWT 58(1):239–246

Ghali Y, Ibrahim N, Gabr S, Aziz H (1979) Modification of corn starch and fine flour by acid and gamma irradiation. Part 1. Chemical investigation of the modified products. Starch-Stärke 31(10):325–328

Gomathi N, Sureshkumar A, Neogi S (2008) RF plasma-treated polymers for biomedical applications. Curr Sci 94(11):1478–1486

Gul K, Singh AK, Sonkawade RG (2016) Physicochemical, thermal and pasting characteristics of gamma irradiated rice starches. Int J Biol Macromol 85:460–466

Hornung PS, Barbi RCT, Teixeira GL, Ávila S et al (2018) Brazilian Amazon white yam (Dioscorea sp.) starch. J Therm Anal Calorim 134(3):2075–2088

Kalaivendan RGT, Mishra A, Eazhumalai G, Annapure US (2022) Effect of atmospheric pressure non-thermal pin to plate plasma on the functional, rheological, thermal, and morphological properties of mango seed kernel starch. Int J Biol Macromol 196:63–71

Khorram S, Zakerhamidi MS, Karimzadeh Z (2015) Polarity functions' characterization and the mechanism of starch modification by DC glow discharge plasma. Carbohydr Polym 127:72–78

Koteswara Reddy C, Vidya PV, Vijina K, Haripriya S (2015) Modification of poovan banana (Musa AAB) starch by γ-irradiation: effect on in vitro digestibility, molecular structure and physicochemical properties. Int J Food Sci Technol 50(8):1778–1784

Lam ND, Quynh TM, Diep TB, Binh PT, Lam TD (2021) Effect of gamma irradiation and pyrolysis on indigestible fraction, physicochemical properties, and molecular structure of rice starch. J Food Proc Preserv 45(10):e15880

Lee J-S, Ee M-L, Chung K-H, Othman Z (2013) Formation of resistant corn starches induced by gamma-irradiation. Carbohydr Polym 97(2):614–617

Li Q, Wu Q-Y, Jiang W, Qian J-Y, Zhang L, Wu M, Rao S-Q, Wu C-S (2019) Effect of pulsed electric field on structural properties and digestibility of starches with different crystalline type in solid state. Carbohydr Polym 207:362–370

Lii C-Y, Liao C-D, Stobinski L, Tomasik P (2002a) Behaviour of granular starches in low-pressure glow plasma. Carbohydr Polym 49(4):499–507

Lii C-Y, Liao C-D, Stobinski L, Tomasik P (2002b) Effects of hydrogen, oxygen, and ammonia low-pressure glow plasma on granular starches. Carbohydr Polym 49(4):449–456

Liu K, Zhang BJ, Chen L, Li XX, Zheng B (2019) Hierarchical structure and physicochemical properties of highland barley starch following heat moisture treatment. Food Chem 271:102–108

Majeed T, Wani IA, Hussain PR (2017) Effect of dual modification of sonication and γ-irradiation on physicochemical and functional properties of lentil (Lens culinaris L.) starch. Int J Biol Macromol 101:358–365

Majeed T, Wani IA, Hamdani AM, Bhat NA (2018) Effect of sonication and γ-irradiation on the properties of pea (Pisum sativum) and vetch (Vicia villosa) starches: a comparative study. Int J Biol Macromol 114:1144–1150

Maniglia BC, Pataro G, Ferrari G, Augusto PED, Le-Bail P, Le-Bail A (2021) Pulsed electric fields (PEF) treatment to enhance starch 3D printing application: effect on structure, properties, and functionality of wheat and cassava starches. Innovative Food Sci Emerg Technol 68:102602

Misra NN, Pankaj SK, Segat A, Ishikawa K (2016) Cold plasma interactions with enzymes in foods and model systems. Trends Food Sci Technol 55:39–47

Mukhtar R, Shah A, Noor N, Gani A, Wani IA, Ashwar BA, Masoodi FA (2017) γ-Irradiation of oat grain–effect on physico-chemical, structural, thermal, and antioxidant properties of extracted starch. Int J Biol Macromol 104:1313–1320

Ocloo FCK, Minnaar A, Emmambux NM (2014) Effects of gamma irradiation and stearic acid, alone and in combination, on functional, structural, and molecular characteristics of high amylose maize starch. Starch-Stärke 66(7–8):624–635

Okyere AY, Boakye PG, Bertoft E, Annor GA (2022) Structural characterization and enzymatic hydrolysis of radio frequency cold plasma treated starches. J Food Sci 87(2):686–698

Othman Z, Hassan O, Hashim K (2015) Physicochemical and thermal properties of gamma-irradiated sago (Metroxylon sagu) starch. Radiat Phys Chem 109:48–53

Pal P, Kaur P, Singh N, Kaur A, Misra NN, Tiwari BK, Cullen PJ, Virdi AS (2016) Effect of nonthermal plasma on physico-chemical, amino acid composition, pasting and protein characteristics of short and long grain rice flour. Food Res Int 81:50–57

Pankaj SK, Wan Z, Keener KM (2018) Effects of cold plasma on food quality: a review. Foods 7(1): 4

Peat S, Bourne EJ, Whelan WJ (1948) Photochemical degradation of starch. Nature 161(4098): 762–763

Polesi LF, Sarmento SBS, Moraes JD, Franco CML, Canniatti-Brazaca SG (2016) Physicochemical and structural characteristics of rice starch modified by irradiation. Food Chem 191:59–66

Reddy CK, Suriya M, Vidya PV, Vijina K, Haripriya S (2015) Effect of γ-irradiation on structure and physico-chemical properties of Amorphophallus paeoniifolius starch. Int J Biol Macromol 79:309–315

Rombo GO, Taylor JRN, Minnaar A (2004) Irradiation of maize and bean flours: effects on starch physicochemical properties. J Sci Food Agric 84(4):350–356

Sarangapani C, Patange A, Bourke P, Keener K, Cullen PJ (2018) Recent advances in the application of cold plasma technology in foods. Annu Rev Food Sci Technol 9(1):609–629

Siow KS, Britcher L, Kumar S, Griesser HJ (2006) Plasma methods for the generation of chemically reactive surfaces for biomolecule immobilization and cell colonization—a review. Plasma Process Polym 3(6–7):392–418

Sofi BA, Wani IA, Masoodi FA, Saba I, Muzaffar S (2013) Effect of gamma irradiation on physicochemical properties of broad bean (Vicia faba L.) starch. LWT Food Sci Technol 54(1):63–72

Sudheesh C, Sunooj KV, George J, Vikas KS, Sajeevkumar VA (2019a) Impact of γ– irradiation on the physico-chemical, rheological properties and in vitro digestibility of kithul (Caryota urens) starch; a new source of nonconventional stem starch. Radiat Phys Chem 162:54–65

Sudheesh C, Sunooj KV, Sinha SK, George J, Kumar S, Murugesan P, Arumugam S, Ashwath Kumar K, Sajeev Kumar VA (2019b) Impact of energetic neutral nitrogen atoms created by glow discharge air plasma on the physico-chemical and rheological properties of kithul starch. Food Chem 294:194–202

Sujka M, Cieśla K, Jamroz J (2015) Structure and selected functional properties of gamma-irradiated potato starch. Starch-Stärke 67(11–12):1002–1010

Sun X, Saleh ASM, Sun Z, Ge X, Shen H, Zhang Q, Yu X, Yuan L, Li W (2022) Modification of multi-scale structure, physicochemical properties, and digestibility of rice starch via microwave and cold plasma treatments. LWT 153:112483

Sunder M, Mumbrekar KD, Mazumder N (2022) Gamma radiation as a modifier of starch—physicochemical perspective. Curr Res Food Sci 5:141–149

Thirumdas R, Kadam D, Annapure US (2017a) Cold plasma: an alternative technology for the starch modification. Food Biophy 12(1):129–139

Thirumdas R, Trimukhe A, Deshmukh RR, Annapure US (2017b) Functional and rheological properties of cold plasma treated rice starch. Carbohydr Polym 157:1723–1731

Verma R, Jan S, Rani S, Jan K, Swer TL, Prakash KS, Dar MZ, Bashir K (2018) Physicochemical and functional properties of gamma irradiated buckwheat and potato starch. Radiat Phys Chem 144:37–42

Wani IA, Jabeen M, Geelani H, Masoodi FA, Saba I, Muzaffar S (2014) Effect of gamma irradiation on physicochemical properties of Indian horse chestnut (Aesculus indica Colebr.) starch. Food Hydrocoll 35:253–263

Wani AA, Wani IA, Hussain PR, Gani A, Wani TA, Masoodi FA (2015) Physicochemical properties of native and γ-irradiated wild arrowhead (Sagittaria sagittifolia L.) tuber starch. Int J Biol Macromol 77:360–368

Whelan WJ, Peat S (1949) The photochemical degradation of starch and allied carbohydrates. J Soc Dye Colour 65(12):748–757

Wongsagonsup R, Deeyai P, Chaiwat W, Horrungsiwat S, Leejariensuk K, Suphantharika M, Fuongfuchat A, Dangtip S (2014) Modification of tapioca starch by non-chemical route using jet atmospheric argon plasma. Carbohydr Polym 102:790–798

Wu C, Wu Q-Y, Wu M, Jiang W, Qian J-Y, Rao S-Q, Zhang L, Li Q, Zhang C (2019) Effect of pulsed electric field on properties and multi-scale structure of japonica rice starch. LWT 116: 108515

Zeng F, Gao Q-Y, Han Z, Zeng X-A, Yu S-J (2016) Structural properties and digestibility of pulsed electric field treated waxy rice starch. Food Chem 194:1313–1319

Zhang B, Chen L, Li X, Li L, Zhang H (2015) Understanding the multi-scale structure and functional properties of starch modulated by glow-plasma: a structure-functionality relationship. Food Hydrocoll 50:228–236

Zhong H, Qian Y, Xin An Z, Dong Hui L, Shu Juan Y, Ben Shan Z, Xiao Dong C (2012) Studies on the microstructure and thermal properties of pulsed electric fields (PEF)-treated maize starch. Int J Food Eng 8(1):1

Zhou Y, Yan Y, Shi M, Liu Y (2019) Effect of an atmospheric pressure plasma jet on the structure and physicochemical properties of waxy and normal maize starch. Polymers 11(1):8

Zhu F (2016) Impact of γ-irradiation on structure, physicochemical properties, and applications of starch. Food Hydrocoll 52:201–212

Zhu F (2018) Modifications of starch by electric field based techniques. Trends Food Sci Technol 75:158–169

Zou J-J, Liu C-J, Eliasson B (2004) Modification of starch by glow discharge plasma. Carbohydr Polym 55(1):23–26

The manufacturer's authorised representative in the EU is Springer Nature Customer Service Centre GmbH, Europaplatz 3, 69115 Heidelberg, Germany. If you have any concerns regarding our products, please contact ProductSafety@springernature.com

Printed and bound by CPI Group (UK) Ltd, Croydon, CR0 4YY
30/09/2026
02225322-0007